深入浅出
.NET框架设计与实现

冯辉 | 著

电子工业出版社
Publishing House of Electronics Industry
北京·BEIJING

内 容 简 介

本书主要介绍.NET 框架的核心部分，不仅阐述了框架的核心设计，还展示了实践代码和运行效果，同时根据不同的功能模块扩展出属于自己的小组件。通过阅读本书，读者可以深入了解.NET 框架的核心设计，掌握.NET 性能调优和 ASP.NET Core 的部署方式。本书包含 18 章。第 1~3 章主要介绍.NET 框架、开发环境和运行模式。第 4~12 章主要介绍依赖注入、配置与选项、后台任务、中间件、缓存、本地化、健康检查、文件系统和日志等常用功能模块。第 13~17 章主要介绍线程、线程同步、内存管理、诊断和调试，以及编译技术等内容。第 18 章主要介绍应用程序的部署方式，包括宿主机的部署，以及 Docker 和 Kubernetes 的部署。

本书可供.NET 开发人员和.NET 初学者阅读与参考，也适合具备其他编程语言基础且想要学习.NET 的开发人员阅读，还可为读者在工作及巩固基础时提供有力支持。

未经许可，不得以任何方式复制或抄袭本书之部分或全部内容。
版权所有，侵权必究。

图书在版编目（CIP）数据

深入浅出：.NET 框架设计与实现 / 冯辉著. —北京：电子工业出版社，2023.3
ISBN 978-7-121-42019-1

Ⅰ. ①深… Ⅱ. ①冯… Ⅲ. ①计算机网络—程序设计 Ⅳ. ①TP393.09

中国版本图书馆 CIP 数据核字（2022）第 219943 号

责任编辑：孙学瑛　　　　　　　　特约编辑：田学清
印　　刷：三河市良远印务有限公司
装　　订：三河市良远印务有限公司
出版发行：电子工业出版社
　　　　　北京市海淀区万寿路 173 信箱　　邮编：100036
开　　本：787×980　1/16　印张：31.75　字数：700 千字
版　　次：2023 年 3 月第 1 版
印　　次：2023 年 3 月第 1 次印刷
定　　价：129.90 元

凡所购买电子工业出版社图书有缺损问题，请向购买书店调换。若书店售缺，请与本社发行部联系，联系及邮购电话：（010）88254888，88258888。
质量投诉请发邮件至 zlts@phei.com.cn，盗版侵权举报请发邮件至 dbqq@phei.com.cn。
本书咨询联系方式：（010）51260888-819，faq@phei.com.cn。

推荐序 Forewords

第一次和冯辉接触是几年前他邀请我参加济南.NET俱乐部的活动，当时我因工作原因未能赴约，到现在仍深感遗憾。在年初的时候得知他正在撰写一本讲解.NET框架的书，又有幸被邀请为该书写序，在忐忑之余，我也想借这个机会介绍一下自己在.NET这条路的心路历程供读者参考。

在编程道路上，有两件事对我产生过很大的影响。第一件事可以追溯到10年前我第一次接触编程时，当时我使用.NET中的XNA框架和WP框架开发了我的第一款手机RPG游戏并在Lumia 820上运行。这次成功开发游戏的经历极大地增强了我对从事编程的信心和对.NET的兴趣。之后因为工作的需要，我对WPF、WCF和ASP.NET等不同方向的.NET技术都有所涉猎。第二件事可以追溯到Microsoft宣布第一个开源的.NET版本ASP.NET VNext时，当时我就对VNext产生了极大的兴趣。在学习和推广VNext时，我结识了Alex LEWIS、He Zhenxi、Xie Yang等好友，之后我们一起创建了NCC（.NET Core Community）社区。

最初，.NET Core提供英文文档作为为数不多的使用参考，所以NCC社区刚刚成立时，我们做的第一件事是翻译ASP.NET Core最初版本的英文文档。得益于Microsoft的开源策略，学习.NET Core的另一个途径就是阅读GitHub官网上的源代码，我也由此养成了阅读开源项目代码的习惯，并且受益至今。受万物皆"Services"并且完全管道化的ASP.NET Core框架的启发，我设计了AspectCore AOP库，也在从事云原生开发之后在Go语言上继续参考ASP.NET Core实现了以依赖注入作为内核的模块化开发框架。

即使对.NET Core已经相对熟悉，我在阅读本书样稿之后还是感觉眼前一亮。本书由

浅入深地介绍了 .NET Core 框架的核心部分，如依赖注入、配置与选项、中间件、缓存、日志、多线程等。我相信，不管是 .NET Core 的初学者，还是想要继续进阶的中高级开发工程师，都能从这本书中获得很大的帮助。

  近几年技术浪潮兴替，从大数据、移动互联网、云计算技术的兴起，再到如今人工智能、云原生技术的流行，.NET Core 完成了从运行时、BCL（Base Class Library，基础类库）到开发框架的一系列蜕变。得益于分层编译、重新实现的集合类、Span、网络/文件 I/O 等诸多细节的优化，.NET Core 不仅在最新几轮的 TechEmpower 性能评测中名列前茅，还可以搭配 C#，使 .NET Core 成为事实上的云原生应用开发的最佳平台之一。谨以此序和同为 .NET Core 的使用者及爱好者共勉之。

Apache SkyWalking PMC、NCC 社区创始人 刘浩杨

2022 年 8 月于杭州

前言 Preface

随着.NET 技术的发展，涌现出众多的设计思想和核心概念。值得开发人员关注的技术点有很多，如 ASP.NET Core 模块的设计、跨平台调试与部署等。

.NET 已经成为一种热门的现代技术体系，从.NET 彻底迈向跨平台和开源开始，已经历经了约 10 个版本。新一代的.NET 平台以拥抱云原生为核心，拥有更小的体积、更少的资源占用和更快的启动速度，并且支持水平扩展。

笔者也算是一个亲历者，从.NET Core 1.0 到现在，是一个从重生到繁荣的成长阶段，.NET 生态更加开放，开源社区越来越活跃，不仅支持传统的 x86 架构体系，还支持 ARM 架构，并且获得了龙芯 LoongArch 架构及诸多新兴架构体系的踊跃支持，同时在工业、IoT、车联网等领域获得了广泛运用。无论是从社区参与度，还是从 NuGet 的下载量，都不难看出.NET 的发展速度。

.NET 具备原生的跨平台部署能力，是一种用于构建多端应用的开放平台。使用.NET 可以构建桌面应用、云服务、嵌入式应用及机器学习应用等，读者可以从 GitHub 官网的 dotnet 组织中获取它所有的源代码。

计算机科学家 Alan J. Perlis 曾说过："不能影响你的编程思维方式的语言不值得学习和使用。"由此可知，"思维"非常重要，只有了解一门编程语言或框架的基础模型与核心设计，才能将其应用到日常的编程中。

框架的设计过程是非常复杂的，笔者偏向于将复杂问题简单化，先研究它的实现方式，再了解它的设计模式，通过这一层层的推导过程，慢慢地了解整体脉络。阅读源代码是一个枯燥但会带来收获的过程。在本书中，笔者将框架设计方法，以及它们的实现（可扩展性）方式毫无保留地写下来。

本书集成了笔者在工作中使用.NET 开发应用程序的编写经验和调试经验，同时结合了笔者关于 Linux 平台和容器云平台的使用经验。通过本书，笔者将介绍每个模块的核心设计与实现，因为要想在生产环境中大规模使用，就需要在这个复杂而庞大的项目中抓到主线，了解内部的实现和调试技术，以便快速定位问题和解决问题。

本书对 ASP.NET Core 的部分核心内容进行了深入解析，在这个基础上延伸内容，以及自定义扩展实例，初学者可以更深入地了解 ASP.NET Core 内部的运作方式。本书也涵盖了很多基础知识，如垃圾回收、调试、线程等，除此之外，添加了部署方面的内容，将应用程序部署到宿主机、Docker 和 Kubernetes 中。

笔者通过对.NET 技术的原理进行剖析及实例的演示，帮助读者快速熟悉框架的核心设计及实现原理。希望读者在阅读完本书后，能够将书中的内容学以致用，使用.NET 构建出高性能的应用程序，同时为开源社区添砖加瓦。

关于勘误

完成本书绝不是一件简单的事情。虽然笔者力争保证内容的准确性，并且花费了很长的时间和大量的精力核对书中的文字和内容，但个人水平有限，书中难免存在一些不足之处，望广大读者批评指正。欢迎发邮件至 hueifeng2020@outlook.com，期待您的反馈。

致谢

感谢邹溪源、严振范、锅美玲、李卫涵、胡心（Azul X）、管生玄、黄新成（一线码农）和周杰等人对本书的审核和校对，同时感谢家人、朋友和同事在笔者编写本书期间给予的支持与鼓励。

感谢符隆美编辑对我的悉心指导，她对本书的审核和建议使我的写作水平有了很大的提高，在此表示感谢！

目录 Contents

第 1 章　.NET 概述和环境安装 ·· 1
　1.1　.NET 框架简介 ··· 1
　1.2　.NET 的开发环境 ··· 6
　1.3　小结 ··· 20

第 2 章　.NET 运行原理概述 ·· 21
　2.1　.NET CLI 概述 ··· 21
　2.2　小结 ··· 36

第 3 章　ASP.NET Core 应用程序的多种运行模式 ···················· 37
　3.1　自宿主 ·· 37
　3.2　IIS 服务承载 ··· 38
　3.3　将 WebAPI 嵌入桌面应用程序中 ·· 39
　3.4　服务承载 ··· 41
　3.5　延伸阅读：WindowsFormsLifetime ··· 53
　3.6　小结 ··· 59

第 4 章 依赖注入 ... 60
- 4.1 .NET 依赖注入 ... 61
- 4.2 实现批量服务注册 ... 80
- 4.3 小结 ... 84

第 5 章 配置与选项 ... 85
- 5.1 配置模式 ... 85
- 5.2 选项模式 ... 101
- 5.3 设计一个简单的配置中心 ... 121
- 5.4 小结 ... 130

第 6 章 使用 IHostedService 和 BackgroundService 实现后台任务 ... 131
- 6.1 IHostedService ... 131
- 6.2 BackgroundService ... 134
- 6.3 任务调度 ... 137
- 6.4 小结 ... 142

第 7 章 中间件 ... 143
- 7.1 中间件的作用 ... 143
- 7.2 中间件的调用过程 ... 144
- 7.3 编写自定义中间件 ... 147
- 7.4 在过滤器中应用中间件 ... 151
- 7.5 制作简单的 API 统一响应格式与自动包装 ... 156
- 7.6 延伸阅读：责任链模式 ... 159
- 7.7 延伸阅读：中间件常见的扩展方法 ... 162
- 7.8 小结 ... 174

第 8 章 缓存 ... 175
- 8.1 内存缓存 ... 175
- 8.2 分布式缓存 ... 180

8.3 HTTP 缓存 · 188

8.4 小结 · 193

第 9 章 本地化 · 194

9.1 内容本地化 · 194

9.2 多样化的数据源 · 203

9.3 小结 · 210

第 10 章 健康检查 · 211

10.1 检查当前应用的健康状态 · 211

10.2 发布健康报告 · 233

10.3 可视化健康检查界面 · 238

10.4 小结 · 240

第 11 章 文件系统 · 241

11.1 ASP.NET Core 静态文件 · 241

11.2 自定义一个简单的文件系统 · 247

11.3 小结 · 253

第 12 章 日志 · 254

12.1 控制台日志 · 254

12.2 调试日志 · 281

12.3 事件日志 · 285

12.4 EventSource 日志 · 292

12.5 TraceSource 日志 · 298

12.6 DiagnosticSource 日志 · 303

12.7 小结 · 309

第 13 章 多线程与任务并行 · 310

13.1 线程简介 · 310

13.2 基于任务的异步编程 · 316

13.3 线程并行 ·· 328
13.4 小结 ··· 329

第 14 章 线程同步机制和锁 ··· 330

14.1 原子操作 ·· 330
14.2 自旋锁 ··· 333
14.3 混合锁 ··· 337
14.4 互斥锁 ··· 340
14.5 信号量 ··· 341
14.6 读写锁 ··· 346
14.7 小结 ··· 348

第 15 章 内存管理 ··· 349

15.1 内存分配 ·· 349
15.2 垃圾回收器 ··· 357
15.3 资源释放 ·· 362
15.4 垃圾回收器的设置 ·· 368
15.5 小结 ··· 374

第 16 章 诊断和调试 ·· 375

16.1 性能诊断工具 ·· 375
16.2 Linux 调试 ·· 416
16.3 小结 ··· 429

第 17 章 编译技术精讲 ··· 430

17.1 IL 解析 ·· 430
17.2 JIT 简介 ··· 435
17.3 JIT 编译 ··· 437
17.4 AOT 编译 ··· 446
17.5 小结 ··· 450

第 18 章　部署 ··········451

- 18.1　发布与部署 ··········451
- 18.2　Docker ··········460
- 18.3　编写 Dockerfile 文件 ··········469
- 18.4　构建.NET 应用镜像 ··········472
- 18.5　Docker Compose ··········477
- 18.6　Docker Swarm ··········480
- 18.7　Kubernetes ··········485
- 18.8　小结 ··········496

第 1 章

.NET 概述和环境安装

随着.NET 技术的发展,从.NET Framework 到.NET Core 和后续的.NET 5、.NET 6、.NET 7 及更高版本,.NET 已经成为一种热门的现代技术体系。从.NET 彻底迈向跨平台和开源开始,已经历经了约 10 个版本。新一代的.NET 平台以拥抱云原生为核心,拥有更小的体积、更少的资源占用和更快的启动速度,并且支持水平扩展。.NET 的许多新特性也都积极贴近时代特征,成为行业发展的风向标。

.NET 已拥有开放的生态体系,不仅支持传统的 x86 架构体系,还支持 ARM 架构,并且获得了龙芯 LoongArch 架构及诸多新兴架构体系的踊跃支持,同时在工业、IoT、车联网等领域获得了广泛应用。.NET 开发人员社区也越来越活跃,每年都有数以万计的开发人员投入.NET 的开发阵营进行上下游应用的开发。

通过学习本章,读者可以初步认识.NET,并了解其发展历史。在这个初识过程中,有的读者可能不太了解.NET 平台的一些术语与缩写,可能会有一头雾水的感觉,笔者会挑选一些重要的内容进行讲解。

1.1 .NET 框架简介

当提到.NET 时,有的读者可能会想到.NET Framework。很多读者隐约记得类似于 WebForms、WCF、Remoting 等传统的.NET Framework(.NET 框架),它们与 Windows 深度绑定,因此,一度跟随 Windows 获得了飞速的发展。

但时代已悄然改变,跨平台已经成为当今开发语言环境的一大基本要求,.NET

Framework 的时代已经一去不复返，昔日只能在 Windows 上运行的.NET Framework 已经成为明日黄花。从 2014 年的.NET Core 开始，.NET 已经彻底走向了跨平台和开源，不仅如此，自.NET Core 开始，.NET 已经越来越强大，发展成目前能够同时在 Windows、Linux、Mac、iOS、Android 等多平台上运行的统一平台。

1.1.1 .NET Core 简介

.NET Core 是.NET 技术走向开源的一个象征，是衍生自.NET Framework 的新一代.NET 框架。与.NET Framework 相比，开源、跨平台是.NET Core 的显著特点。Microsoft 也将整个.NET Core 交给.NET 基金会进行管理，.NET Core 已经成为该基金会管理的主要开源项目之一。

.NET Core 于 2014 年 11 月 12 日公布，并于 2016 年发布.NET Core 1.0 版本。.NET Core 是.NET Framework 的下一个版本。

.NET Core 支持的语言包括 C#、F#、Visual Basic .NET 和 C++/CLI，具有 Windows 和 macOS 安装程序，以及 Linux 安装包和 Docker 镜像。

.NET Core 具有许多优点，下面列举其中的几个。

- 跨平台：可以在 Windows、Linux 和 macOS 上运行。
- 跨体系架构一致性：支持多种架构体系（如 x64、x86 和 ARM）。
- 命令行工具：.NET Core 所有的运行脚本都可以通过命令行工具执行。
- 部署灵活：多样化与灵活性的部署方式。
- 兼容：.NET Core 通过.NET Standard 与.NET Framework、Xamarin 和 Mono 兼容。
- 开源：.NET Core 是开源的，使用 MIT 许可证。另外，.NET Core 是.NET Foundation（.NET 基金会）项目。
- 扩展性：.NET Core 是一个模块化、轻量级，以及具有灵活性和深度扩展性的框架。

.NET Core 包括.NET 编译器平台 Roslyn、.NET Core 运行时 CoreCLR、.NET Core 框架 CoreFX 和 ASP.NET Core。由于 ASP.NET Core 是.NET Core SDK 的一部分，因此无须单独安装。.NET Core 组件如图 1-1 所示。

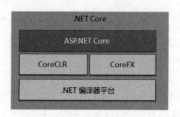

图 1-1 .NET Core 组件

.NET Core 3.1 之前的版本一直采用.NET Core 这个名称，而.NET Core 3.1 后续的版本直接被命名为.NET 5。为了避免与.NET Framework 的版本号发生冲突，它跳过了版本号 4，而.NET Core 也完成了它的使命，成为一个历史名词。

.NET 又分为.NET 运行时（.NET Runtime）和.NET SDK（Software Development Kit，软件开发工具包）。.NET 运行时的安装包的体积比较小，仅包含运行时的相关资源，可以作为部署生产环境时的基础组件，.NET 应用程序必须依赖它才能运行。而.NET SDK 不仅包含.NET 运行时的相关资源，还包含开发应用程序时所依赖的相关资源。如果要开发.NET 应用程序，则需要安装.NET SDK。

.NET 的产品路线图如图 1-2 所示，由此可知，未来每年都会发布新版本的.NET。

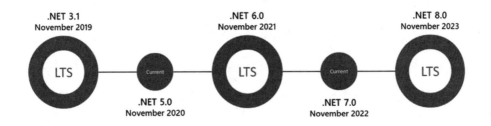

图 1-2　.NET 的产品路线图

1.1.2　.NET Standard 简介

.NET Standard（.NET 标准）的原称为.NET 平台标准（.NET Platform Standard）。.NET Standard 适用于所有.NET 实现的 API，主要用于统一平台，是一种可用于.NET Core、Xamarin 和.NET Framework 的可移植库。也就是说，.NET Standard 是一个基础类库。.NET Standard 支持的技术广泛，如支持.NET Framework、.NET Core 及后续的.NET 版本、MAUI（前身为由 Mono 基金会推出的 Xamarin，后被 Microsoft 收购，并且 MAUI 支持使用 C#开发 Android、iOS 等移动端的原生应用）和 Unity 等，并且包含基础类库，如集合类、I/O、网络、数据访问 API、线程 API 等。

.NET Standard，顾名思义，是一个标准，不包含具体的实现。.NET Standard 定义了一套标准的接口，通过这一套标准的接口的设计，反而促成了一个意想不到的结果，即.NET Core 的诞生。

.NET 平台由.NET Framework、.NET Core 和 Xamarin 这三大技术分支组成，如图 1-3 所示。

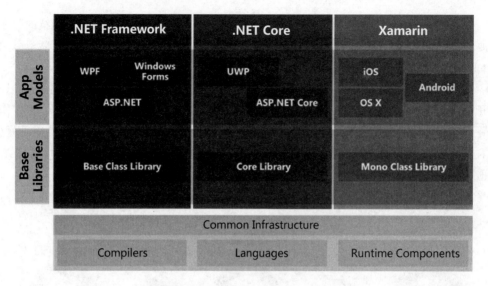

图 1-3 .NET 平台的三大技术分支

如图 1-3 所示,每个技术分支都具有一个基础库(Base Library)。为了解决代码复用的问题,推出了一个统一的基础类库,这个统一的基础类库称为.NET Standard。图 1-4 所示为.NET Standard 的布局,.NET Standard 为.NET Framework、.NET Core 和 Xamarin 提供了统一的 API,这个统一的 API 可以被所有类型的应用程序做代码复用。

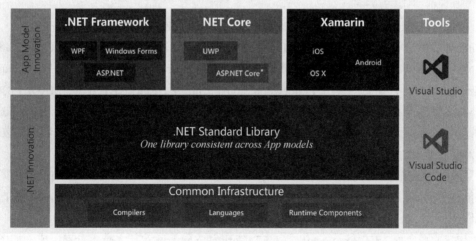

图 1-4 .NET Standard 布局

.NET Standard 提供的 API 随着版本而丰富。另外,.NET Standard 也是 PCL(Portable Class Library,可移植类库)的替代品。开发人员可以通过.NET Standard 来开发支持多平台的应用程序,如图 1-5 所示,不同的框架版本对应不同的.NET Standard 版本。

.NET Standard	1.0	1.1	1.2	1.3	1.4	1.5	1.6	2.0	2.1
.NET	5.0	5.0	5.0	5.0	5.0	5.0	5.0	5.0	5.0
.NET Core	1.0	1.0	1.0	1.0	1.0	1.0	1.0	2.0	3.0
.NET Framework	4.5	4.5	4.5.1	4.6	4.6.1	4.6.1[2]	4.6.1[2]	4.6.1[2]	N/A[3]
Mono	4.6	4.6	4.6	4.6	4.6	4.6	4.6	5.4	6.4
Xamarin.iOS	10.0	10.0	10.0	10.0	10.0	10.0	10.0	10.14	12.16
Xamarin.Mac	3.0	3.0	3.0	3.0	3.0	3.0	3.0	3.8	5.16
Xamarin.Android	7.0	7.0	7.0	7.0	7.0	7.0	7.0	8.0	10.0
Universal Windows Platform	10.0	10.0	10.0	10.0	10.0	10.0.16299	10.0.16299	10.0.16299	TBD
Unity	2018.1	2018.1	2018.1	2018.1	2018.1	2018.1	2018.1	2018.1	TBD

图 1-5　不同的框架版本对应不同的.NET Standard 版本

注意:.NET Standard 具有多个版本,每个新的版本支持的 API 都会增加一些。.NET Standard 的版本越新支持的 API 就越多(但这也意味着旧版本的.NET Standard 支持的 API 少),同时可以兼容的旧版本也就越多。

开发人员可以创建基于.NET Standard 的类库项目,并且将该类库项目的目标框架(TargetFramework 属性)定义为 netstandard{version},如代码 1-1 所示。

代码1-1

```
<Project Sdk="Microsoft.NET.Sdk">
  <PropertyGroup>
    <TargetFramework>netstandard2.1</TargetFramework>
  </PropertyGroup>
</Project>
```

目前,.NET Standard 的最新版本为 2.1,并且已经存档,未来将不再发布新版本的.NET

Standard，但.NET 5、.NET 6 及所有将来的版本将继续支持.NET Standard 2.1 及更早的版本。.NET Standard 不仅不会成为历史名词，还将继续成为众多.NET 家族成员之间赖以共享的底层标准。

1.2 .NET 的开发环境

受到.NET 开发人员广泛欢迎的集成开发环境（Integrated Development Environment，IDE）是被称为"宇宙最强 IDE"的 Visual Studio，它包括整个软件生命周期中所需要的大部分工具，如 UML 工具、代码管控工具、IDE 等。Visual Studio 提供了社区版、专业版和企业版，社区版是完全免费的，专业版和企业版需要付费购买。Microsoft 这款为开发人员精心打造的开发工具，不仅能用于常规的.NET 应用程序开发，还能用来开发 Python、Java、前端等多种语言的应用。目前，Visual Studio 不仅能在 Windows 平台上使用，还能在 mac OS 平台上使用，开发人员使用 mac OS 版本能够获得与 Windows 版本相同的操作功能。

图 1-6 所示为 Visual Studio 在 mac OS 中的界面，可以看到，该界面与 Visual Studio 在 Windows 平台上的界面类似。

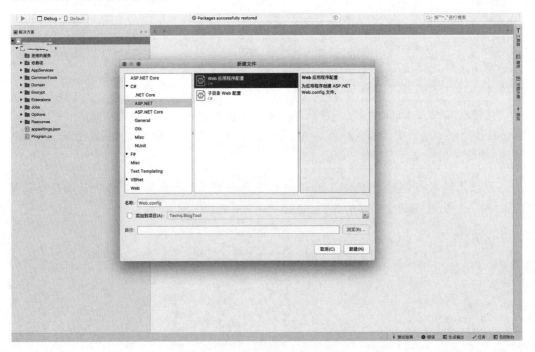

图 1-6　Visual Studio 在 mac OS 中的界面

除了 Visual Studio，还可以使用 Visual Studio Code 和 Rider。Visual Studio Code 是免费且提供全平台（Windows、macOS 和 Linux）支持的 IDE，支持语法高亮、代码自动补全、代码重构功能，并且内置了命令行工具和 Git 版本控制系统。用户不仅可以通过更改主题和键盘快捷方式实现个性化设置，还可以通过内置的应用商店安装扩展插件以拓展软件功能，如 Visual Studio Code 代码自动补全，如图 1-7 所示。

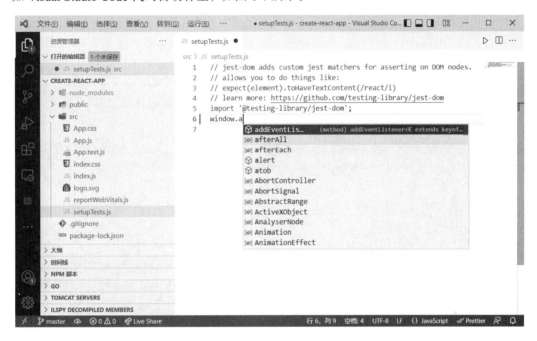

图 1-7　Visual Studio Code 代码自动补全

Rider 是 JetBrains 公司开发的专注于.NET 的 IDE，并且是跨平台的 IDE。开发人员可以在 Windows、macOS 和 Linux 平台上使用 Rider。Rider 提供了 30 天的试用权，并且支持学校、开源项目、社区组织等申请激活码，当然，也可以直接购买它的授权码。

除了上述 3 种开发工具，还有一种备受社区开发人员支持的开发工具——LINQPad。LINQPad 是面向.NET Framework 和.NET Core 开发的实用工具，用于 LINQ 交互式查询 SQL 数据库，以及交互式编写 C#代码，而无须 IDE。LINQPad 可以作为通用的"测试工作台"，利用该工作台，开发人员可以在 Visual Studio 外部快速为 C#代码创建原型。图 1-8 所示为利用 LINQ 在 LINQPad 中查询数据。

下面先介绍 Visual Studio 和 Visual Studio Code 的安装过程，读者可以根据自己的喜好选择感兴趣的工具自行安装。

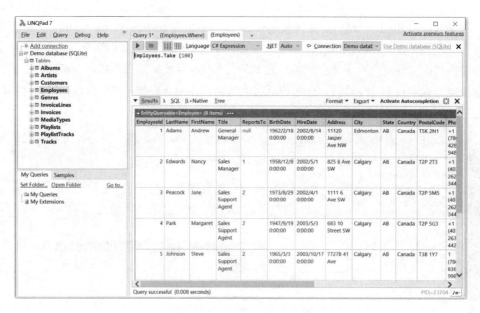

图 1-8　利用 LINQ 在 LINQPad 中查询数据

1.2.1　安装 Visual Studio

打开 Visual Studio 官网，单击 Visual Studio 板块的 "下载 Visual Studio" 下拉按钮，在弹出的下拉列表中任意选择一个版本即可下载 Visual Studio，如图 1-9 所示。

图 1-9　Visual Studio 板块

单击如图 1-9 所示的"下载 Visual Studio"下拉按钮，即可下载 Visual Studio 2022，而 Visual Studio 2022 包含以下几个版本。

- 社区版：完全免费的 IDE，适合学生和个人开发人员使用。
- 专业版：适合小型开发团队使用。
- 企业版：适合中型和大型企业使用。

关于上述 3 个版本的详细功能，读者可以通过 Visual Studio 官网的资料进行对比。另外，上述 3 个版本可同时安装在同一台计算机上，因为这 3 个版本是相互独立的。

下载完 Visual Studio 安装程序后，运行安装程序，如图 1-10 所示。在 Visual Studio 安装程序对话框中单击"工作负荷"选项卡，开发人员可以根据需要，按需安装 ASP.NET 和 Web 开发、桌面应用开发、游戏开发等环境。

图 1-10　Visual Studio 2022 安装程序的运行

例如，创建一个 C#应用程序，启动 Visual Studio，在 Visual Studio 2022 窗口中可以看到"开始使用"栏，如图 1-11 所示，单击"创建新项目"链接。

打开如图 1-12 所示的"创建新项目"窗口，选择的语言为 C#，选中"控制台应用"项目模板，单击"下一步"按钮。

图 1-11　Visual Studio 2022 窗口

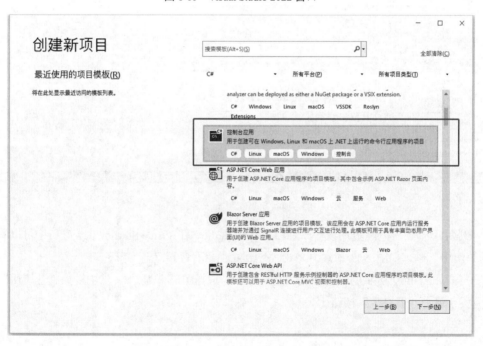

图 1-12　"创建新项目"窗口

打开如图 1-13 所示的"配置新项目"窗口,在"项目名称"文本框中输入 ConsoleApp1,单击"下一步"按钮。

图 1-13 "配置新项目"窗口

打开如图 1-14 所示的"其他信息"窗口,选择目标框架的版本作为项目的版本,单击"创建"按钮。

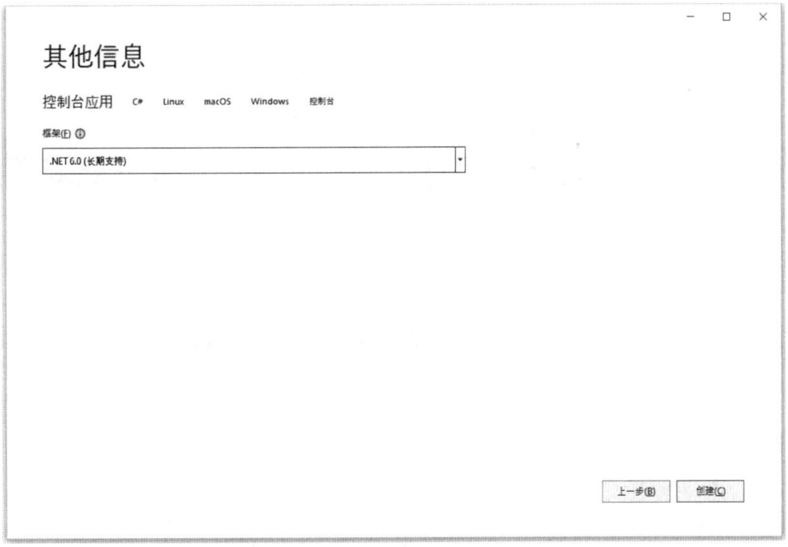

图 1-14 "其他信息"窗口

先在"解决方案资源管理器"面板中选择 ConsoleApp1 文件，再在代码编辑器中添加一个实例（整数运算），如图 1-15 所示。

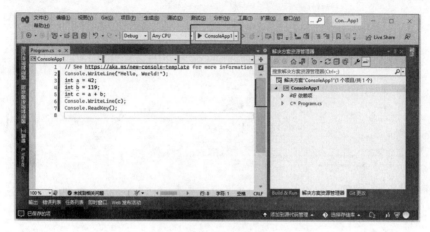

图 1-15　在代码编辑器中添加一个实例

当需要运行应用程序时，按 F5 键，随后开启一个控制台窗口，如图 1-16 所示，输出运行结果。

图 1-16　控制台窗口

当需要断点调试时，可以在指定行的左侧单击并插入断点，如图 1-17 所示。

插入断点后，可以进一步运行并调试应用程序，如图 1-18 所示，调试并查看变量。

图 1-17　插入断点　　　　　　　　　图 1-18　调试并查看变量

1.2.2　安装 Visual Studio Code

打开 Visual Studio Code 官网，如图 1-19 所示，可以看到，Visual Studio Code 支持 macOS、Windows 和 Linux，选择对应的版本下载即可。

第 1 章 .NET 概述和环境安装

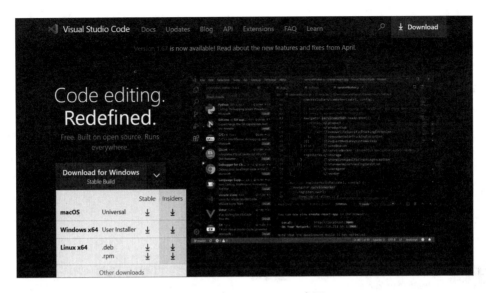

图 1-19 Visual Studio Code 官网

Visual Studio Code 安装包下载完成后，双击安装包可以直接安装，安装成功后还需要下载并安装.NET SDK。

打开.NET SDK 的下载页面，单击页面中的 Download .NET SDK x64 按钮，下载.NET SDK 安装包。.NET SDK 的下载页面如图 1-20 所示。

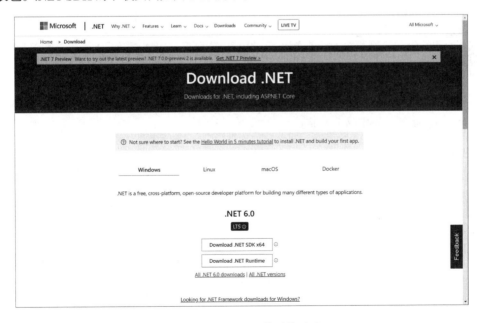

图 1-20 .NET SDK 的下载页面

13

下载完成后，运行安装包，将.NET SDK 安装到本地环境中，在如图 1-20 所示的页面中不仅可以下载.NET SDK 安装包，还可以下载.NET 运行时（.NET Runtime）安装包。

1.2.3 设置 Visual Studio Code 环境

在安装完.NET SDK 后，还需要安装开发.NET 应用程序所需的 C#扩展插件，在 Visual Studio Code 编辑器左侧的"扩展工具栏"中搜索 C#安装即可，如图 1-21 所示。

图 1-21　安装 C#扩展插件

在插件安装完成后，重启编辑器，并使用 dotnet CLI 工具在 Visual Studio Code 终端中创建一个控制台实例，如代码 1-2 所示。

代码1-2

```
mkdir dotnet
cd dotnet
dotnet new console
code .
```

在执行完上述命令后，不仅创建了一个控制台实例，还通过 code.命令打开了 Visual Studio Code 编辑器，并且会加载当前创建的目录，如图 1-22 所示。

第 1 章 .NET 概述和环境安装

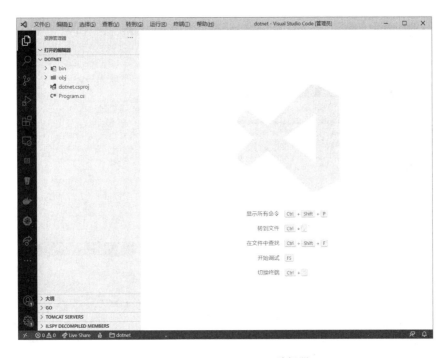

图 1-22　Visual Studio Code 编辑器

1.2.4　Visual Studio Code 调试

使用 Visual Studio Code 打开 .NET 项目后，右下角会出现一个通知提示，如图 1-23 所示，提示是否为该项目创建编译和调试文件。

图 1-23　通知提示

单击 Yes 按钮，可以看到项目文件中会多出一个名为 .vscode 的文件夹，如图 1-24 所示。.vscode 文件夹中包含两个文件：tasks.json 是任务配置文件，用于放置命令任务；launch.json 是调试器配置文件。

在这些条件必备的情况下，就可以在代码中进行打断点和调试。首先，打开 .cs 文件，在某行代码的左侧单击，此时会显示"小红点"，这说明已经成功添加了断点；然后，按 F5 键调试控制台实例，如图 1-25 所示。

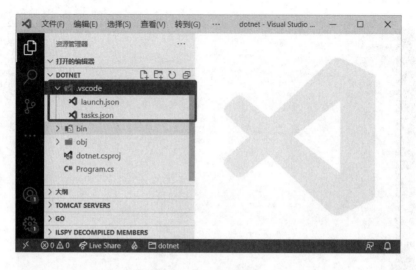

图 1-24　.vscode 文件夹

图 1-25　调试控制台实例

1.2.5　.NET CLI

事实上，CLI（Command-Line Interface，命令行界面）本身也可以作为一种开发手段，开发人员可以在不方便安装开发工具的情况下，使用 CLI 进行.NET 项目的创建、编译、运行等操作。

在安装.NET SDK 时，.NET CLI 也会随着工具链一并安装。开发人员可以通过在控制台窗口中输入 dotnet-version 命令来查看计算机安装的.NET 的版本。

可以通过 CLI 快速创建项目，如 WebAPI 等项目。在控制台窗口中输入 dotnet new list 命令后显示的相关界面提示如图 1-26 所示。

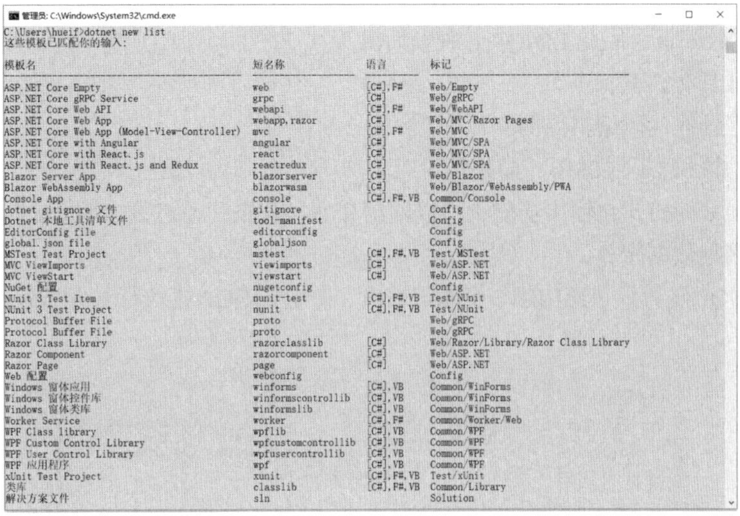

图 1-26　在控制台窗口中输入 dotnet new list 命令后显示的相关界面提示

通过 dotnet CLI 控制台窗口，可以看到丰富多样的 dotnet 项目模板，因此可以选择所需的项目模块开发项目。图 1-27 所示为.NET 常用项目模板列表。

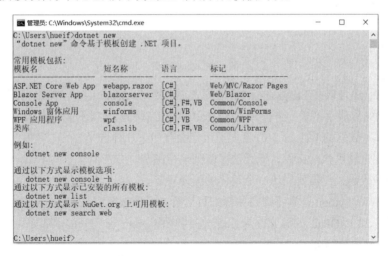

图 1-27　.NET 常用项目模板列表

当然，也可以通过执行 dotnet new --help 命令查阅更多的信息。

1.2.6　LINQPad

LINQPad 是一个功能强大且轻量级的工具。虽然 Visual Studio 可以满足大多数场景的要求，但是它比较笨重，因此 LINQPad 脱颖而出。

LINQPad 具有如下特点。
- 具有简约的代码编辑界面。
- 占用的体积不到 20MB，即超轻量级。
- 具有强大的格式化输出功能，可以输出文字、表格和动态数据。
- 支持多种数据库等。

使用 LINQPad 可以快速构建一个测试输出。使用 LINQPad 执行 C#代码并输出结果，如图 1-28 所示。

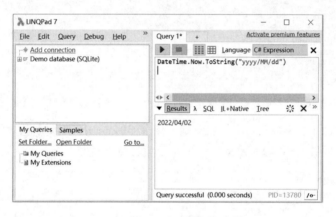

图 1-28　输出结果

除此之外，还可以在 Language 的下拉列表中选择 C# Statement(s)选项，并通过 Dump 扩展方法将实例的值展示出来，并使用 LINQPad 执行 C#代码，如图 1-29 所示。

在需要一些数据的 ETL 下，开发人员可能会有如下几种选择。
- 使用二进制可执行文件，但无法了解到运行细节。
- 使用源代码，但需要经历一个具体的编译环节。
- Node.js 或 Python 脚本不需要编译即可运行，但依赖要定义在 package.json 文件中。

综上所述，LINQPad 的源文件为.linq，它可以像 Node.js、Python 那样，不需要单独编译也可以了解到运行细节。

如代码 1-3 所示，创建一个名为 test 的.linq 文件。

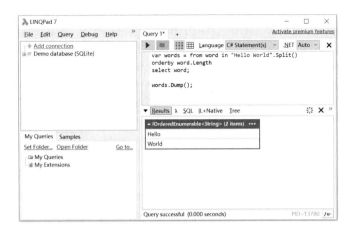

图 1-29　展示实例的值

代码1-3

```
<Query Kind="Statements">
  <NuGetReference>Newtonsoft.Json</NuGetReference>
  <Namespace>System.Net.Http</Namespace>
  <Namespace>Newtonsoft.Json.Linq</Namespace>
</Query>

using var http = new HttpClient();
string url = "https://******.com/manifest.json";
string json = await http.GetStringAsync(url);
JToken.Parse(json)["icons"].Select(x => (string)x["src"]).Dump();
```

在 LINQPad 工具下，有一个名为 lprun7 的命令行工具，可以通过该工具运行。图 1-30 所示为通过 lprun7 命令行工具执行.linq 文件。

图 1-30　通过 lprun7 命令行工具执行.linq 文件

1.3 小结

通过学习本章的内容，读者可以大致了解.NET 的发展、.NET 的基本结构和.NET 开发可使用的 IDE。通过这些工具，读者将开启新世界的大门，真正走向专业.NET 开发人员的成长之路。

第 2 章

.NET 运行原理概述

C#是一门现代化的编程语言,开发人员不需要清楚计算机的每个细节就可以进行高效的编码。C#编译器会将 C#代码编译成中间语言,同时在.NET 运行时通过 JIT 编译器将中间语言编译成机器可以理解的机器语言。这套机制涉及众多的技术点和名词,本章将对部分技术点展开介绍。

2.1 .NET CLI 概述

.NET CLI 是一个跨平台工具,用于创建、构建、运行和发布.NET 应用程序。.NET CLI 不需要开发人员主动安装,在安装.NET SDK 时会默认安装.NET CLI。因此,不需要在计算机中特意安装.NET CLI。

为了验证是否安装了.NET CLI,可在 Windows 中打开命令提示符窗口输入 dotnet 命令,如果是 Linux,则可在终端 Bash 窗口中直接输入 dotnet 命令,按 Enter 键执行该命令。执行 dotnet 命令的过程如图 2-1 所示,图中展示了 Windows 命令提示符窗口中的输出结果。

.NET CLI 提供了一些命令,将.NET 应用程序开发的常规操作进行集成管理。熟悉这些命令,有助于开发人员在一些非 IDE 的编辑器工具中进行.NET 应用程序的开发、构建、测试和发布。

.NET CLI 最方便的应用场景可能是在 DevOps 流程中进行持续集成(Continuous Integration,CI),开发人员可以将命令灵活地运用到持续集成的工具链中,从而对.NET 应用程序进行生成和构建操作。

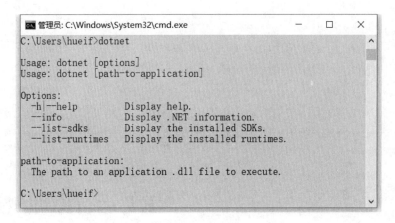

图 2-1　执行 dotnet 命令的过程

dotnet 命令说明如表 2-1 所示，表中列举的是一些基本命令。

表 2-1　dotnet 命令说明

命令	描述/说明
dotnet add	将包或引用添加到 .NET 项目中
dotnet build	构建 .NET 项目，并编译为 IL 二进制文件
dotnet clean	清理 .NET 项目的构建（build）输出
dotnet help	显示命令行帮助
dotnet list	列出 .NET 项目的项目引用
dotnet msbuild	运行 Microsoft 生成引擎（MSBuild）命令
dotnet new	创建新的 .NET 项目或文件
dotnet pack	创建 NuGet 包
dotnet publish	发布 .NET 项目，用于部署
dotnet restore	还原 .NET 项目中指定的依赖项
dotnet run	构建并运行 .NET 项目
dotnet sln	修改 Visual Studio 解决方案文件
dotnet store	将指定的程序集存储在运行时包存储中
dotnet test	指定单元测试项目，运行单元测试
dotnet vstest	用于指定程序集，运行单元测试

下面使用 CLI 创建一个控制台应用程序，以加深读者对 .NET CLI 的理解。

如代码 2-1 所示，创建一个控制台应用程序。

代码2-1

```
C:\App>dotnet new console
```

生成的控制台应用程序的目录如图 2-2 所示。

图 2-2　生成的控制台应用程序的目录

使用 dotnet build 命令可以构建.NET 应用程序，如代码 2-2 所示。

代码2-2

```
C:\App >dotnet build
```

运行.NET 应用程序，如代码 2-3 所示。

代码2-3

```
C:\App>dotnet run
Hello World!
```

也可以使用 dotnet 命令指定.NET 应用程序的.dll 文件，运行控制台应用程序，输出结果如图 2-3 所示。

图 2-3　输出结果

通过上述内容，读者可以基本上了解.NET CLI 的使用。有的读者可能想深入了解.NET CLI 内部的工作机制，在命令背后.NET CLI 执行了什么操作，以及.NET CLI 是如何演化的。

代码 2-4 展示了 RestoreCommand 类的实现，开发人员可以通过对 CLI 的源代码进行探索来查看其运行机制（关于 CLI 的源代码可以从 GitHub 官网的 dotnet/sdk 仓库中查看）。

代码2-4

```
public class RestoreCommand : MSBuildForwardingApp
{
    //...
    public static RestoreCommand FromArgs(string[] args,
                        string msbuildPath = null,
                        bool noLogo = true)
    {
        var parser = Parser.Instance;
        var result = parser.ParseFrom("dotnet restore", args);
        result.ShowHelpOrErrorIfAppropriate();
        var parsedRestore = result["dotnet"]["restore"];
        var msbuildArgs = new List<string>();
        if(noLogo)
        {
            msbuildArgs.Add("-nologo");
        }
        msbuildArgs.Add("-target:Restore");
        msbuildArgs.AddRange(parsedRestore.OptionValuesToBeForwarded());
        msbuildArgs.AddRange(parsedRestore.Arguments);
        return new RestoreCommand(msbuildArgs, msbuildPath);
    }

    public static int Run(string[] args)
    {
        return FromArgs(args).Execute();
    }
}
```

上述代码来源于 .NET CLI 工具，用于实现 dotnet restore 命令。在 dotnet restore 命令的背后，该命令及参数会生成对应的 MSBuild 命令，如代码 2-5 所示。

代码2-5

```
msbuild -target:Restore
```

代码 2-6 展示了其他命令，它们与 dotnet build 命令等效。

代码2-6

```
dotnet build
dotnet msbuild /t:Build
msbuild /t:Build
```

关于 MSBuild，本章不展开介绍，感兴趣的读者可以查阅 MSBuild 的相关资料进行了解。

通过学习前面的内容，读者基本了解了.NET CLI。读者也可以自行创建一个.NET CLI 工具。

为了帮助读者更深入地理解.NET CLI，笔者通过一个简单的实例来演示创建一个控制台。该实例是一个日期提供程序，如代码 2-7 所示，使用 Windows 命令创建一个名为 dotnet-date-tool 的文件夹。

代码2-7

```
mkdir dotnet-date-tool
cd dotnet-date-tool
```

在文件夹 dotnet-date-tool 创建完成后，使用 cd 命令切换到指定的目录下。

在目录切换完成后，输入 dotnet new console 命令，创建控制台实例，如代码 2-8 所示。

代码2-8

```
C:\dotnet-date-tool>dotnet new console
The template "Console Application" was created successfully.

Processing post-creation actions...
Running 'dotnet restore' on C:\dotnet-date-tool\dotnet-date-tool.csproj...
  正在确定要还原的项目…
  已还原 C:\dotnet-date-tool\dotnet-date-tool.csproj (用时 71 ms)。
Restore succeeded.


C:\dotnet-date-tool>
```

在控制台实例创建完成后，可以通过 Visual Studio 或 Visual Studio Code 等编辑器打开项目文件。

图 2-4 所示为 System.CommandLine 包的安装过程。System.CommandLine 是一个命令行解析器，提供了规范化的 API，使开发人员可以快速创建命令行工具。

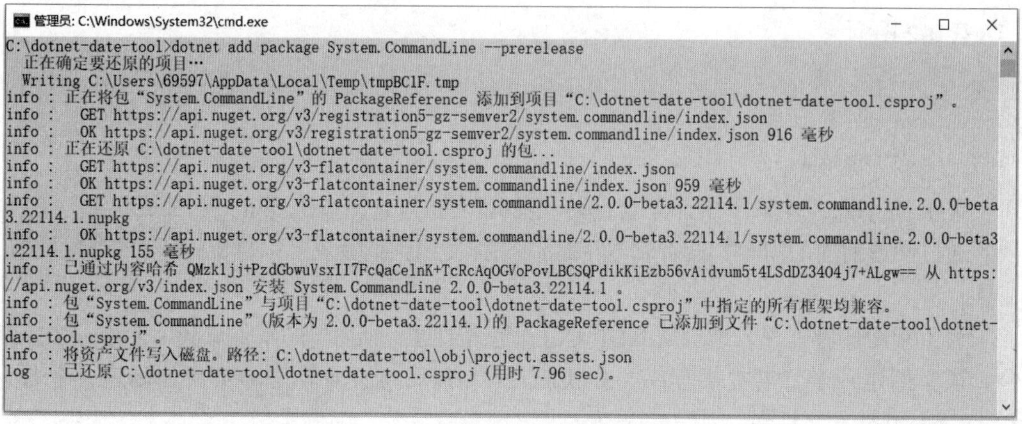

图 2-4 System.CommandLine 包的安装过程

在如图 2-4 所示的安装过程中，--prerelease 参数表示 System.CommandLine 包目前正处于预发布状态，但是作为 dotnet 背后的命令行引擎，它是安全的。

在 Program 类中编写 dotnet-date-tool 命令行工具，主要用于输出具有格式化的日期字符串。如代码 2-9 所示，将 System.CommandLine 包作为命令行引擎，创建 RootCommand 对象，RootCommand 对象表示程序本身的命令（如 dotnet、docker），创建 HandleCmd 方法用于处理命令行参数。

代码2-9

```
class Program
{
    static async Task<int> Main(string[] args)
    {
        var cmd = new RootCommand
        {
            new Option<string>("--name", "请输入执行者名称"),
            new Option<string>("--format", "获取指定格式的时间字符串")
            {
                IsRequired = true
            }
        };
        cmd.Name = "dotnet-date-tool";
        cmd.Description = "日期获取工具";
        cmd.SetHandler<string, string, IConsole>(HandleCmd,
                            cmd.Options[0],cmd.Options[1]);
```

```
        return await cmd.InvokeAsync(args);
    }
    static void HandleCmd(string name, string format, IConsole console)
    {
        if(!string.IsNullOrWhiteSpace(format))
        {
            var date = DateTime.Now.ToString(format);
            if (!string.IsNullOrWhiteSpace(name))
            {
                console.Out.WriteLine(
                            $"你好，{name}，日期根据指定格式转换后为{date}");
                return;
            }
            console.Out.WriteLine(date);
        }
    }
}
```

在 RootCommand 对象中创建两个 Option 对象，dotnet-date-tool 命令可以接收一个或多个 Option 对象，Option 对象表示命令的参数，分别创建参数--name 和--format。在 Option 对象的构造函数中可以添加命令行参数的描述信息，在--format 参数中将它对应的 Option 对象的 IsRequired 属性设置为 true，表示该参数为必须项。也可以通过 RootCommand 对象设置该命令行程序的基础信息，如调用 RootCommand 对象的 Name 属性，添加命令行工具的名称，调用 Description 属性，添加命令行程序的描述信息等。

基础信息的配置编写完后，先调用 SetHandler 扩展方法添加命令处理方法，再调用 RootCommand 对象的 InvokeAsync 扩展方法对参数进行解析。

在 HandleCmd 方法中，接收参数--name 和--format 的信息，并判断--format 参数的值是否为空，在--format 参数的值不为空的情况下判断--name 参数的值，如果不为空则输出包含 name 名称的字符串，否则只输出格式化日期。图 2-5 所示为执行 dotnet-date-tool 命令输出的帮助信息。

执行 dotnet-date-tool 命令后，提示 Option '--format' is required.，这证明 Option 对象的 IsRequired 属性的设置已生效，--format 参数为必填项。在控制台窗口输出的信息中可以看到对该命令及其参数的描述。接下来通过命令的参数执行，如图 2-6 所示为 dotnet- date-tool 参数的验证信息，通过多次输入，对 HandleCmd 方法内部的判断逻辑进行验证。

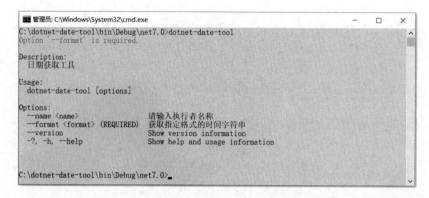

图 2-5　执行 dotnet-date-tool 命令输出的帮助信息

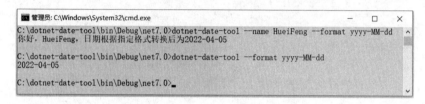

图 2-6　dotnet-date-tool 参数的验证信息

2.1.1　将 C#编译成机器代码

代码编译是程序语言设计的基石，IDE 是开发人员日常开发过程中最重要的生产力工具，这些开发工具是如何将代码编译成目标语言和可执行文件的？

Visual Studio 集成了 C#编译器，用于将 C#代码转换为机器语言（CPU 可以理解的语言），并以.dll 和.exe 的形式返回输出文件。将 C#代码转换为机器语言的过程可以划分为两个阶段，分别是编译时过程和运行时过程。

在计算机编程的初期阶段，开发人员用机器代码编写应用程序，它看起来与代码 2-10 中的代码类似，这些实际上是一条条计算机指令。开发人员通过一条条指令直接与硬件打交道，同样可以实现许多复杂的功能。

代码2-10

```
0000002e
0000002f
00000031
00000032
```

机器语言的执行效率通常比较高，但可读性很差。众所周知，在计算机科学中有一句名言，

"计算机科学中没有什么是不能通过增加一层抽象来解决的"，这就促成了高级语言的诞生。

高级语言是对机器语言进行抽象，在编写过程中允许开发人员无须像机器语言那样输入复杂的指令，也不需要花费大量的时间编写诸如内存管理、硬件的兼容性等与实际业务无关的底层代码。

注意：开发人员不需要清楚计算机的每个细节就可以进行高效的编码，这意味着有一套机制用来实现高级语言到机器语言的转换，同样，C#也必须转换为处理器可以理解的语言，因为处理器不知道 C#，只知道机器语言。

C#编译器在编译时会将代码作为输入，并以中间语言（Intermediate Language，IL）的形式输出，该代码保存在*.exe 文件或*.dll 文件中。将 C#代码编译为 IL 代码的过程如图 2-7 所示。

图 2-7　将 C#代码编译为 IL 代码的过程

如图 2-7 所示，这并不是一个完整的编译流程，没有生成处理器能够处理的指令，也就是缺少机器语言的生成过程，因此，需要一个过程将 IL 代码转换为机器代码。而处理这个过程的正是公共语言运行时，即 CLR。

CLR 在计算机上运行，可以管理 IL 代码的执行。简单来说，它知道如何执行通过 IL 代码编写的应用程序，并且使用 JIT 编译器将 IL 代码转换为机器代码，有时候也被称为本机代码（Machine Code）。代码的执行过程如图 2-8 所示。

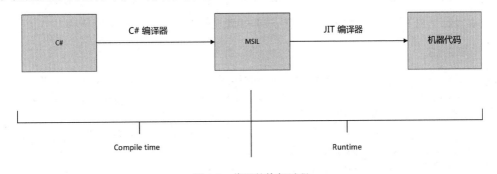

图 2-8　代码的执行过程

由上述内容可知，编写 C# 代码既不需要关注硬件设备的组成，也不需要考虑兼容性，因为 CLR 和 JIT 将负责这个过程，它们可以将 IL 代码编译为计算机使用的代码。

上面介绍的是 C# 编译器，下面介绍 Roslyn 编译器。Roslyn 是 C# 和 Visual Basic.NET 的开源编译器，是一个完全使用 C# 托管代码开发的编译器和分析器。如代码 2-11 所示，创建 C# 代码，通过 C# Roslyn 编译器进行编译。

代码2-11

```
using System;

namespace Program
{
    class Program
    {
        static void Main(string[] args)
        {
            Console.WriteLine(
                $"{ (System.Runtime.InteropServices
                        .RuntimeInformation.FrameworkDescription)}");
            Console.ReadLine();
        }
    }
}
```

对源文件进行编译后，可以打开 .NET SDK 中的 C# Roslyn 编译器。C# Roslyn 编译器可以直接通过 dotnet 命令进行调用，如代码 2-12 所示。

代码2-12

```
dotnet "C:\Program Files\dotnet\sdk\6.0.201\Roslyn\bincore\csc.dll" -help
```

如代码 2-13 所示，调用 csc.dll 文件，将 C#源文件编译为 Program.dll 文件。

代码2-13

```
dotnet "C:\Program Files\dotnet\sdk\6.0.201\Roslyn\bincore\csc.dll"
 -reference:"C:\Program     Files\dotnet\shared\Microsoft.NETCore.App\6.0.3\
System.Private.CoreLib.dll"
-reference:"C:\Program     Files\dotnet\shared\Microsoft.NETCore.App\6.0.3\
System.Console.dll"  -reference:"C:\Program    Files\dotnet\shared\Microsoft.
NETCore.App\6.0.3\System.Runtime.dll"
-reference:"C:\Program     Files\dotnet\shared\Microsoft.NETCore.App\6.0.3\
System.Runtime.InteropServices.RuntimeInformation.dll" *.cs -out:Program.dll
```

需要注意的是，-reference 参数用于指定外部程序集。*.cs 表示编译当前目录下的所有 .cs 文件，当然也可以编译指定的单个文件。-out 参数用于指定输出的文件名，这不是一个必选参数，若不指定该参数，则默认生成 Program.exe 文件。

图 2-9 所示为执行 csc 命令编译后的 Program.dll 文件，返回了错误信息。

```
C:\app>dotnet Program.dll
Cannot use file stream for [C:\app\Program.deps.json]: No such file or directory
A fatal error was encountered. The library 'hostpolicy.dll' required to execute the application was not found in
'C:\app\'.
Failed to run as a self-contained app.
  - The application was run as a self-contained app because 'C:\app\Program.runtimeconfig.json' was not found.
  - If this should be a framework-dependent app, add the 'C:\app\Program.runtimeconfig.json' file and specify th
e appropriate framework.

C:\app>_
```

图 2-9　执行 csc 编译后的 Program.dll 文件

如图 2-9 所示，需要创建 Program.runtimeconfig.json 文件，用于指定运行时信息。如代码 2-14 所示，定义运行时信息。

代码2-14

```json
{
  "runtimeOptions": {
    "tfm": "net6.0",
    "framework": {
      "name": "Microsoft.NETCore.App",
      "version": "6.0.3"
    }
  }
}
```

使用 dotnet 命令调用 Program.dll 文件，输出结果如图 2-10 所示。

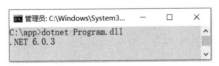

图 2-10　输出结果

代码从编写到执行的流程如图 2-11 所示。

学习一门语言第一个经典的例子就是编写 Hello World 应用程序。下面介绍 Hello World 应用程序从 C# 代码到机器代码的编译过程。

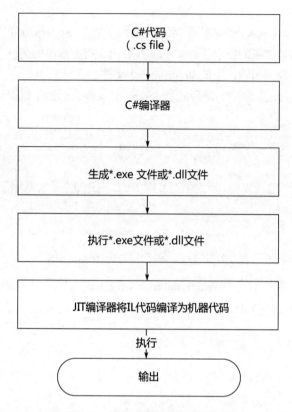

图 2-11　代码从编写到执行的流程

C#代码

```
class Program
{
    static void Main()
    {
        Console.WriteLine("Hello World!");
    }
}
```

IL代码

```
.method private hidebysig static void
```

```
Main() cil managed
{
  .entrypoint
  .maxstack 8

  IL_0000: ldstr      "Hello World!"
  IL_0005: call  void [System.Console]System.Console:: WriteLine (string)

  IL_000a: ret

} //end of method Program::Main
```

机器代码

```
00007FF7A2BF8220   push    rbp
00007FF7A2BF8221   push    rdi
00007FF7A2BF8222   push    rsi
00007FF7A2BF8223   sub     rsp,20h
00007FF7A2BF8227   mov     rbp,rsp
00007FF7A2BF822A   cmp     dword ptr [00007FF7A2A00E80h],0
00007FF7A2BF8231   je      00007FF7A2BF8238
00007FF7A2BF8233   call    00007FF8025BC9C0
00007FF7A2BF8238   nop
00007FF7A2BF8239   mov     rcx,2D41A855B90h
00007FF7A2BF8243   mov     rcx,qword ptr [rcx]
00007FF7A2BF8246   call    00007FF7A2BF8110
00007FF7A2BF824B   nop
00007FF7A2BF824C   nop
00007FF7A2BF824D   lea     rsp,[rbp+20h]
00007FF7A2BF8251   pop     rsi
00007FF7A2BF8252   pop     rdi
```

```
00007FF7A2BF8253    pop         rbp
00007FF7A2BF8254    ret
```

2.1.2 运行时

.NET 提供了 CLR，运行时会将中间语言代码转换为当前 CPU 平台支持的机器代码并执行这些机器代码。在 CLR 下托管的代码称为托管代码（Managed Code），运行时是托管代码的执行环境。

CLR 逐渐成为运行时的代名词，而在技术上更准确的虚拟执行系统（VES）则很少在 CLI 规范之外的地方提到。事实上，CLR 提供了类型安全、内存安全、异常处理、垃圾回收、多线程等机制，这些机制为 .NET 应用程序提供了一个更安全、更高效的运行环境。

2.1.3 程序集和清单

在众多计算机语言中，若按编译特点划分，可以分为编译型语言、解释型语言和混合型语言。

编译型语言是需要通过编译器将源代码编译为机器代码才能执行的高级语言，如 C、C++ 等。解释型语言则不需要预先编译，在执行时逐行编译。混合型语言也需要编译，但不直接编译成机器代码，而是编译成中间语言，通过运行时执行中间语言，将中间语言解释为机器代码来执行。

C# 可视为混合型语言，这意味着当开发人员创建源文件时，这些文件需要先编译才能运行。C# 不像 JavaScript 和 PHP 这种动态类型语言（或脚本语言）一样能直接运行。程序集的产生正是为了解决该问题，程序集内会保存相关的中间语言和其他资源文件（如 TXT、JPG、Excel、XML 等，若是作为嵌入程序集的资源，则作为嵌入文件保存到程序集中，本章不对其展开介绍）。

程序集是由一个或多个源代码文件生成的输出文件。程序集是 .NET 应用程序在资源管理器中基本的文件单元，具有 .exe 扩展名和 .dll 扩展名两种类型，扩展名为 .exe 的程序集是可执行文件（Executable File），扩展名为 .dll 的程序集是动态链接库（Dynamic-Link Library）。

程序集清单从本质上来说是程序集的一个标头（Header），提供程序集运行所需的描述信息和程序集唯一性标识信息，包含程序集的版本信息、范围信息，以及与程序集相关的其他信息，清单内容如图 2-12 所示。

.NET 程序集中包含描述程序集自身的元数据（清单，Manifest），而清单内容页则包含所需要的外部程序集、程序集版本号、模块名称等其他信息。

```
// Metadata version: v4.0.30319
.assembly extern System.Private.CoreLib
{
  .publickeytoken = (7C EC 85 D7 BE A7 79 8E )                    // |.....y.
  .ver 5:0:0:0
}
.assembly extern System.Runtime.InteropServices.RuntimeInformation
{
  .publickeytoken = (B0 3F 5F 7F 11 D5 0A 3A )                    // .?_....:
  .ver 5:0:0:0
}
.assembly extern System.Console
{
  .publickeytoken = (B0 3F 5F 7F 11 D5 0A 3A )                    // .?_....:
  .ver 5:0:0:0
}
.assembly Program
{
  .custom instance void [System.Private.CoreLib]System.Runtime.CompilerServices.CompilationRelaxationsAttribute::.
  .custom instance void [System.Private.CoreLib]System.Runtime.CompilerServices.RuntimeCompatibilityAttribute::.ct

  // --- 下列自定义特性会自动添加,不要取消注释 ------
  // .custom instance void [System.Private.CoreLib]System.Diagnostics.DebuggableAttribute::.ctor(valuetype [Syste

  .hash algorithm 0x00008004
  .ver 0:0:0:0
}
.module Program.dll
// MVID: {ECFC7F20-744A-49CC-A074-CD5B0A518825}
.imagebase 0x00400000
.file alignment 0x00000200
.stackreserve 0x00100000
.subsystem 0x0003       // WINDOWS_CUI
.corflags 0x00000001    // ILONLY
// Image base: 0x09C00000
```

图 2-12 清单内容

程序集中可以包含多个不同类型的资源文件。单个文件和多个文件的存储方式如图 2-13 所示,程序集中包括.jpg 文件和.bmp 文件,或者其他格式的文件。

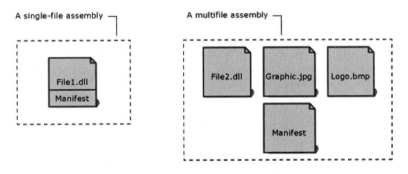

图 2-13 单个文件和多个文件的存储方式

多文件程序集不是将多个文件物理链接起来,而是通过程序集清单进行链接,CLR 将它们作为一个单元进行管理。

2.1.4 公共中间语言

公共中间语言（Common Intermediate Language，CIL），简称中间语言，包括 Microsoft 中间语言（MSIL）和中间语言（IL），由这个名字可以得知 CIL 的一个重要特点，即它支持多种语言在同一个应用程序中进行交互。CIL 不只是 C#的中间语言，还是其他许多.NET 大家族编程语言的中间语言，如 Visual Basic、F#等。

2.1.5 .NET Native

.NET Native 是一项预编译技术，用于创建平台特定的可执行文件。通常，.NET 应用程序会编译为中间语言，在运行期间利用 JIT 编译器将 IL 代码翻译为机器代码。相比之下，.NET Native 则将应用程序直接编译为机器代码，这意味着用这种方式编译的应用程序具有机器代码的性能。

通常，.NET 应用程序首先编译为 IL 代码，然后由 JIT 编译器编译为 Native 代码。利用.NET Native 可以不需要.NET 运行时，也不需要 JIT 编译器，而是直接运行机器代码。

2.2 小结

通过学习本章，读者可以初步了解.NET 涉及的一些技术概念。在.NET 平台中，中间语言具有关键作用。中间语言也是一种面向对象的编程语言，会在应用程序运行时被编译成机器代码。

第 3 章

ASP.NET Core 应用程序的多种运行模式

ASP.NET Core 应用程序可以在多种模式下运行,包括自宿主、IIS 服务承载、桌面应用程序、服务承载,开发人员可以根据实际需求选择对应的运行模式。例如,在 Windows 环境下部署,打算通过 Kestrel(Kestrel 是 ASP.NET Core 中内置的高性能服务器)服务器直接承载外部请求,同时想"开机自启",在这种情况下可以选择注册 Windows 服务。ASP.NET Core 应用程序的运行模式具有多样化特点,因此选择合适的模式很重要。

3.1 自宿主

通常,开发人员会使用自宿主的运行模式,这种模式实际上支持以控制台的方式运行 ASP.NET Core 应用程序。

以 Web API 实例为例,在 Windows 中打开命令提示符窗口并输入 dotnet new webapi 命令,随后按 Enter 键,创建一个 Web API 应用程序。如图 3-1 所示,通过 dotnet 命令创建 Web API 应用程序。

应用程序创建完成后,通过 dotnet 命令启动项目,如图 3-2 所示。

运行后,应用程序会直接对端口进行监听,由此可以访问应用程序。除此之外,还可以通过--urls 参数指定 URL 地址进行监听,如图 3-3 所示。

图 3-1　通过 dotnet 命令创建 Web API 应用程序　　　图 3-2　通过 dotnet 命令启动项目

图 3-3　通过 --urls 参数指定 URL 地址进行监听

3.2　IIS 服务承载

执行 dotnet publish 命令，代码会随之发布，打开 IIS 管理器，右击"网站"选项，在弹出的快捷菜单中选择"添加网站"命令，在打开的对话框的"网站名称"文本框中输入 webapi，"物理路径"需要设置为当前应用程序所在的目录，将"端口"设置为 8081，如图 3-4 所示，通过 IIS 管理器添加网站。

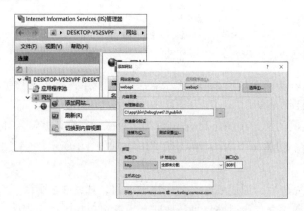

图 3-4　通过 IIS 管理器添加网站

如图 3-5 所示，先单击 IIS 管理器左侧窗口中的"应用程序池"，再双击 webapi 选项，打开"编辑应用程序池"对话框，将".NET CLR 版本"设置为"无托管代码"。

随后，使用 curl 命令访问网站，如图 3-6 所示。

图 3-5　编辑应用程序池　　　　图 3-6　使用 curl 命令访问网站

如图 3-6 所示，在输出结果中可以看到 Server 的值为 Microsoft-IIS/10.0，这充分说明当前应用程序的 Web 服务器为 IIS。

3.3　将 WebAPI 嵌入桌面应用程序中

在生产环境中，通常采用自宿主模式或 IIS 服务承载模式，但有时开发桌面应用程序需要提供一种便捷的方式来实现其他应用与其进行主机间的通信，那么将 WebAPI 放置于 WinForm 应用程序中进行承载或许是最快捷的方式。

如代码 3-1 所示，创建一个 WinForm 应用程序，首先调用 WebApplication 对象的 CreateBuilder 扩展方法，获取 WebApplicationBuilder 对象，然后调用 WebApplicationBuilder 对象的 Build 方法，创建 WebApplication 对象实例，通过调用 MapGet 扩展方法来处理 Get 请求，最后调用 RunAsync 扩展方法启动该服务。

代码3-1

```
internal static class Program
{
    [STAThread]
    static void Main(string[] args)
    {
        var builder = WebApplication.CreateBuilder(args);
        var app = builder.Build();
        app.MapGet("/", () => "Hello World!");
        app.RunAsync();

        var urls = app.Urls;
```

```
        ApplicationConfiguration.Initialize();
        Application.Run(new Form1(string.Join(",", urls)));
    }
}
```

通过调用 WebApplication 对象的 Urls 属性，获取当前 Web 服务开启的 URL 地址集合，并通过 Form1 窗体进行展示。

创建 Web 服务需要在.csproj 文件中引入名为 Microsoft.AspNetCore.App 的 SDK，如代码 3-2 所示。值得注意的是，从.NET Core 3.1 开始，该依赖包已经被内置在.NET 运行时中，无须通过安装 NuGet 包的方式来引入，如果通过 NuGet 管理器进行搜索，则只能找到低版本（.NET Core 2.2）的 Microsoft.AspNetCore 组件，可能无法使用最新版.NET 的特性。

代码3-2

```
<ItemGroup>
    <FrameworkReference Include="Microsoft.AspNetCore.App" />
</ItemGroup>
```

在 Form1 窗体中定义一个 label 控件，并且将该控件命名为 txt_urls，用于展示当前 Web 服务的 URL 地址，如代码 3-3 所示，通过 txt_urls 设置 URL 地址。当前服务的 URL 地址如图 3-7 所示。

代码3-3

```
public partial class Form1 : Form
{
    public Form1(string urls)
    {
        InitializeComponent();
        this.txt_urls.Text = urls;
    }
}
```

运行应用程序后，Web 服务的 URL 地址会显示在应用程序界面中，如图 3-8 所示，通过 curl 命令行工具请求 Web 服务。

图 3-7 当前服务的 URL 地址

图 3-8 通过 curl 命令行工具请求 Web 服务

3.4 服务承载

自.NET Core 2.0 起，新加了一个名为 IHostedService 的接口，使用该接口有助于开发人员在日常开发中轻松地实现托管服务，实现多个 IHostedService 接口注册托管服务，托管服务可以非常轻松地将其注册到 Windows 服务或 Linux 服务中进行守护。

3.4.1 使用 Worker Service 项目模板

Worker Service 是一个.NET 项目模板，可以通过 Visual Studio 或.NET CLI 工具创建，允许开发人员创建基于操作系统级别的应用程序服务，这些服务既可以部署为 Windows 服务，也可以部署为 Linux Systemd 服务。

先通过 Visual Studio 工具栏选择"文件"→"新建"→"项目"选项，再在搜索栏中搜索 Worker Service 模板，这样就可以通过该模板创建项目，如图 3-9 所示。

图 3-9 搜索 Worker Service 模板

项目创建完成后，如代码 3-4 所示，在 Program 类中，首先通过 Host 对象调用 CreateDefaultBuilder 方法，然后调用 ConfigureServices 扩展方法，通过 services 实例对象注册托管服务，调用 Build 方法创建 WebApplication 对象实例，最后调用 RunAsync 扩展方法启动该服务。

代码3-4

```
IHost host = Host.CreateDefaultBuilder(args)
    .ConfigureServices(services =>
    {
        services.AddHostedService<Worker>();
    })
    .Build();

await host.RunAsync();
```

在模板中定义 Worker 类,该类在代码 3-4 中通过 AddHostedService 扩展方法注册,如代码 3-5 所示,Worker 类继承了 BackgroundService 类,实现了 ExecuteAsync 方法,在该方法中持续循环并输出日志。

代码3-5

```
public class Worker : BackgroundService
{
    private readonly ILogger<Worker> _logger;

    public Worker(ILogger<Worker> logger)
    {
        _logger = logger;
    }

    protected override async Task ExecuteAsync(CancellationToken stoppingToken)
    {
        while (!stoppingToken.IsCancellationRequested)
        {
            _logger.LogInformation(
                    "Worker running at: {time}", DateTimeOffset.Now);
            await Task.Delay(1000, stoppingToken);
        }
    }
}
```

3.4.2 Windows 服务注册

目前,可以轻松地将应用程序以 Windows 服务的方式进行部署,但是,为了让 Worker Service 能够以 Windows 服务的方式运行,还需要引入相关的 NuGet 包。Windows 服务的 NuGet

包为 Microsoft.Extensions.Hosting.WindowsServices，在创建的 Worker Service 项目实例中，将 UseWindowsService 扩展方法添加到 CreateDefaultBuilder 方法中，如代码 3-6 所示。

代码3-6

```
IHost host = Host.CreateDefaultBuilder(args)
    .UseWindowsService()
    .ConfigureServices(services =>
    {
        services.AddHostedService<Worker>();
    })
    .Build();

await host.RunAsync();
```

如代码 3-7 所示，演示 UseWindowsService 扩展方法的内部实现，在 Windows 环境下部署，主要执行以下几个操作：设置 Lifetime 为 WindowsServiceLifetime，设置 ContentRoot 为 BaseDirectory，并且开启 EventLog 事件日志。

代码3-7

```
public static class WindowsServiceLifetimeHostBuilderExtensions
{
    public static IHostBuilder UseWindowsService(
                                    this IHostBuilder hostBuilder)
    {
        return UseWindowsService(hostBuilder, _ => { });
    }
    public static IHostBuilder UseWindowsService(
                                    this IHostBuilder hostBuilder,
            Action<WindowsServiceLifetimeOptions> configure)
    {
        if (WindowsServiceHelpers.IsWindowsService())
        {
            hostBuilder.UseContentRoot(AppContext.BaseDirectory);
            hostBuilder.ConfigureLogging((hostingContext, logging) =>
            {
                Debug.Assert(RuntimeInformation.IsOSPlatform(
                                        OSPlatform.Windows));
                logging.AddEventLog();
```

```
            })
         .ConfigureServices((hostContext, services) =>
         {

            Debug.Assert(
               RuntimeInformation.IsOSPlatform(OSPlatform.Windows));
            services.AddSingleton<IHostLifetime,
                           WindowsServiceLifetime>();
            services.Configure<EventLogSettings>(settings =>
            {
               Debug.Assert(
                  RuntimeInformation.IsOSPlatform(OSPlatform. Windows));
               if (string.IsNullOrEmpty(settings.SourceName))
               {
                  settings.SourceName =
                     hostContext.HostingEnvironment.ApplicationName;
               }
            });
            services.Configure(configure);
         });
   }

   return hostBuilder;
}
```

如代码 3-8 所示，WindowsServiceLifetime 类继承了 IHostLifetime 接口，使其在 Windows 服务中进行生命周期的管理，首先在 WaitForStartAsync 方法中进行启动，并注册应用程序等，然后实例化一个线程，为 Run 方法开辟一个新线程，并将其设置为后台线程。另外，在 OnStop 方法和 OnShutdown 方法中调用了 ApplicationLifetime.StopApplication 方法，WindowsServiceLifetime 的基类为 ServiceBase，当服务停止时会调用 OnStop 方法和 OnShutdown 方法。

代码3-8

```
[SupportedOSPlatform("windows")]
public class WindowsServiceLifetime : ServiceBase, IHostLifetime
{
    private readonly TaskCompletionSource<object> delayStart =
        new TaskCompletionSource<object>
            (TaskCreationOptions.RunContinuationsAsynchronously);
```

```csharp
private readonly ManualResetEventSlim _delayStop =
                                    new ManualResetEventSlim();
private readonly HostOptions _hostOptions;
private IHostApplicationLifetime ApplicationLifetime { get; }
private IHostEnvironment Environment { get; }
private ILogger Logger { get; }
public Task WaitForStartAsync(CancellationToken cancellationToken)
{
    cancellationToken.Register(delegate
    {
        _delayStart.TrySetCanceled();
    });
    ApplicationLifetime.ApplicationStarted.Register(delegate
    {
    });
    ApplicationLifetime.ApplicationStopping.Register(delegate
    {
    });
    ApplicationLifetime.ApplicationStopped.Register(delegate
    {
        _delayStop.Set();
    });
    Thread thread = new Thread(new ThreadStart(Run));
    thread.IsBackground = true;
    thread.Start();
    return _delayStart.Task;
}
private void Run()
{
    try
    {
        ServiceBase.Run(this);
        _delayStart.TrySetException(
            new InvalidOperationException("Stopped without starting"));
    }
    catch (Exception exception)
    {
        _delayStart.TrySetException(exception);
    }
}
public Task StopAsync(CancellationToken cancellationToken)
```

```
    {
        Task.Run(new Action(base.Stop), CancellationToken.None);
        return Task.CompletedTask;
    }
    protected override void OnStart(string[] args)
    {
        delayStart.TrySetResult(null);
        base.OnStart(args);
    }
    protected override void OnStop()
    {
        ApplicationLifetime.StopApplication();
        delayStop.Wait(_hostOptions.ShutdownTimeout);
        base.OnStop();
    }
    protected override void OnShutdown()
    {
        ApplicationLifetime.StopApplication();
        delayStop.Wait(_hostOptions.ShutdownTimeout);
        base.OnShutdown();
    }
    protected override void Dispose(bool disposing)
    {
        if (disposing)
        {
            delayStop.Set();
        }
        base.Dispose(disposing);
    }
}
```

现在，可以通过 Windows 操作系统中的 Service Create（sc）命令行工具，将应用程序注册为 Windows 服务。

Service Create（sc）是用于服务管理的命令行工具，可以用于添加新服务，也可以用于查询、修改、启动、停止和删除现有服务。

通过 dotnet 命令行工具，可以还原应用程序的依赖项，还原后执行 publish 命令进行项目的发布，如代码 3-9 所示。

代码3-9

```
dotnet restore
```

```
dotnet publish
```

如代码3-10所示，使用sc.exe命令行工具部署，输入sc.exe create SERVICE NAME binpath= SERVICE FULL PATH。需要注意的是，binpath需要指定.exe文件的完整路径，当服务创建完成后，使用sc.exe start SERVICE NAME可以启动服务。

代码3-10

```
sc.exe create MyWorker binpath=publish\app.exe
sc.exe start MyWorker
```

使用sc.exe命令行工具可以停止和删除Windows服务，如代码3-11所示。

代码3-11

```
sc.exe stop MyWorker
sc.exe delete MyWorker
```

注册Windows服务后，可以在系统中按Win+R键，在"运行"窗口中输入services.msc，单击"确定"按钮，即可查看Windows服务，如图3-10所示。

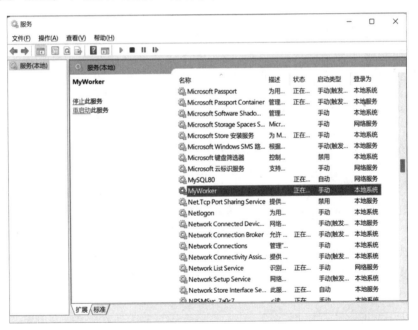

图3-10 查看Windows服务

使用sc.exe命令行工具可以查看Windows服务的状态信息，如图3-11所示。

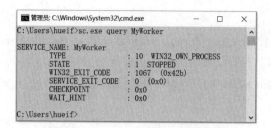

图 3-11 查看 Windows 服务的状态信息

3.4.3 Linux 服务注册

.NET 在 Linux 上运行也具有非常好的支持，采用 Worker Service 项目模板创建的应用程序也可以在 Linux 上进行托管，作为 Systemd Service 执行。

Linux 服务的 NuGet 包为 Microsoft.Extensions.Hosting.Systemd，代码 3-12 展示了 Worker Service 项目实例，将 UseSystemd 扩展方法添加到 CreateDefaultBuilder 方法中。

代码3-12

```
IHost host = Host.CreateDefaultBuilder(args)
   .UseSystemd()
   .ConfigureServices(services =>
   {
       services.AddHostedService<Worker>();
   })
   .Build();

await host.RunAsync();
```

代码 3-13 展示了 UseSystemd 扩展方法的内部实现，在 Linux 环境下部署需要执行如下几个主要操作：设置 Lifetime 为 SystemdLifetime；控制台记录器更改为 ConsoleFormatterNames.Systemd；注册 ISystemdNotifier 对象，用于通知 systemd 服务何时启动和何时停止，使用它不仅可以为 systemd 看门狗功能定期发送心跳，还可以实现 SDNotify 协议。

代码3-13

```
public static class SystemdHostBuilderExtensions
{
    public static IHostBuilder UseSystemd(this IHostBuilder hostBuilder)
    {
        if (SystemdHelpers.IsSystemdService())
```

```
    {
        hostBuilder.ConfigureServices((hostContext, services) =>
        {
            services.Configure<ConsoleLoggerOptions>(options =>
            {
                options.FormatterName = ConsoleFormatterNames.Systemd;
            });

            services.AddSingleton<ISystemdNotifier, SystemdNotifier>();
            services.AddSingleton<IHostLifetime, SystemdLifetime>();
        });
    }
    return hostBuilder;
}
```

在 Linux 环境下新建一个后缀为.service 的服务配置文件，该文件主要放在/usr/lib/systemd/system 目录下，也可以存放在用户配置目录/etc/systemd/system 下，如代码 3-14 所示。

代码3-14

```
[Unit]
Description=myworker

[Service]
Type=notify
WorkingDirectory=/home/user/myworker/
ExecStart=/home/user/myworker/03

[Install]
WantedBy=multi-user.target
```

在.service 文件中定义的服务主要分为[Unit]、[Service]和[Install]这 3 个部分。[Unit]中仅添加了 Description 属性，用于描述 Unit，但还有更多的可选项。[Service]中定义了有关应用的详细信息，对于.NET 应用程序，可以将 Type 设置为 notify，以便在主机启动或停止时通知 Systemd；[Service]中还定义了 WorkingDirectory 属性，用于设置工作目录；ExecStart 属性用于定义执行启动的脚本，需要有完整的路径，/home/user/myworker/03 路径中的 03 为项目名称。[Install]中定义了如何启动，以及开机是否启动，在此处使用了 WantedBy 属性，它的值可以为一个或多个 Target，在激活 Unit 时，符号链接会放在/etc/systemd/system 目录下以

49

Target 名+.wants 后缀构成的子目录中。

如代码 3-15 所示，发布.NET 应用程序，生成独立的应用程序，有关.NET 运行时和所需的依赖项都绑定在一个可执行文件中。

代码3-15

```
dotnet publish -c Release -r linux-x64
            --self-contained=true -p:PublishSingleFile=true
            -p:GenerateRuntimeConfigurationFiles=true -o artifacts
```

最后，在 Linux 环境下安装和运行服务，将.service 文件命名为 myworker.service，并且将它保存到/etc/systemd/system/目录下。

运行如代码 3-16 所示的命令可以加载新配置文件。

代码3-16

```
sudo systemctl daemon-reload
```

如代码 3-17 所示，启动和停止服务。

代码3-17

```
sudo systemctl start myworker.service
sudo systemctl stop myworker.service
sudo systemctl restart myworker.service
```

启动后可以使用如图 3-12 所示的命令查看服务的状态。

```
root@VM-16-6-ubuntu:/# systemctl status myworker.service
● myworker.service - myworker
     Loaded: loaded (/etc/systemd/system/myworker.service; disabled; vendor preset: enabled)
     Active: active (running) since Thu 2022-06-30 00:03:00 CST; 7s ago
   Main PID: 21868 (03)
      Tasks: 15 (limit: 8668)
     Memory: 10.1M
     CGroup: /system.slice/myworker.service
             └─21868 /home/user/myworker/03

Jun 30 00:03:00 VM-16-6-ubuntu 03[21868]: _01.Worker[0] Worker running at: 06/30/2022 00:03:00 +08:00
Jun 30 00:03:00 VM-16-6-ubuntu 03[21868]: Microsoft.Hosting.Lifetime[0] Application started. Hosting environmen
t: Production; Content root path: /home/user/myworker
Jun 30 00:03:00 VM-16-6-ubuntu systemd[1]: Started myworker.
Jun 30 00:03:01 VM-16-6-ubuntu 03[21868]: _01.Worker[0] Worker running at: 06/30/2022 00:03:01 +08:00
Jun 30 00:03:02 VM-16-6-ubuntu 03[21868]: _01.Worker[0] Worker running at: 06/30/2022 00:03:02 +08:00
Jun 30 00:03:03 VM-16-6-ubuntu 03[21868]: _01.Worker[0] Worker running at: 06/30/2022 00:03:03 +08:00
Jun 30 00:03:04 VM-16-6-ubuntu 03[21868]: _01.Worker[0] Worker running at: 06/30/2022 00:03:04 +08:00
Jun 30 00:03:05 VM-16-6-ubuntu 03[21868]: _01.Worker[0] Worker running at: 06/30/2022 00:03:05 +08:00
Jun 30 00:03:06 VM-16-6-ubuntu 03[21868]: _01.Worker[0] Worker running at: 06/30/2022 00:03:06 +08:00
Jun 30 00:03:07 VM-16-6-ubuntu 03[21868]: _01.Worker[0] Worker running at: 06/30/2022 00:03:07 +08:00
root@VM-16-6-ubuntu:/#
```

图 3-12 查看服务的状态

3.4.4 将 WebAPI 托管为 Windows 服务

将 WebAPI 通过 Windows 服务进行托管是一种什么体验呢？首先可以明确的是，通过服务托管即使计算机重启，当服务模式为"自动启动"时也会开机自启。

创建一个 WebAPI 项目，在 Program.cs 文件中创建一个 WebApplicationOptions 对象，并将其 ContentRootPath 属性设置为 AppContext.BaseDirectory，将 WebApplicationOptions 对象传递给 CreateBuilder 方法，如代码 3-18 所示。

代码3-18

```
WebApplicationOptions options = new()
{
   ContentRootPath = AppContext.BaseDirectory,
   Args = args
};
var builder = WebApplication.CreateBuilder(options);
builder.Host.UseWindowsService();

var app = builder.Build();

app.MapGet("/", () =>
{
   return "Hello World!";
});

app.Run();
```

当 WebAPI 项目部署为 Windows 服务后，需要考虑日志的输出位置，最简单的方式是输出 Windows 事件日志，如代码 3-19 所示，可以通过 appsettings.json 文件设置事件日志的级别。

代码3-19

```
{
  "Logging": {
    "LogLevel": {
      "Default": "Information",
      "Microsoft.AspNetCore": "Warning"
    },
    "EventLog": {
```

```
    "LogLevel": {
      "Default": "Information",
      "Program": "Information"
    }
  }
},
"AllowedHosts": "*"
}
```

然后使用 dotnet publish 命令将应用程序发布到指定的目录下，如代码 3-20 所示。

代码3-20

```
dotnet publish -c Release -o c:\webapiservice
```

通过 sc.exe 命令行工具创建 Windows 服务，如代码 3-21 所示。

代码3-21

```
sc.exe create WebAPIService binpath=c:\webapiservice
```

启动 Windows 服务后，如图 3-13 所示，可以使用 Windows 事件查看器查看日志信息，目前的"来源"列对应的是项目名称 WebAPIService，可以进行日志筛选。

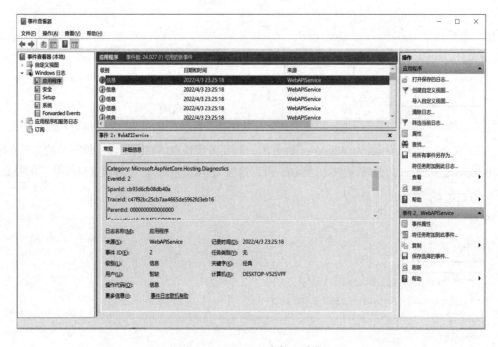

图 3-13　Windows 事件查看器

3.5 延伸阅读：WindowsFormsLifetime

下面对 3.3 节创建的 WinForm 应用程序进行优化，但是需要对它做出一些改动。虽然可以在 WinForm 应用程序中托管并运行 WebAPI，但无法结合 WinForm 应用程序的生命周期管理。本节主要介绍使用 WinForm 应用程序对其生命周期进行托管。

如代码 3-22 所示，修改 Program.cs 文件，先调用 Host 属性返回一个 IHostBuilder 实例，再调用 UseWindowsForms 扩展方法注册核心服务。IHostBuilder 实例承载了应用程序的核心功能，如配置系统、依赖注入、Host 的承载等。

代码3-22

```
var builder = WebApplication.CreateBuilder(args);
builder.Host.UseWindowsForms<Form1>();

var app = builder.Build();
app.MapGet("/", () => "Hello World!");
app.RunAsync();
```

创建一个名为 WindowsFormsLifetime 的类，如代码 3-23 所示，实现 IHostLifetime 接口，并注入 IHostApplicationLifetime 对象，用于管理应用程序的生命周期。

代码3-23

```
public class WindowsFormsLifetime : IHostLifetime, IDisposable
{
    private readonly IHostApplicationLifetime _applicationLifetime;

    public WindowsFormsLifetime(IHostApplicationLifetime applicationLifetime)
    {
        _applicationLifetime = applicationLifetime;
    }
    public Task WaitForStartAsync(CancellationToken cancellationToken)
    {
        Application.ApplicationExit += OnExit;
        return Task.CompletedTask;
    }
    public Task StopAsync(CancellationToken cancellationToken)
    {
```

```
        return Task.CompletedTask;
    }

    public void Dispose()
    {
        Application.ApplicationExit -= OnExit;
    }
    private void OnExit(object sender, EventArgs e)
    {
        _applicationLifetime.StopApplication();
    }
}
```

如代码 3-24 所示,创建 WindowsFormsApplicationHostedService 类,继承 IHostedService 接口用于承载服务,该接口定义了两个方法用于启动服务和关闭服务。在启动服务时会调用 StartAsync 方法,并开启一个 UI 线程用于管理 WinForm 应用程序;在关闭服务时会调用 StopAsync 方法。

代码3-24

```
public class WindowsFormsApplicationHostedService : IHostedService
{
    private readonly IServiceProvider _serviceProvider;
    private readonly Thread _thread;

    public WindowsFormsApplicationHostedService(
                        IServiceProvider serviceProvider)
    {
        _serviceProvider = serviceProvider;
        //创建一个 STA 线程
        _thread = new Thread(UIThreadStart);
        _thread.SetApartmentState(ApartmentState.STA);
    }

    public Task StartAsync(CancellationToken cancellationToken)
    {
        _thread.Start();
        return Task.CompletedTask;
    }
```

```csharp
public Task StopAsync(CancellationToken cancellationToken)
{
    Application.Exit();
    return Task.CompletedTask;
}

private void UIThreadStart()
{
    var applicationContext =
        serviceProvider.GetRequiredService<ApplicationContext>();
    Application.Run(applicationContext);
}
}
```

如代码 3-25 所示，创建 UseWindowsForms 扩展方法，首先注册一个主窗体，并且将其注册为 Singleton 服务实例，然后注册 IHostLifetime 服务和 WindowsFormsApplicationHostedService 承载服务，同时创建一个 ApplicationContext 对象作为 WinForm 应用程序的上下文对象。

代码3-25

```csharp
public static class WinFormLifetimeHostBuilderExtensions
{
    public static IHostBuilder UseWindowsForms<TMainForm>(
                this IHostBuilder builder) where TMainForm : Form
    {
        return builder.ConfigureServices((context, services) =>
        {
            services
                .AddSingleton<TMainForm>()
                .AddSingleton<IHostLifetime, WindowsFormsLifetime>()
                .AddSingleton(c =>
                    new ApplicationContext(c.GetRequiredService<TMainForm>()))
                .AddHostedService<WindowsFormsApplicationHostedService>();
        });
    }
}
```

IHost 接口表示一个宿主服务的定义，通常应用程序的启动和停止都由 IHost 接口管理。另外，IHost 接口也声明了 StartAsync 方法和 StopAsync 方法，如代码 3-26 所示。

代码3-26

```
public interface IHost : IDisposable
{
    IServiceProvider Services { get; }
    Task StartAsync(CancellationToken cancellationToken = default);
    Task StopAsync(CancellationToken cancellationToken = default);
}
```

IHostApplicationLifetime 接口用于对应用程序的生命周期进行管理，如代码 3-27 所示。

代码3-27

```
public interface IHostApplicationLifetime
{
    CancellationToken ApplicationStarted { get; }
    CancellationToken ApplicationStopped { get; }
    CancellationToken ApplicationStopping { get; }
    void StopApplication();
}
```

代码 3-28 展示了 IHostLifetime 接口的定义，当启动 Host 时会调用 IHostLifetime 对象的 WaitForStartAsync 方法，当关闭 Host 时会调用 StopAsync 方法。

代码3-28

```
public interface IHostLifetime
{
    Task StopAsync(CancellationToken cancellationToken);
    Task WaitForStartAsync(CancellationToken cancellationToken);
}
```

不难发现，Host 对象承载了应用程序的托管工作，下面简化 Host 类的实现。如代码 3-29 所示，在 StartAsync 方法中，首先调用了 IHostLifetime 对象的 WaitForStartAsync 方法，也就是说，笔者编写的 WindowsFormsLifetime 对象会被第一时间调用，用于注册 Host 本身的生命周期管理，接下来就是获取所有注册的 IHostedService 服务，它们都会在此处被集中启动，最后通过调用 ApplicationLifetime 对象的 NotifyStarted 方法用于通知应用程序。同样，StopAsync 方法首先通过 ApplicationLifetime 对象来停止，然后调用 IHostedService 对象停止所有服务，最后通过 ApplicationLifetime 对象通知应用程序。

代码3-29

```
internal sealed class Host : IHost, IAsyncDisposable
```

```csharp
{
    private readonly ILogger<Host> _logger;
    private readonly IHostLifetime _hostLifetime;
    private readonly ApplicationLifetime _applicationLifetime;
    private readonly HostOptions _options;
    private readonly IHostEnvironment _hostEnvironment;
    private readonly PhysicalFileProvider _defaultProvider;
    private IEnumerable<IHostedService> _hostedServices;
    private volatile bool _stopCalled;

    public IServiceProvider Services { get; }
    public async Task StartAsync(CancellationToken cancellationToken
                                                         = default)
    {
        await _hostLifetime.WaitForStartAsync(combinedCancellationToken)
                                                .ConfigureAwait(false);
        combinedCancellationToken.ThrowIfCancellationRequested();
        _hostedServices = Services.GetService<IEnumerable <IHostedService>>();
        foreach (IHostedService hostedService in _hostedServices)
        {
            await hostedService.StartAsync(
                    combinedCancellationToken).ConfigureAwait(false);
            if (hostedService is BackgroundService backgroundService)
            {
                _ = TryExecuteBackgroundServiceAsync(backgroundService);
            }
        }
        _applicationLifetime.NotifyStarted();
    }
    private async Task TryExecuteBackgroundServiceAsync(
                            BackgroundService backgroundService)
    {
        Task backgroundTask = backgroundService.ExecuteTask;
        if (backgroundTask == null)
            return;
        try
        {
            await backgroundTask.ConfigureAwait(false);
```

```csharp
        }
        catch (Exception ex)
        {
            if (options.BackgroundServiceExceptionBehavior
                    == BackgroundServiceExceptionBehavior.StopHost)
            {
                _applicationLifetime.StopApplication();
            }
        }
    }

    public async Task StopAsync(CancellationToken cancellationToken
                                                    = default)
    {
        using(var cts = new CancellationTokenSource(
                                        options.ShutdownTimeout))
        using(var linkedCts = CancellationTokenSource
                .CreateLinkedTokenSource(cts.Token, cancellationToken))
        {
            CancellationToken token = linkedCts.Token;
            applicationLifetime.StopApplication();
            IList<Exception> exceptions = new List<Exception>();
            if (hostedServices != null)
            {
                foreach (IHostedService hostedService in
                                            _hostedServices.Reverse())
                {
                    try
                    {
                        awaithostedService.StopAsync(token)
                                            .ConfigureAwait(false);
                    }
                    catch (Exception ex)
                    {
                        exceptions.Add(ex);
                    }
                }
            }
```

```
        applicationLifetime.NotifyStopped();
        try
        {
            await hostLifetime.StopAsync(token).ConfigureAwait(false);
        }
        catch (Exception ex)
        {
            exceptions.Add(ex);
        }
    }
}
public void Dispose() => DisposeAsync().AsTask()
                                       .GetAwaiter().GetResult();
public async ValueTask DisposeAsync(){ }
}
```

3.6 小结

通过学习本章,读者可以基本了解.NET 多样化的运行方式。读者可以根据自己的实际需求选择不同类型的运行方式。

第 4 章

依赖注入

依赖注入（Dependency Injection，DI），又称为依赖关系注入，是一种软件设计模式，也是依赖倒置原则（Dependence Inversion Principle，DIP）的一种体现。依赖倒置原则的含义如下：一是高层模块不依赖低层模块，二者都依赖抽象；二是抽象不应依赖细节；三是细节应依赖抽象。在业务代码各处使用关键字 new 来创建对象也是细节实现，一旦广泛使用了关键字 new，就可能导致对象间的关系变得非常复杂。

依赖注入原则有别于传统的通过关键字 new 直接依赖低层模块的形式，以第三方容器注入的形式进行依赖项的管理。依赖注入是实现控制反转的一种手段，而用来实现依赖注入的技术框架又被称为 IoC 框架。

控制反转（Inversion of Control，IoC）最早是由 Martin Fowler 提出的一种概念。Martin Fowler 指出，由于对象间存在紧密的相互依赖关系，每个对象都需要管理依赖对象的引用，因此应用程序代码变得高度耦合且难以拆分。而依赖注入则改变了对象原本的依赖形式，可以实现对依赖关系的控制反转。

在传统的.NET Framework 时代，许多有经验的开发人员已经普遍使用 IoC 框架来管理依赖项的关系，如 Spring.NET、Unity（不是游戏框架 Unity 或 Unity3D，而是由 Microsoft 企业库提供的一种 IoC 框架）就是常用的几种框架。但是由于这些框架不是.NET Framework 的一部分，需要由开发人员自行引入，因此会提高学习成本。

而从.NET Core 开始，.NET 平台就原生提供了一种 IoC 框架（其命名空间为 Microsoft.Extension.DependencyInjection），该框架具有轻量级和易用性的特点，因此，可以很轻松地在应用程序中使用。依赖注入在.NET 中无处不在，已经成为.NET 应用程序开发的理论基石。

4.1 .NET 依赖注入

"高内聚，低耦合"是软件开发人员一直在追求的目标，而依赖项的管理则是耦合问题的直接体现，将依赖注入作为手段是.NET 应用程序开发过程中最常见的方式。例如，配置、日志和选项等模块其实都是开发应用程序过程中必不可少的基础设施,当采用关键字new时，如果这些模块的实现发生了一些变化，就会不可避免地影响业务代码。

依赖注入也体现了面向对象多态的特征，将直接管理对象改成管理对象的引用关系，是面向对象编程的一大魅力。通过使用依赖注入，采用面向接口编程的方式，将常用的实现抽象成接口，通过调用接口减少创建（new）对象的过程。当面对不同的实现类时，开发人员不需要修改之前的代码，只需要创建新的实现类并改变引用关系即可。用来管理引用关系的基础设施就是依赖注入框架。容器管理对象的引用关系如图 4-1 所示。有人将依赖注入框架称为管理对象的容器，在这个容器中，人们无须手动创建和销毁对象，容器会自动帮助开发人员管理这些过程。

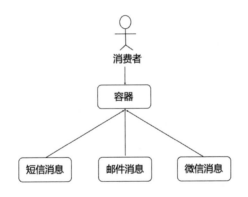

图 4-1 容器管理对象的引用关系

4.1.1 服务的注册

依赖注入的理论比较复杂，为了更好地帮助读者理解，笔者从代码实现层面介绍.NET中的依赖注入。

假设有一个消息打印类 MessagePrinter，需要调用 MessagePrinter 类来输出相关消息，按照传统的开发习惯，实现过程如代码 4-1 所示，有两处使用了关键字 new 的代码段。

代码4-1

```csharp
class Program
{
    static void Main(string[] args)
    {
        MessagePrinter messagePrinter = new MessagePrinter();
        messagePrinter.Print();
    }
}
///<summary>
///打印类
///</summary>
public class MessagePrinter
{
    private MessageWriter writer;
    public MessagePrinter()
    {
        writer = new MessageWriter();
    }
    public void Print()
    {
        writer.Write("Hello World");
    }
}

///<summary>
///消息输出
///</summary>
public class MessageWriter
{
    public void Write(string msg)
    {
        Console.Out.WriteLine(msg);
    }
}
```

MessagePrinter 相当于高层模块，MessageWriter 相当于低层模块，在这段代码中，MessagePrinter 通过 new 的形式直接依赖 MessageWriter，如果后续 MessageWriter 的实现方

式发生了变化,如新增了实现、类改名或方法改名,就会对 MessagePrinter 造成影响。这种典型的对象耦合行为在大型项目中很常见,低层模块的实现变化总是为高层模块的开发人员带来许多困扰。

如果使用接口编程的方式,则会显著改善这种状况,如代码 4-2 所示,将原本的 MessageWriter 抽象为接口 IWriter,并在构造函数中管理该接口的实现细节,如果后续 MessageWriter 需要改成 ConsoleWriter,则只需对一行代码进行修改。

代码4-2

```
///<summary>
///打印类
///</summary>
public class MessagePrinter
{
    private IWriter writer;
    public MessagePrinter()
    {
        //后续对象行为发生重点变更或改名,只需修改此处
        writer = new MessageWriter();
        //writer = new ConsoleWriter();
    }
    public void Print()
    {
        writer.Write("Hello World");
    }
}


///<summary>
///消息输出
///</summary>
public class MessageWriter:IWriter
{
    public void Write(string msg)
    {
        Console.Out.WriteLine(msg);
    }
}
///<summary>
```

```csharp
///控制台输出
///</summary>
public class ConsoleWriter:IWriter
{
    public void Write(string msg)
    {
        Console.Out.WriteLine(msg);
    }
}
```

上述代码采用接口编程的形式看似已经使高层模块更加稳定，但.NET 的依赖注入则更进一步，可以将整个对象的创建过程容器化，开发人员可以通过构造函数注入的形式，在创建 MessagePrinter 对象时自动创建其依赖的底层对象。

先创建控制台实例，创建完成后，再引入相应的 NuGet 包，依赖注入的 NuGet 包为 Microsoft.Extensions.DependencyInjection，如代码 4-3 所示。

代码4-3

```csharp
///<summary>
///打印类
///</summary>
public class MessagePrinter
{
    private IWriter _writer;
    ///<summary>
    ///注入构造函数，由依赖注入框架管理对象的创建过程，开发人员无须使用 new 来创建
    ///</summary>
    ///<param name="writer"></param>
    public MessagePrinter(IWriter writer)
    {
        //writer = new MessageWriter();
        _writer = writer;
    }
    public void Print()
    {
        _writer.Write("Hello World");
    }
}
///<summary>
///消息输出
```

```
///</summary>
public class MessageWriter:IWriter
{
    public void Write(string msg)
    {
        Console.Out.WriteLine(msg);
    }
}
```

因为依赖注入具有传染性,所以还需要对入口程序进行修改。如代码 4-4 所示,在 Main 函数中,首先创建一个 ServiceCollection 对象,用这个对象来管理对象的引用关系,然后调用 AddSingleton 扩展方法,由方法名称可知,该方法针对服务注册采用不同的生命周期(Transient、Scoped 和 Singleton,本节不对其展开介绍,读者可以在 4.1.2 节中对服务的生命周期进行学习和了解),将 IMessageWriter 接口注册相应的服务,接着调用 BuildServiceProvider 扩展方法,返回 ServiceProvider 对象,并调用该对象的 GetService<T>扩展方法获取相应的服务实例,最后调用实例的 Print 方法。

代码4-4

```
class Program
{
    static void Main(string[] args)
    {
        var provider = new ServiceCollection()
            .AddSingleton<MessagePrinter>()
            .AddSingleton<IWriter, MessageWriter>()
            .BuildServiceProvider();
        var messageWriter = provider.GetService<MessagePrinter>();
        messageWriter.Print();
    }
}
```

注意:控制台应用中不包含依赖注入框架,所以需要手动添加的 NuGet 包为 Microsoft. Extension.DependencyInjection,而 ASP.NET Core 应用不需要单独引入,该 SDK 已经被自动添加基础 NuGet 包的依赖。

扩展延伸

根据上述内容继续做延伸性理解,可以发现 IServiceProvider 是依赖注入的核心抽象接口。代码 4-5 展示了 IServiceProvider 接口的定义。

代码4-5

```
public interface IServiceProvider
{
    object? GetService(Type serviceType);
}
```

一般来说，当服务注册到 DI 容器（IServiceCollection）后，后续的服务实例的获取可以直接使用 GetService 扩展方法。除此之外，获取服务实例还可以使用 GetRequiredService<T>扩展方法，该方法被定义在一个名为 ServiceProviderServiceExtensions 的静态类中，在该类中还实现了扩展方法 GetService<T>和 GetRequiredService<T>，如代码 4-6 所示。笔者简化了该代码，让代码更加精简一些。如果读者需要阅读完整的代码，则可以从 GitHub 官网的 dotnet/runtime 仓库中阅读源代码。

代码4-6

```
public static class ServiceProviderServiceExtensions
{
    public static T? GetService<T>(this IServiceProvider provider)
    {
        return (T?)provider.GetService(typeof(T));
    }
    public static T GetRequiredService<T>(this IServiceProvider provider)
    {
        return (T)provider.GetRequiredService(typeof(T));
    }
}
```

正如代码 4-6 所示，获取服务实例可以使用扩展方法 GetRequiredService<T>和 GetService<T>。而扩展方法 GetRequiredService<T>和 GetService<T>的区别，也是值得读者关注的内容。另外，读者必须了解选择每个方法原因。对于 GetRequiredService<T>扩展方法来说，如果服务不可用，就会抛出异常；GetService<T>扩展方法的返回类型是空的，所以是有返回空的可能性的，需要做非空检查。也就是说，如果开发人员不进行判断，则可能会返回 NullReferenceException 异常对象。而 GetRequiredService<T>扩展方法则直接抛出 InvalidOperationException 异常对象。

4.1.2 生命周期

通过学习 4.1.1 节，读者可以基本认识依赖注入。本节将介绍.NET 依赖注入的生命周期。.NET 依赖注入提供了以下 3 种生命周期。

（1）单例，.NET 中使用的 Singleton 与设计模式中的单例模式类似，表示一个应用程序的生命周期只会创建一次对象，后续每次使用都将复用该对象。

（2）作用域，有人将使用的 Scoped 称为"请求单例"，表示在每个 HTTP 请求期间创建一次。

（3）瞬时，使用 Transient，每个方法调用周期内只创建一次对象。

正如代码 4-7 所示，依赖注入的 3 种生命周期如图 4-2 所示。

代码4-7

```
public enum ServiceLifetime
{
    Singleton,
    Scoped,
    Transient,
}
```

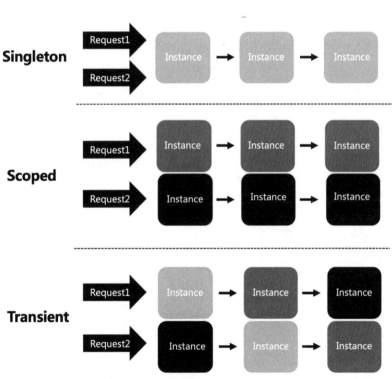

图 4-2　依赖注入的 3 种生命周期

为了更好地帮助读者了解服务生命周期，笔者通过一个 Web 应用程序来演示。再创建一个 ASP.NET Core WebAPI 应用程序，在该应用程序中先创建 ISample 接口，再创建 ISampleSingleton 接口、ISampleScoped 接口和 ISampleTransient 接口，创建完成后需要创建一个名为 Sample 的实现类，用于记录对象的生成次数，不是某个服务的累加次数，而是统计类实例化对象的个数，如代码 4-8 所示。

代码4-8

```csharp
public interface ISample
{
    int Id { get; }
}
public interface ISampleSingleton : ISample
{
}
public interface ISampleScoped : ISample
{
}
public interface ISampleTransient : ISample
{
}
public class Sample : ISampleSingleton, ISampleScoped, ISampleTransient
{
    private static int _counter;
    private int _id;
    public Sample()
    {
        _id = ++_counter;
    }
    public int Id => _id;
}
```

上面创建了接口和实现类，接下来需要注册服务。可以使用 3 种生命周期分别创建并注册 Sample 类的实例，在 Startup 类中，通过调用 WebApplicationBuilder 对象的 Services 属性，分别调用不同生命周期的扩展方法注册服务，如代码 4-9 所示。

代码4-9

```csharp
var builder = WebApplication.CreateBuilder(args);
//注册服务
```

```csharp
builder.Services.AddTransient<ISampleTransient, Sample>();
builder.Services.AddScoped<ISampleScoped, Sample>();
builder.Services.AddSingleton<ISampleSingleton, Sample>();

builder.Services.AddControllers();

var app = builder.Build();

app.UseHttpsRedirection();

app.UseAuthorization();

app.MapControllers();

app.Run();
```

服务注册完成后,需要对 Controller 注入 3 个服务。在之后的使用中不仅可以在 Controller 中注入服务,还可以在 ASP.NET Core MVC 视图中注入服务。在本实例中,输出代码 4-8 中统计类实例化的次数,通过服务实例的 HashCode 进行观察比较,在代码 4-10 中对其进行演示。

代码4-10

```csharp
using System;

public class WeatherForecastController : ControllerBase
{
    private readonly ISampleSingleton _sampleSingleton;
    private readonly ISampleScoped _sampleScoped;
    private readonly ISampleTransient _sampleTransient;

    public WeatherForecastController(ISampleSingleton sampleSingleton
            ,ISampleScoped sampleScoped,ISampleTransient sampleTransient)
    {
        _sampleSingleton = sampleSingleton;
        _sampleScoped = sampleScoped;
        _sampleTransient = sampleTransient;
    }
```

```
[HttpGet]
public OkResult Get()
{
    Console.WriteLine(
        $"name: sampleScoped,Id: {_sampleScoped.Id},hashCode
                             : {_sampleScoped.GetHashCode()},\n" +
        $"name: sampleTransient,Id: {_sampleTransient.Id},hashCode
                             : {_sampleTransient.GetHashCode()},\n" +
        $"name: sampleSingleton, Id: {_sampleSingleton.Id},hashCode
                             : {_sampleSingleton.GetHashCode()}\n");
    return Ok();
}
```

运行上述应用程序,并访问指定的地址,请求两次记录请求的内容。图 4-3 展示了 3 种不同生命周期对象的 HashCode,可以发现,Singleton 实例的第一次请求和第二次请求输出的内容是一致的。因此,对于 Singleton 来说,在应用程序生命周期内都是相同的实例。

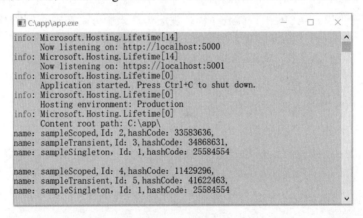

图 4-3　3 种不同生命周期对象的 HashCode

下面对 Scoped 生命周期进行验证。在 Controller 中注入 IServiceProvider 对象,如代码 4-11 所示,通过 IServiceProvider 对象进行实例的检索和解析操作,该对象可以在运行时快速检索到对象实例,Controller 的 Get 方法通过 IServiceProvider 对象检索应用程序中定义的 3 个服务实例,主要目的是在同一个请求中对比对应实例的 HashCode 是否一致。

代码4-11

```
public OkResult Get()
{
```

```
    var sampleSingleton = _serviceProvider.GetService<ISampleSingleton>();
    var sampleScoped = _serviceProvider.GetService<ISampleScoped>();
    var sampleTransient = _serviceProvider.GetService<ISampleTransient>();

    Console.WriteLine(
        $"name：sampleScoped,Id：{ _sampleScoped.Id },hashCode
                            ：{_sampleScoped.GetHashCode()},\n" +
        $"name：sampleTransient,Id：{ _sampleTransient.Id},hashCode
                            ：{_sampleTransient.GetHashCode()},\n" +
        $"name：sampleSingleton, Id：{_sampleSingleton.Id},hashCode
                            ：{_sampleSingleton.GetHashCode()}\n");

    Console.WriteLine(
        $"name：sampleScoped,Id：{ sampleScoped.Id },hashCode
                            ：{sampleScoped.GetHashCode()},\n" +
        $"name：sampleTransient,Id：{sampleTransient.Id},hashCode
                            ：{sampleTransient.GetHashCode()},\n" +
        $"name：sampleSingleton, Id：{sampleSingleton.Id},hashCode
                            ：{sampleSingleton.GetHashCode()}\n");
    return Ok();
}
```

图 4-4 展示了对实例对象的生命周期进行多次验证的结果，Scoped 作用域可以在同一个请求内使用同一个实例，而 Transient（瞬时）是"即用即走"，所以每次获取的都是不同的实例。

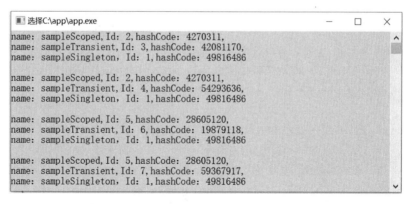

图 4-4　对实例对象生命周期进行多次验证的结果

在.NET 中，IServiceProvider 对象提供了 CreateScope 扩展方法，使用该方法可以获取一个全新 ServiceProvider 实例对象。顾名思义，使用 CreateScope 扩展方法会创建一个范围性的实例，而对于实例来说，ServiceProvider 的 Scope 相当于一个单例，通过 IServiceProvider 对象，调用它的 CreateScope 扩展方法创建出一个相应的服务范围，这个服务范围代表子容器，而这个子容器获取相应的 IServiceProvider 对象。获取上面注册的 3 个不同生命周期的服务实例，如代码 4-12 所示。

代码4-12

```
using (var scope = _serviceProvider.CreateScope())
{
    var p = scope.ServiceProvider;

    var scopeobj1 = p.GetService<ISampleScoped>();
    var transient1 = p.GetService<ISampleTransient>();
    var singleton1 = p.GetService<ISampleSingleton>();

    var scopeobj2 = p.GetService<ISampleScoped>();
    var transient2 = p.GetService<ISampleTransient>();
    var singleton2 = p.GetService<ISampleSingleton>();

    Console.WriteLine(
        $"name：scope1,Id：{ scopeobj1.Id },
                    hashCode：{scopeobj1. GetHashCode()},\n" +
        $"name：transient1,Id：{transient1.Id},
                    hashCode：{transient1. GetHashCode()},\n" +
        $"name：singleton1, Id：{singleton1.Id},
                    hashCode：{singleton1. GetHashCode()}\n");
    Console.WriteLine($"name：scope2,Id：{ scopeobj2.Id },
                    hashCode：{scopeobj2.GetHashCode()},\n " +
        $"name：transient2,Id：{transient2.Id},
                    hashCode：{transient2.GetHashCode()}, \n" +
        $"name：singleton2,Id：{singleton2.Id},
                    hashCode：{singleton2.GetHashCode()}\n");
}
```

获取根容器的 IServiceProvider 对象，通过该对象创建服务范围。需要注意的是，服务范围对象是在 using 代码块中进行的，所以它们始终会被 Dispose（释放）。子容器的生命周期如图 4-5 所示，将不同生命周期的服务实例的 HashCode 输出到控制台中。

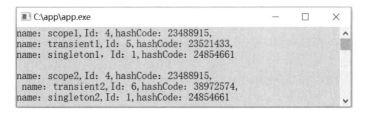

图 4-5　子容器的生命周期

上面的服务范围输出的 HashCode 与预期一致，接下来，需要对 CreateScope 扩展方法进行二次验证。对于二次验证，就是复制一个 using 代码块，通过两个子容器对比生命周期，如图 4-6 所示。

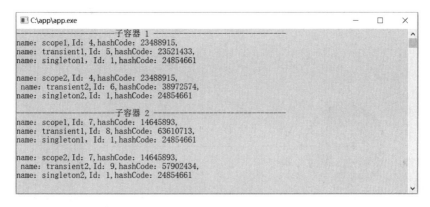

图 4-6　通过两个子容器对比生命周期

IServiceProvider 有两种定义：一种是根容器（Root Scope）中的 IServiceProvider 对象，是位于应用程序顶端的容器，一般被称为 ApplicationServices；另一种是通过 IServiceScopeFactory 服务创建的带有服务范围的 IServiceScope 对象，而对于 IServiceScope 对象来说，拥有的是一个"范围"性的 IServiceProvider 对象。代码 4-13 展示了内部相应的源代码。

代码4-13

```
public interface IServiceScope : IDisposable
{
    IServiceProvider ServiceProvider { get; }
}
public interface IServiceScopeFactory
{
```

```
    IServiceScope CreateScope();
}
public static class ServiceProviderServiceExtensions
{
    public static IServiceScope CreateScope(this IServiceProvider provider)
    {
        return provider.GetRequiredService<IServiceScopeFactory>()
                                            .CreateScope();
    }
}
```

可以通过依赖的一个容器创建子容器，需要注意的是，只可以拥有一个根容器。根容器与子容器的关系如图 4-7 所示。

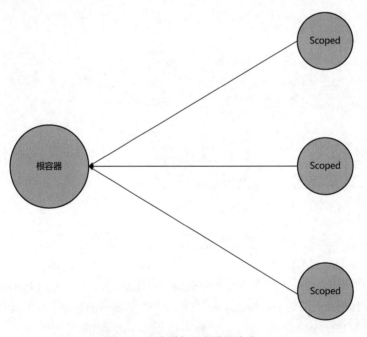

图 4-7　根容器与子容器的关系

图 4-8 所示为 ASP.NET Core 的作用域。在 ASP.NET Core 中，每个请求都会创建一个全新的 Scope（作用域）服务，在这个请求过程中创建的服务实例都会保存在当前的 IServiceProvider 对象上，所以，在当前范围内提供的实例都是单例的。如图 4-8 所示，在 ASP.NET Core 中有两种容器，一种是单例的容器，另一种是 HTTP 请求中的作用域容器。

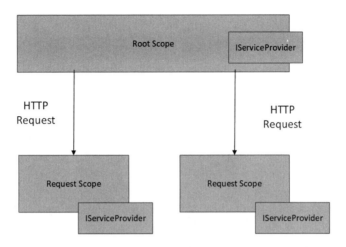

图 4-8　ASP.NET Core 的作用域

4.1.3　服务描述

当 ServiceDescriptor 对象用于注册指定的服务时，对服务注册项进行描述，而依赖注入容器则利用这种描述为开发人员提供服务实例。

如代码 4-14 所示，创建一个 ServiceDescriptor 实例，通过调用其构造函数来创建实例对象。

代码4-14

```
var singletonDescriptor = new ServiceDescriptor(typeof (ISampleSingleton),
            typeof(Sample), ServiceLifetime. Singleton);
```

除了上面的方式，还可以通过调用 ServiceDescriptor 类型中的 Describe 静态方法创建实例对象，如代码 4-15 所示。

代码4-15

```
var scopedDescriptor = ServiceDescriptor.Describe
 (typeof(ISampleScoped),typeof(Sample), ServiceLifetime.Scoped);
```

正如上述内容所示，每个方法调用中都传递了服务的生命周期参数，ServiceDescriptor 类型还为不同的生命周期提供了对应的静态方法，可以参考代码 4-16 来使用。

代码4-16

```
var transientDescriptor = ServiceDescriptor.Transient
            (typeof (ISampleTransient), typeof(Sample));
```

ServiceDescriptor 对象提供了 3 个构造方法，同时为 3 种不同的生命周期提供了 3 个方法，用于对服务实例进行描述。如代码 4-17 所示，ServiceType 属性对应要注册的服务的类型，Lifetime 属性表示服务的生命周期。以 Implementation 开头的属性代表不同方式传递的服务实现：ImplementationType 属性表示服务实现的类型，通过传递指定类型，调用相应的构造函数来创建服务实例；ImplementationInstance 属性表示直接传递一个现成的对象，而该对象正是最终通过依赖注入容器获取的服务实例；ImplementationFactory 属性提供了创建服务实例的委托对象。

代码4-17

```
public class ServiceDescriptor
{
    public ServiceLifetime Lifetime { get; }
    public Type ServiceType { get; }

    public Type? ImplementationType { get; }
    public object? ImplementationInstance { get; }
    public Func<IServiceProvider, object>? ImplementationFactory { get; }

    public ServiceDescriptor(Type serviceType,
        Type implementationType, ServiceLifetime lifetime);
    public ServiceDescriptor(Type serviceType, object instance);
    public ServiceDescriptor(Type serviceType,
        Func<IServiceProvider, object> factory, ServiceLifetime lifetime);
}
```

依赖注入容器对象，正是利用 ServiceDescriptor 对象为应用程序提供所需的服务实例的。代码 4-18 展示了 IServiceCollection 对象的扩展方法 AddSingleton，该方法用来注册 Singleton 服务。在 AddSingleton 方法内部可以看到，通过 AddSingleton 扩展方法将类型和实现对象传递过来以后，先通过服务描述类 ServiceDescriptor 创建一个实例，再通过 IServiceCollection 对象注册 ServiceDescriptor 实例，并将其添加到 IServiceCollection 集合中。

代码4-18

```
public static IServiceCollection AddSingleton(
    this IServiceCollection services,
    Type serviceType,
    object implementationInstance)
{
```

```
    var serviceDescriptor = new
            ServiceDescriptor(serviceType, implementationInstance);
    services.Add(serviceDescriptor);
    return services;
}
```

4.1.4 作用域验证

ASP.NET Core 容器可以分为两种，一种是单例的根容器，另一种是请求作用域容器。例如，应用程序有一个 Context 对象，开发人员通常会将其注册为 Scoped 生命周期，而在请求作用域中，可以从一个请求的开始到结束友好地进行释放，如对数据连接的释放，还可以保证 Context 对象在当前上下文中保持重复使用。

如代码 4-19 所示，以控制台实例为例，创建 ServiceCollection 对象，调用 AddScoped 扩展方法，用于注册作用域服务，调用 IServiceCollection 的 BuildServiceProvider 扩展方法得到依赖注入容器的 IServiceProvider 对象之后，调用 GetRequiredService<T>扩展方法来获取服务实例，最终通过 Console.WriteLine 方法将其输出。

代码4-19

```
var provider = new ServiceCollection()
   .AddScoped<ISampleScoped, Sample>()
   .BuildServiceProvider(validateScopes: true);
var sampleScoped = provider.GetRequiredService<ISampleScoped>();
Console.WriteLine(sampleScoped.Id);
```

运行该实例后，抛出异常信息 Cannot resolve scoped service，表示无法解析作用域服务，如图 4-9 所示。

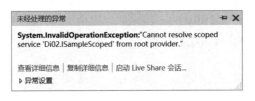

图 4-9 无法解析作用域服务

这意味着应该从 IServiceScope 对象中解析，而不是从 IServiceProvider 对象中解析。在 ASP.NET Core 请求中，框架会为开发人员自动创建 ServiceProviderServiceExtensions.CreateScope 作用域容器，并使用该作用域容器和其他服务。如果应用程序不是 ASP.NET Core，则可以自行创建一个作用域，如代码 4-20 所示。

代码4-20

```
using (var scope = ServiceProviderServiceExtensions.CreateScope (provider))
{
    var sampleScoped =
        scope.ServiceProvider.GetRequiredService<ISampleScoped>();
    Console.WriteLine(sampleScoped.Id);
}
```

ServiceProviderOptions 对象中还有一个名为 ValidateOnBuild 的属性，如果将该属性设置为 true，则意味着在构建 IServiceProvider 对象时会验证服务的有效性，即确保每个服务的实例都是可用的。在默认情况下，ValidateOnBuild 属性的值为 false。ValidateOnBuild 属性只有在通过 IServiceProvider 对象获取服务实例的时候，才会触发相应的异常。

修改控制台实例，如代码 4-21 所示，将 ServiceProviderOptions 对象的 ValidateOnBuild 属性设置为 true，在 Sample 类中添加一个有参构造方法。

代码4-21

```
new ServiceCollection()
    .AddTransient<ISample, Sample>()
    .BuildServiceProvider(new ServiceProviderOptions
    {
        ValidateOnBuild = true,
    });

public class Sample : ISampleSingleton, ISampleScoped, ISampleTransient
{
    private static int _counter;
    private int _id;

    public Sample(ISample sample)
    {
        _id = ++_counter;
    }
    public int Id => _id;
}
```

控制台实例修改完成后，运行应用程序可以发现，在启动时就会对服务进行检查，对于 ValidateOnBuild 属性，可以在应用程序启动时检查服务的有效性，并在服务无效的情况下抛出异常，防止开发人员遗漏服务问题。如图 4-10 所示，将 ValidateOnBuild 属性设置为 true

后，如果服务无效就会报错。

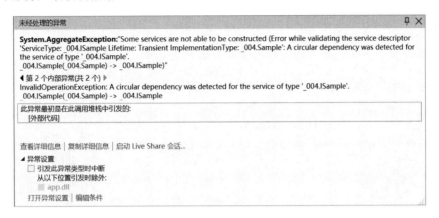

图 4-10　报错信息

代码 4-22 展示了 ServiceProviderOptions 类的定义，该类包括 ValidateScopes 属性和 ValidateOnBuild 属性，这些属性在默认情况下都为 false。

代码4-22

```
public class ServiceProviderOptions
{
    public bool ValidateScopes { get; set; }
    public bool ValidateOnBuild { get; set; }
}
```

ServiceProviderOptions 对象用于在构建 ServiceProvider 对象时检查服务的有效性，如果将 ValidateOnBuild 属性设置为 true，如代码 4-23 所示，则在 ServiceProvider 构造方法内进行处理，先获取 ServiceDescriptor 集合，并循环遍历该集合，逐一判断它的有效性，如果无效，则将其添加到错误集合中，最后将异常对象的集合抛出。

代码4-23

```
internal ServiceProvider(
        IEnumerable<ServiceDescriptor> serviceDescriptors,
        IServiceProviderEngine engine,
        ServiceProviderOptions options)
{
    _engine = engine;

    if(options.ValidateScopes)
```

```
{
    _engine.InitializeCallback(this);
    _callSiteValidator = new CallSiteValidator();
}

if(options.ValidateOnBuild)
{
    List<Exception> exceptions = null;
    foreach (ServiceDescriptor serviceDescriptor in serviceDescriptors)
    {
        try
        {
            _engine.ValidateService(serviceDescriptor);
        }
        catch(Exception e)
        {
            exceptions = exceptions ?? new List<Exception>();
            exceptions.Add(e);
        }
    }

    if(exceptions != null)
    {
        throw new AggregateException(
                "Some services are not able to be constructed",
                                        exceptions.ToArray());
    }
}
}
```

4.2 实现批量服务注册

当使用依赖注入时，如果应用程序需要注册的服务很多，那么手动为每个服务编写代码进行注册将是一场灾难。一方面是开发人员需要维护的代码比较繁杂，另一方面是不方便对其进行管理，而采用批量注册则有望减轻负担。开发人员可以根据某个指定的类的名称进行注册，或者采用其他规则进行批量注册，这样可以降低维护代码的成本。

4.2.1 根据名称匹配并注册

如代码 4-24 所示，首先创建一个 ServiceCollection 对象，通过 Assembly 对象的 Load 扩展方法加载指定的程序集，可以针对某个程序集中的类进行扫描，调用 Assembly 对象的 GetTypes 方法获取所有的类型，并进行筛选和过滤，而命名则通过 t.Name.EndsWith 方法匹配后缀是否一致。然后创建一个字典（Dictionary）保存接口和实例，通过循环遍历 IEnumerable<Type>集合，获取指定对象的接口类型，同时将实现类和接口保存到字典中。最后通过循环遍历字典调用 AddSingleton 方法逐一进行服务注册。

代码4-24

```
static void Main(string[] args)
{
    var services = new ServiceCollection();
    var assembly = Assembly.Load("Di05");
    IEnumerable<Type> typeList = assembly.GetTypes()
                .Where(t => !t.IsInterface && !t.IsAbstract &&
                                    t.Name.EndsWith ("Sample"));
    var dic = new Dictionary<Type, Type[]>();
    foreach(var type in typeList)
    {
        var interfaces = type.GetInterfaces();
        dic.Add(type, interfaces);
    }
    if(dic.Keys.Count > 0)
    {
        foreach(var instanceType in dic.Keys)
        {
            foreach(var interfaceType in dic[instanceType])
            {
                services.AddSingleton(interfaceType, instanceType);
            }
        }
    }

    var serviceProvider = services.BuildServiceProvider();
    var sample = serviceProvider.GetRequiredService<ISample>();
```

```
    Console.WriteLine(sample.Id);
}
```

注册完成之后,先调用 BuildServiceProvider 方法获取 ServiceProvider 依赖注入容器对象,再调用 GetRequiredService 方法获取指定的实例。

4.2.2 根据标记注册

对于批量注册来说,只要制定一套规则,就可以灵活地实现一个批量注册器。本节通过标记的形式实现批量注册,由于篇幅有限,本节只做思维扩展,至于全面的实现,读者可以根据该实例进行扩展。首先创建一个 TransientServiceAttribute 特性类,用于标注当前类型要注册的类型,如代码 4-25 所示。

代码4-25

```
[AttributeUsage(AttributeTargets.Class, Inherited = false)]
public class TransientServiceAttribute : Attribute
{
}
```

创建完标记类之后,修改 4.1 节的控制台实例,如代码 4-26 所示,应用程序只需要检索具有自定义特性的类即可,如果匹配,那么进行后续的服务注册操作。

代码4-26

```
var services = new ServiceCollection();
var assembly = Assembly.Load("app");
var typeList = assembly.GetTypes().Where(t =>
t.GetCustomAttribute<TransientServiceAttribute>()?
              .GetType() == typeof(TransientServiceAttribute));

var dic = new Dictionary<Type, Type[]>();
foreach(var type in typeList)
{
    var interfaces = type.GetInterfaces();
    dic.Add(type, interfaces);
}
if(dic.Keys.Count > 0)
{
    Foreach(var instanceType in dic.Keys)
    {
```

```
            foreach(var interfaceType in dic[instanceType])
            {
                services.AddTransient(interfaceType, instanceType);
            }
        }
}
var serviceProvider = services.BuildServiceProvider();
var sample = serviceProvider.GetRequiredService<ISample>();
Console.WriteLine(sample.Id);
```

4.2.3 配置扩展方法

至此，相信读者对批量注册已经有了一定的了解，在这种操作模式下，可以通过标记类型或其他方式制定一套规则来进行注册。可以将这套规则定义为扩展方法，简化注册时的代码量，同时方便统一化管理。如代码 4-27 所示，对控制台实例进行模拟，首先创建 ServiceCollection 对象，然后调用自定义的 BatchRegisterService 扩展方法进行服务批量注册操作。

代码4-27

```
var services = new ServiceCollection();
var assembly = Assembly.Load("app");
services.BatchRegisterService(assembly, "Sample", ServiceLifetime.Singleton);
var serviceProvider = services.BuildServiceProvider();
var sample = serviceProvider.GetRequiredService<ISample>();
Console.WriteLine(sample.Id);
```

如代码 4-28 所示，笔者编写了一个简单的扩展方法，使用该方法可以传递指定的条件，在方法的最后定义了生命周期参数，可以指定服务的生命周期。

代码4-28

```
public static IServiceCollection BatchRegisterService(
    this IServiceCollection services, Assembly assembly, string endWith,
        ServiceLifetime serviceLifetime= ServiceLifetime.Singleton)
{
    IEnumerable<Type> typeList = assembly.GetTypes()
        .Where(t => !t.IsInterface && !t.IsSealed &&
            !t.IsAbstract && t.Name.EndsWith(endWith));
    var dic = new Dictionary<Type, Type[]>();
    foreach(var type in typeList)
    {
```

```
        var interfaces = type.GetInterfaces();
        dic.Add(type, interfaces);
    }
    if(dic.Keys.Count > 0)
    {
        foreach(var instanceType in dic.Keys)
        {
            foreach(var interfaceType in dic[instanceType])
            {
                switch(serviceLifetime)
                {
                    case ServiceLifetime.Singleton:
                        services.AddSingleton(interfaceType, instanceType);
                        break;
                    case ServiceLifetime.Scoped:
                        services.AddScoped(interfaceType, instanceType);
                        break;
                    case ServiceLifetime.Transient:
                        services.AddTransient(interfaceType, instanceType);
                        break;
                }
            }
        }
    }
    return services;
}
```

4.3 小结

至此，相信读者对依赖注入框架已经有了基本的了解。在.NET 中，依赖注入无处不在，依托依赖注入框架提供的操作机制，开发人员可以高效地管理代码中的依赖关系，从而实现代码耦合度的进一步降低。

第 5 章

配置与选项

配置与选项是两种完全独立的模式。自公布.NET Core 之后，就出现了一套全新的配置（Configuration）模式及选项（Options）模式。通过使用依赖注入机制，可以将配置作为服务传递给各种中间件组件和其他应用程序类。本章主要介绍配置与选项的使用和设计，并设计了一个简单的配置中心。

5.1 配置模式

配置，对于开发人员来说并不陌生。早期的.NET 开发人员使用了大量的 System.Configuration 和 XML 配置文件（如 web.config 文件、app.config 文件），而在.NET Core 之后，配置系统发生了翻天覆地的变化，新的类型系统变得更加轻量级、更加灵活，支持丰富多样的数据源的配置类型，如文件、环境变量或其他存储方式（如 Azure Key Vault、Redis、数据库），还可以使用内存存储和命令行参数。配置模式的使用可以通过依赖注入的形式被业务代码调用，并且支持动态更新。

5.1.1 获取和设置配置

在配置系统中，一般采用键值对结构。在.NET 中，通过 Configuration 对象的 Providers 集合属性提供配置的获取和设置功能，开发人员可以通过定义标准化的配置格式被业务代码调用。配置的格式规则大致包括以下几点。

- 键不区分大小写。
- 值是字符串/布尔型/整型的。
- 层次结构以":"隔开。

如代码5-1所示,以appsettings.json文件为例,展示一个JSON配置文件。

代码5-1

```json
{
  "ConnectionStrings": {
    "DefaultConnection": "connection-string"
  },
  "Logging": {
    "LogLevel": {
      "Default": "Information",
      "Microsoft": "Warning",
      "Microsoft.Hosting.Lifetime": "Information"
    }
  }
}
```

假设要在应用程序中读取配置文件,需要设置"复制到输出目录"为"始终复制",并且引入的 NuGet 包为 Microsoft.Extensions.Configuration.Binder 和 Microsoft.Extensions.Configuration.Json,创建并使用 ConfigurationBuilder 对象,通过调用 AddJsonFile 扩展方法注册配置项(一个或多个文件)。代码 5-2 展示了调用 ConfigurationBuilder 对象的 Build 方法创建并返回 Configuration 对象,以及调用 Configuration 对象进行配置的读/写操作。

代码5-2

```csharp
IConfiguration configuration = new ConfigurationBuilder()
        .AddJsonFile("appsettings.json")
        .Build();

var defaultConnection = configuration
        .GetValue<string>("ConnectionStrings:DefaultConnection");
var connectionString =
        configuration.GetConnectionString ("DefaultConnection");
configuration["ConnectionStrings:DefaultConnection"] =
                        "new connectionn-string";
var newConnectionString = configuration
        .GetValue<string>("ConnectionStrings:DefaultConnection");
```

```
Console.WriteLine($"GetValue: {defaultConnection}");
Console.WriteLine($"GetConnectionString: {connectionString}");
Console.WriteLine($"newConnectionString: { newConnectionString}");
```

读取配置文件的信息，并将结果输出到控制台中，如图 5-1 所示。

图 5-1　读取配置文件的信息

在 ConfigurationBuilder 对象中有一个名为 Properties 的字典（IDictionary<Key,Value>）属性，在注册配置资源时会使用该属性。

在每次调用扩展方法（如 AddJsonFile 和 AddXmlFile）进行注册时，都会增加一个 IConfigurationSource 对象并保存到 ConfigurationBuilder 对象的 Sources 集合属性中，而 ConfigurationBuilder 对象会通过调用 Build 方法处理注册的配置资源。

每个 IConfigurationSource 对象都会创建一个 IConfigurationProvider 对象。当调用 ConfigurationBuilder 对象的 Build 方法时，也会依次调用每个 IConfiguartionSource 对象的 Build 方法并进行构建配置。

在 IConfigurationProvider 对象中实际上是读取并解析配置的，并在配置上提供标准化的视图，因此，系统可以查询键值对。

在调用 IConfigurationBuilder 对象的 Build 方法时会创建并返回 IConfiguration 对象，IConfiguration 对象可以通过调用如表 5-1 所示的方法/属性进行配置。

表 5-1　IConfiguration 对象的方法/属性

属性/方法	说明
Item[String]	获取或设置配置值
GetChildren()	获取子配置节
GetReloadToken()	返回一个 IChangeToken 对象，接收配置源被重新加载的通知
GetSection(String)	获取指定键的子配置节
Bind(IConfiguration,Object)	将配置根据匹配名称绑定到给定的对象实例中
Bind(IConfiguration,Object,Action<BinderOptions>)	将配置根据匹配名称绑定到给定的对象实例中
Bind(IConfiguration,String,Object)	将指定的子配置节绑定到预先创建的对象中
Get(IConfiguration,Type)	将指定的 IConfiguration 对象转换成指定类型的 POCO 对象

续表

属性/方法	说明
Get(IConfiguration,Type,Action<BinderOptions>)	将指定的 IConfiguration 对象转换成指定类型的 POCO 对象
Get<T>(IConfiguration)	将指定的 IConfiguration 对象转换成指定类型的 POCO 对象
Get<T>(IConfiguration,Action<BinderOptions>)	将指定的 IConfiguration 对象转换成指定类型的 POCO 对象
GetValue(IConfiguration,Type,String)	根据指定的键将 IConfiguration 对象转换为指定的类型
GetValue(IConfiguration,Type,String,Object)	根据指定的键将 IConfiguration 对象转换为指定的类型
GetValue<T>(IConfiguration,String)	根据指定的键将 IConfiguration 对象转换为 T 类型
GetValue<T>(IConfiguration,String,T)	根据指定的键将 IConfiguration 对象转换为 T 类型
AsEnumerable(IConfiguration)	根据 IConfiguration 对象获取 IEnumerable 对象集合
AsEnumerable(IConfiguration,Boolean)	根据 IConfiguration 对象获取 IEnumerable 对象集合
GetConnectionString(IConfiguration,String)	GetSection("ConnectionString")[name]的简化形式

当 ConfigurationBuilder 对象的 Build 方法被调用后，如代码 5-3 所示，会循环调用每个 IConfigurationSource 对象的 Build 方法并依次构建 IConfigurationProvider 对象，同时将构建的 IConfigurationProvider 对象保存到 List<IConfigurationProvider>集合中。

代码5-3

```
public IConfigurationRoot Build()
{
    var providers = new List<IConfigurationProvider>();
    foreach (IConfigurationSource source in Sources)
    {
        IConfigurationProvider provider = source.Build(this);
        providers.Add(provider);
    }
    return new ConfigurationRoot(providers);
}
```

5.1.2 使用强类型对象承载配置

在实际的开发过程中，如果逐条读取所有的内容，那么这将是一件非常烦琐的事情。因此，可以将配置直接转换为一个 POCO 对象（Plain Old C# Object，是指只有属性而没有行为的简单对象）。

在配置系统中，当 IConfiguration 对象与对应的 POCO 对象具有兼容性的数据结构时，就会利用其配置自动绑定对象，并将 IConfiguration 对象指定的值直接转换为 POCO 对象。

代码 5-4 展示了将 IConfiguration 对象转换为 POCO 对象的操作。

代码5-4

```
var systemSettings = new ConfigurationBuilder()
   .AddJsonFile("appsettings.json")
   .Build()
   .GetSection("ConnectionStrings")
   .Get<SystemSettings>();
```

5.1.3 使用环境变量配置源

在.NET 全新的配置系统中，支持将内存变量、命令行参数、环境变量和物理文件作为配置源数据。例如，物理文件支持 JSON、XML、INI 等格式。如果这些配置源无法满足需求，那么还可以注册自定义 IConfigurationSource 对象，并以此作为数据源的扩展。

假设有如代码 5-5 所示的环境变量。

代码5-5

```
set DATABASE__CONNECTION=connection-string
set LOGGING__ENABLED=True
set LOGGING__LEVEL=Debug
```

调用 AddEnvironmentVariables 扩展方法注册 IConfigurationSource 对象，如代码 5-6 所示。

代码5-6

```
var config = new ConfigurationBuilder()
   .AddEnvironmentVariables()
.Build();
```

读取的配置内容如代码 5-7 所示。

代码5-7

```
database:connection = "connection-string"
logging:enabled = "True"
logging:level = "Debug"
```

当使用环境变量作为配置源时，会检索出系统中所有的环境变量作为配置，为了避免出现这种情况，可以使用具有特定前缀的变量进行标识，如代码 5-8 所示。

代码5-8

```
set CONFIGURATION_DATABASE__CONNECTION = connection-string
```

```
set LOGGING__ENABLED = True
set LOGGING__LEVEL = Debug
```

在配置注册时,同样指定一个前缀作为过滤条件,如代码 5-9 所示。

代码5-9

```
var config = new ConfigurationBuilder()
    .AddEnvironmentVariables("CONFIGURATION_")
    .Build();
```

过滤后的数据结果如代码 5-10 所示。

代码5-10

```
database:connection = "connection-string"
```

注意:当使用环境变量时,需要使用"__"隔开而不是使用":"隔开。

在本节实例中,默认获取的是系统级别的环境变量。在 Windows 操作系统中,可以使用"系统属性"命令查看或设置系统级别与用户级别的环境变量,具体的步骤为选择"此电脑"→"属性"→"高级系统设置"→"环境变量"命令。在 Visual Studio 中,可以在"项目属性"选项卡中找到"启动配置文件"界面进行设置或查看环境变量,具体的步骤为选择"属性"→"调试"→"常规"→"打开调试启动配置文件 UI"命令。设置的环境变量如图 5-2 所示。

图 5-2 设置的环境变量

使用 Visual Studio 设置的环境变量会被添加并保存到 launchSettings.json 文件中，如代码 5-11 所示。

代码5-11

```
{
  "profiles": {
    "Configuration01": {
      "commandName": "Project",
      "environmentVariables": {
        "Key": "Value",
        "Key1": "Value1"
      }
    }
  }
}
```

如代码 5-12 所示，环境变量的配置源是通过 EnvironmentVariablesConfigurationSource 类来表示的，该类继承了 IConfigurationSource 接口，用于定义环境变量配置源的实现。在 EnvironmentVariablesConfigurationSource 类中定义了一个名为 Prefix 的字符串类型的属性，用于标识环境变量的前缀。如果设置了 Prefix 属性，那么只会匹配名称（Prefix 属性值）作为前缀的环境变量。

代码5-12

```
public class EnvironmentVariablesConfigurationSource : IConfigurationSource
{
    public string Prefix { get; set; }

    public IConfigurationProvider Build(IConfigurationBuilder builder)
    {
        return new EnvironmentVariablesConfigurationProvider(Prefix);
    }
}
```

EnvironmentVariablesConfigurationSource 配置源会利用 EnvironmentVariablesConfigurationProvider 对象来读取环境变量，如代码 5-13 所示，由 Load 方法可以看出，通过_prefix 字段对集合数据进行筛选，从而将筛选出来的数据添加到配置字典中。

代码5-13

```
public class EnvironmentVariablesConfigurationProvider : ConfigurationProvider
```

```csharp
{
    private const string MySqlServerPrefix = "MYSQLCONNSTR_";
    private const string SqlAzureServerPrefix = "SQLAZURECONNSTR_";
    private const string SqlServerPrefix = "SQLCONNSTR_";
    private const string CustomPrefix = "CUSTOMCONNSTR_";
    private readonly string _prefix;

    public EnvironmentVariablesConfigurationProvider() =>
                                        _prefix = string.Empty;
    public EnvironmentVariablesConfigurationProvider(string? prefix) =>
                                        _prefix = prefix ?? string.Empty;

    ///<summary>
    ///加载环境变量
    ///</summary>
    public override void Load() =>
            Load(Environment.GetEnvironmentVariables());

    public override string ToString() =>
                    $"{GetType().Name} Prefix: '{_prefix}'";

    internal void Load(IDictionary envVariables)
    {
        var data = new Dictionary<string, string?>(
                            StringComparer.OrdinalIgnoreCase);
        IDictionaryEnumerator e = envVariables.GetEnumerator();
        try
        {
            while (e.MoveNext())
            {
                DictionaryEntry entry = e.Entry;
                string key = (string)entry.Key;
                string? provider = null;
                string prefix;
                if(key.StartsWith(MySqlServerPrefix,
                            StringComparison.OrdinalIgnoreCase))
                {
                    prefix = MySqlServerPrefix;
                    provider = "MySql.Data.MySqlClient";
```

```csharp
        }
        else if(key.StartsWith(SqlAzureServerPrefix,
                    StringComparison.OrdinalIgnoreCase))
        {
            prefix = SqlAzureServerPrefix;
            provider = "System.Data.SqlClient";
        }
        Else if(key.StartsWith(SqlServerPrefix,
                    StringComparison.OrdinalIgnoreCase))
        {
            prefix = SqlServerPrefix;
            provider = "System.Data.SqlClient";
        }
        else if(key.StartsWith(CustomPrefix,
                    StringComparison.OrdinalIgnoreCase))
        {
            prefix = CustomPrefix;
        }
        else if(key.StartsWith(prefix,
                    StringComparison.OrdinalIgnoreCase))
        {
            key = NormalizeKey(key.Substring(_prefix.Length));
            //指定 Key 值并进行复制
            data[key] = entry.Value as string;
            continue;
        }
        else
        {
            continue;
        }
        //将 key 中的 "__" 替换为 ":"
        key = NormalizeKey(key.Substring(prefix.Length));
        //如果存在 prefix 不为空,则从 key 中移除 prefix 部分
        AddIfPrefixed(data, $"ConnectionStrings:{key}",
                                    (string?) entry.Value);
        if (provider != null)
        {
            AddIfPrefixed(data, $"ConnectionStrings:{key}_ProviderName",
                                            provider);
```

```
            }
        }
    }
    finally
    {
        (e as IDisposable)?.Dispose();
    }
    Data = data;
}

private void AddIfPrefixed(Dictionary<string, string?> data,
                                     string key, string? value)
{
    if(key.StartsWith(_prefix, StringComparison.OrdinalIgnoreCase))
    {
        key = key.Substring(_prefix.Length);
        data[key] = value;
    }
}
private static string NormalizeKey(string key) => key.Replace("__",
                                    ConfigurationPath.KeyDelimiter);
```

在 ASP.NET Core 中，前缀 ASPNETCORE_ 中也预定义了一些环境变量，比较常见的有以下几个。

- ASPNETCORE_ENVIRONMENT：定义环境名称，一般为 Development、Staging 和 Production。
- ASPNETCORE_URLS：定义应用程序监听的地址/端口。
- ASPNETCORE_WEBROOT：定义应用程序静态资源的路径，默认为 Content Root/wwwroot 目录。

5.1.4 使用命令行配置源

应用程序在启动时，使用命令行开关（Switch）作为应用程序的控制行为也是一个不错的选择。至此，读者已经对环境变量配置源有了基本的了解，接下来介绍命令行配置源。

命令行配置源与 5.1.3 节中的 EnvironmentVariablesConfigurationSource 类一样，都具有一个数据解析器。应用程序接收命令行参数，命令行参数通常是一串字符串，所以，

CommandLineConfigurationSource 类的目的在于将命令行参数由字符串数组转换为配置字典。

需要注意的是，参数必须以单横杠（-）或双横杠（--）开头，映射字典不能包含重复的 Key。命令行配合的几种格式命令如下。

- {key}={value}。
- {prefix}{key}={value}。
- {key} {value}。
- {prefix}{key} {value}。

命令行参数会在参数名称前加一个前缀，目前支持"/"、"--"和"-"这 3 种格式。

下面先介绍一个实例，如代码 5-14 所示。

代码5-14

```
dotnet run CommandLineKey1=value1 --CommandLineKey2=value2
                                              /CommandLineKey3= value3
dotnet run -CommandLineKey1 value1 /CommandLineKey2 value2
dotnet run CommandLineKey1=CommandLine Key2=value2
```

假设应用程序需要根据命令行参数进行启动，开发人员应当首先引入的 NuGet 包为 Microsoft.Extensions.Configuration.CommandLine，如代码 5-15 所示，通过调用 AddCommandLine 扩展方法进行注册，当然，这需要一个参数，而这个参数可以直接填入 args，args 接收用户输入的参数。

代码5-15

```
IConfiguration configuration = new ConfigurationBuilder()
        .AddCommandLine(args)
        .Build();
Console.WriteLine($"os: {configuration["os"]}");
Console.WriteLine($"framework: {configuration["f"]}");
Console.ReadLine();
```

图 5-3 所示为读取的命令行参数。

如代码 5-16 所示，命令行参数的配置源可以通过 CommandLineConfigurationSource 类来表示，在该类中定义了 Args 属性和 SwitchMappings 属性，Args 属性主要用于承载命令行参数集合，SwitchMappings 属性则保存了命令行开关的缩写和全称之间的映射关系。CommandLineConfigurationSource 类通过继承 IConfigurationSource 接口实现 Build 方法，根据 SwitchMappings 属性和 Args 属性创建并返回了 CommandLineConfigurationProvider 对象。

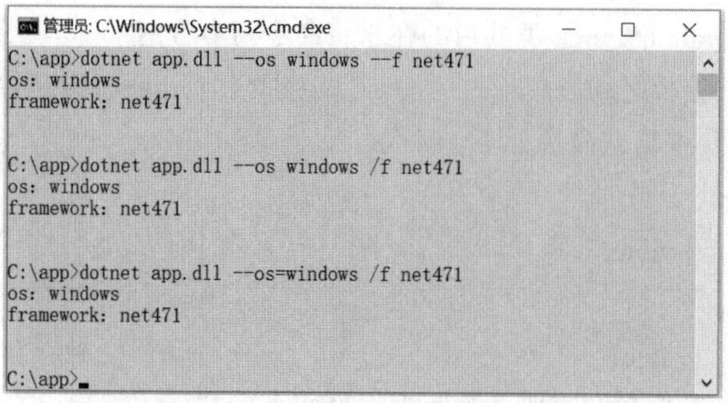

图 5-3 读取的命令行参数

代码5-16

```
public class CommandLineConfigurationSource : IConfigurationSource
{
    public IDictionary<string, string>? SwitchMappings { get; set; }
    public IEnumerable<string> Args { get; set; } = Array.Empty<string>();

    public IConfigurationProvider Build(IConfigurationBuilder builder)
    {
        return new CommandLineConfigurationProvider(Args, SwitchMappings);
    }
}
```

CommandLineConfigurationProvider 对象和 EnvironmentVariablesConfigurationProvider 对象的设计是一样的，都是维护数据解析器，主要用于对命令行参数进行数据解析，并将解析后的参数名称和值保存到配置字典中，如代码 5-17 所示。

代码5-17

```
public class CommandLineConfigurationProvider : ConfigurationProvider
{
    private readonly Dictionary<string, string> _switchMappings;

    public CommandLineConfigurationProvider(
        IEnumerable<string> args,
        IDictionary<string, string> switchMappings = null)
    {
        this.Args = args ?? throw new ArgumentNullException(nameof(args));
        if(switchMappings == null)
            return;
```

```csharp
        this.switchMappings =
                this.GetValidatedSwitchMappingsCopy (switchMappings);
}

protected IEnumerable<string> Args { get; private set; }

public override void Load()
{
    Dictionary<string, string> dictionary =
     new Dictionary <string,string>(StringComparer.OrdinalIgnoreCase);
    using(IEnumerator<string> enumerator =
                            this.Args. GetEnumerator())
    {
        while(enumerator.MoveNext())
        {
            string key1 = enumerator.Current;
            int startIndex = 0;
            if(key1.StartsWith("--"))
                startIndex = 2;
            else if(key1.StartsWith("-"))
                startIndex = 1;
            else if(key1.StartsWith("/"))
            {
                key1 = string.Format("--{0}", (object)key1.Substring(1));
                startIndex = 2;
            }
            int length = key1.IndexOf('=');
            string key2;
            string str1;
            //...more
            dictionary[key2] = str1;
        }
    }
    this.Data = (IDictionary<string, string>)dictionary;
}
```

如代码 5-18 所示，通过默认的输入参数可以直接将配置信息放在字典中，同时以默认的方式对参数进行绑定。

代码5-18

```
var mapping = new Dictionary<string, string>
{
    ["-os"] = "windows",
    ["-f"] = "net471"
};
var configuration = new ConfigurationBuilder()
    .AddCommandLine(args, mapping)
    .Build();
Console.WriteLine($"os：{configuration["os"]}");
Console.WriteLine($"framework：{configuration["f"]}");
```

命令行配置源始终保持以键值对的设计模型为基础，从而可以联想到在配置系统中，配置系统的底层设计无法脱离键值对数据模型，但是开发人员可以根据需要，通过一定的规则模式，实现一套属于自己的配置源。

5.1.5 配置管理器（ConfigurationManager）

在配置系统中，除了上面介绍的，还可以采用 ConfigurationManager 对象。先引入的 NuGet 包为 Microsoft.Extensions.Configuration，如代码 5-19 所示，笔者以控制台实例为例，首先创建 ConfigurationManager 对象，然后通过指定的 Key 读取配置，接着调用 AddJsonFile 扩展方法注册 JSON 配置文件，最后进行第二次读取操作。

代码5-19

```
using Microsoft.Extensions.Configuration;

class Program
{
    static void Main(string[] args)
    {
        var key"= "ConnectionStrings:DefaultConnect"on";
        //创建 ConfigurationManager 对象
        ConfigurationManager configurationManager = new ConfigurationManager();
        //通过指定的 Key 读取配置的值
        Console.WriteLine($"第一次输出：{configurationManager[key]}");
```

```
        //调用AddJsonFile扩展方法，添加config.json配置文件
        configurationManager.AddJsonFile("config.json", true, true);
        Console.WriteLine($"第二次输出：{configurationManager[key]}");
        Console.ReadKey();
    }
}
```

输出结果如图 5-4 所示。

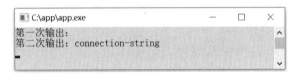

图 5-4　以键值对的形式读取配置

由图 5-4 可以看到，ConfigurationManager 对象第一次输出为空，第二次输出时配置项已经保存到配置系统中，从而可以取值并输出。

ConfigurationManager 对象的产生是为了向 Minimal 出发，从而简化开发人员的使用及代码量。为此，它实现了 IConfigurationBuilder 接口、IConfigurationRoot 接口和 IDisposable 接口，将两个对象的特性集成在 ConfigurationManager 对象中，如代码 5-20 所示。

代码5-20

```
public sealed class ConfigurationManager :
            IConfigurationBuilder, IConfigurationRoot, IDisposable
{
    private readonly ConfigurationSources _sources;
    private readonly ConfigurationBuilderProperties _properties;
    private readonly ReferenceCountedProviderManager
                                                _providerManager = new();
    private readonly List<IDisposable> changeTokenRegistrations = new();
    private ConfigurationReloadToken changeToken = new();

    public ConfigurationManager()
    {
        _sources = new ConfigurationSources(this);
        _properties = new ConfigurationBuilderProperties(this);
        _sources.Add(new MemoryConfigurationSource());
    }
```

```
//...
}
```

如代码 5-21 所示,在实例化 ConfigurationManager 对象时会创建一个 ConfigurationSources 对象。ConfigurationSources 类继承了 IList<IConfigurationSource> 接口,用于维护一个 ConfigurationSource 集合对象。

代码5-21

```
private sealed class ConfigurationSources : IList<IConfigurationSource>
{
    private readonly List<IConfigurationSource> _sources = new();
    private readonly ConfigurationManager _config;

    public ConfigurationSources(ConfigurationManager config)
    {
        _config = config;
    }

    public IConfigurationSource this[int index]
    {
        get => _sources[index];
        set
        {
            _sources[index] = value;
            _config.ReloadSources();
        }
    }

    public void Add(IConfigurationSource source)
    {
        _sources.Add(source);
        _config.AddSource(source);
    }
    //...
}
```

实际上,在添加 ConfigurationSource 配置源对象时,也会注册一个 IConfigurationProvider 对象,随后会加载配置源的数据并注册一个更新事件令牌,所以,开发人员在配置注册完之后才会读取到配置信息,如代码 5-22 所示。

代码5-22

```csharp
public sealed class ConfigurationManager :
            IConfigurationBuilder, IConfigurationRoot, IDisposable
{
    private readonly ConfigurationSources sources;
    private readonly ConfigurationBuilderProperties properties;
    private readonly ReferenceCountedProviderManager providerManager
                                                                = new();
    private readonly List<IDisposable> changeTokenRegistrations = new();
    private ConfigurationReloadToken changeToken = new();

    private void RaiseChanged()
    {
        var previousToken = Interlocked.Exchange(ref changeToken,
                                    new ConfigurationReloadToken());
        previousToken.OnReload();
    }

    private void AddSource(IConfigurationSource source)
    {
        IConfigurationProvider provider = source.Build(this);
        provider.Load();
        changeTokenRegistrations.Add(ChangeToken.OnChange(()=>
                    provider.GetReloadToken(), () => RaiseChanged()));

        providerManager.AddProvider(provider);
        RaiseChanged();
    }
}
```

5.2 选项模式

在.NET Core 时期就已经引入了选项模式，开发人员可以轻松地将配置绑定到 POCO 对象上。选项模式可以采用依赖注入的方式，使开发人员可以直接使用绑定配置信息的 POCO

对象，这个 POCO 对象为特定的 Options 对象，也被称为配置选项。在选项模式中有 3 种使用方式，分别是 IOptions、IOptionsSnapshot 和 IOptionsMonitor，接下来介绍如何使用这 3 种方式，从而帮助读者了解其基本运作。

5.2.1 将配置绑定 Options 对象

选项模式使用依赖注入注册到容器中，但依赖注入框架并不会提供 Options 对象，而是通过依赖注入框架创建一个 IOptions<TOptions>对象，泛型参数 TOptions 对应定义的目标配置对象。下面演示如何利用 IOptions<TOptions>对象获取所需的 Options 对象。

笔者以控制台项目为例，在项目中定义一个名为 person.json 的配置文件，如代码 5-23 所示。为了使 person.json 文件能在编译后自动复制到输出目录下，需要将文件属性中的"复制到输出目录"设置为"始终复制"。

代码5-23

```
{
  "age": 18,
  "languages": [
    "English",
    "Chinese"
  ],
  "company": {
    "title": "Microsoft",
    "country": "USA"
  }
}
```

该文件中的信息实际上是对人员的一些描述，如果希望将这些数据自动映射为一个对象，就需要创建一个 Person 类，如代码 5-24 所示。

代码5-24

```
public class Person
{
    public string Name { get; set; }
    public int Age { get; set; }
    public List<string> Languages { get; set; }
    public Company Company { get; set; }
}
```

```csharp
public class Company
{
    public string Title { get; set; }
    public string Country { get; set; }
}
```

如代码 5-24 所示，定义 Company 类对应公司信息，Person 类的定义与 person.json 文件的结构是完全相同的。

代码 5-25 展示了如何通过选项模式来获取配置信息并绑定到 Person 对象中。首先创建一个 ConfigurationBuilder 对象，然后通过调用 AddJsonFile 扩展方法注册 JSON 配置文件（person.json）的配置源，并通过它创建对应的 IConfiguration 对象。

代码5-25

```csharp
IConfiguration configuration = new ConfigurationBuilder()
        .AddJsonFile("person.json")
        .Build();
var person = new ServiceCollection()
   .Configure<Person>(configuration.GetSection(""))
   .BuildServiceProvider()
   .GetRequiredService<IOptions<Person>>()
   .Value;
Console.WriteLine($"Age：{person.Age}");
Console.WriteLine($"Company：{person.Company.Title}");
Console.WriteLine("Languages：");
foreach (var item in person.Languages)
{
   Console.WriteLine(item);
}
```

接下来创建一个 ServiceCollection 对象，并调用 Configure<Person>扩展方法。该扩展方法被定义在 Microsoft.Extensions.Options.ConfigurationExtensions 命名空间下，而 Configure<TOptions>扩展方法内部默认已经调用了 AddOptions 扩展方法，注册选项模式的核心服务。代码 5-26 展示了 Configure<TOptions>扩展方法内部的实现，将创建的 IConfiguration 对象作为参数传递。Configure<TOptions>扩展方法将 IConfiguration 对象与指定的 TOptions 对象进行映射，在需要提供对应的 TOptions 对象时，IConfiguration 对象承载的配置数据会被提取出来并绑定到 TOptions 对象中。

代码5-26

```csharp
public static class OptionsConfigurationServiceCollectionExtensions
{
    public static IServiceCollection Configure<TOptions>(this
      IServiceCollection services,IConfiguration config) where
      TOptions : class =>
        services.Configure<TOptions>(Options.Options.DefaultName, config);

    public static IServiceCollection Configure<TOptions>(this
       IServiceCollection services, stringname, IConfiguration config)
        where TOptions : class =>
          services.Configure<TOptions>(name, config, => { });

    public static IServiceCollection Configure<TOptions>(this
       IServiceCollection services,IConfiguration config,
       Action<BinderOptions> configureBinder) where TOptions : class =>
       services.Configure<TOptions>(Options.Options.DefaultName,
                                          config, configureBinder);

    public static IServiceCollection Configure<TOptions>(this
       IServiceCollection services!!, string name,
        IConfiguration config!!, Action<BinderOptions> configureBinder)
       where TOptions : class
    {
      services.AddOptions();
      services.AddSingleton<IOptionsChangeTokenSource<TOptions>>(
          new ConfigurationChangeTokenSource<TOptions>(name, config));
      return services.AddSingleton<IConfigureOptions<TOptions>>(
         new NamedConfigureFromConfigurationOptions<TOptions>(name, config,
                                          configureBinder));
    }
}
```

首先调用 IServiceCollection 对象的 BuildServiceProvider 扩展方法，获取依赖注入容器的 IServiceProvider 对象。然后调用 GetRequiredService<T>扩展方法获取 IOptions<Person>对象，该对象的 Value 属性返回的就是 IConfiguration 对象绑定生成的 Person 对象。最后将 Person 对象相关的数据直接输出到控制台中。Person 对象值的输出结果如图 5-5 所示。

图 5-5　Person 对象值的输出结果

如代码 5-27 所示，TOptions 对应的是 Value 属性的类型，在 Value 属性被调用时，Value 属性先执行 Get 方法，并为 Value 属性赋值，调用 Get 方法时会传递一个 Options.DefaultName 参数值，Get 方法在 OptionsCache<TOptions>对象中读取名称为 name 的值，如果值存在则通过 out 关键字进行传递，如果值不存在则执行 if 语句内的操作，将值缓存起来。

注意：OptionsCache<TOptions>对象中维护了一个线程安全的集合 ConcurrentDictionary<string, Lazy<TOptions>>。

代码5-27

```
public class OptionsManager<TOptions> :
    IOptions<TOptions>,
    IOptionsSnapshot<TOptions>
    where TOptions : class
{
    private readonly IOptionsFactory<TOptions> _factory;

    //通过 IOptionsFactory<TOptions>对象创建的 Options 对象会被其缓存起来
    private readonly OptionsCache<TOptions> _cache =
                                    new OptionsCache<TOptions>();

    public TOptions Value => Get(Options.DefaultName);

    public virtual TOptions Get(string name)
    {
        name = name ?? Options.DefaultName;

        if(!_cache.TryGetValue(name, out TOptions options))
        {
            IOptionsFactory<TOptions> localFactory = _factory;
            string localName = name;
            options =
            _cache.GetOrAdd(name, () => localFactory.Create(localName));
```

```
        }
        return options;
    }
}
```

5.2.2 命名选项的使用

命名选项可以使开发人员在配置系统中拥有相同的强类型对象和不同的实例对象，它们之间通过名称来区分。在需要的时候，可以按名称对它们进行检索。下面通过控制台实例进行演示。

创建一个配置文件 company.json，其结构如代码 5-28 所示。

代码5-28

```
{
  "company1": {
    "title": "Company1",
    "country": "USA"
  },
  "company2": {
    "title": "Company2",
    "country": "USA"
  }
}
```

使用命名选项，通过一个类获取多个独立的配置实例，可以按照如代码 5-28 所示的 company.json 文件创建一个相应的配置类，如代码 5-29 所示。

代码5-29

```
public class Company
{
    public string Title { get; set; }
    public string Country { get; set; }
}
```

如代码 5-30 所示，演示如何做到多个独立的配置实例，通过调用 AddJsonFile 扩展方法将 JSON 配置文件（company.json）的配置源创建到 ConfigurationBuilder 对象上，并利用它创建对应的 IConfiguration 对象。

代码5-30

```
IConfiguration configuration = new ConfigurationBuilder()
        .AddJsonFile("company.json")
        .Build();
var services = new ServiceCollection()
    .Configure<Company>("Company1", configuration.GetSection("Company1"))
    .Configure<Company>("Company2", configuration.GetSection("Company2"));
var options = services.BuildServiceProvider()
    .GetRequiredService<IOptionsSnapshot<Company>>();
var company1 = options.Get("Company1");
var company2 = options.Get("Company2");
Console.WriteLine($"Company1: {company1.Title}");
Console.WriteLine($"Company2: {company2.Title}");
```

创建 ServiceCollection 对象，调用 Configure<Company>扩展方法，同时根据指定的名称对 Options 进行初始化，第二个参数用于指定相应的配置项。调用 IServiceCollection 对象的 BuildServiceProvider 扩展方法得到依赖注入容器的 IServiceProvider 对象后，先调用 GetRequiredService<T>扩展方法得到 IOptionsSnapshot<Company>对象，再通过 Get 方法获取指定名称的配置对象。当按照指定名称读取后，创建两个 Company 对象，将两个对象的 Title 属性值打印在控制台中。如图 5-6 所示，采用命名选项方式读取并输出结果。

图 5-6 采用命名选项方式读取并输出结果

如代码 5-31 所示，关于命名选项的处理，需要先在缓存中获取对应的配置信息，本节不再重复介绍，读者可以参考 5.2.1 节的内容。

代码5-31

```
public virtual TOptions Get(string? name)
{
    name = name ?? Options.DefaultName;
    if (!_cache.TryGetValue(name, out var options))
    {
        IOptionsFactory<TOptions> localFactory = _factory;
        string localName = name;
        return _
```

```
        cache.GetOrAdd(name, () => localFactory.Create(localName));
    }
    return options;
}
```

5.2.3 Options 内容验证

上面介绍了 IOptions，简单来说就是将配置注入服务中。但在此之前并没有提过它允许配置验证，IOptions 对象对配置项并不具有很强的约束性，如果配置文件中没有值，就使用默认值进行填充，这可能会在运行时给应用程序造成严重的隐式错误。

验证配置内容可以使用 System.ComponentModel.DataAnnotations 命名空间下的数据验证特性类，如代码 5-32 所示，创建一个名为 Person 的 POCO 类，在 Age 属性中添加 RangeAttribute 特性。

代码5-32

```
public class Person
{
    public string Name { get; set; }
    [RangeAttribute(minimum: 20, maximum: int.MaxValue, ErrorMessage =
                        "填写信息有误，年龄必须大于19岁。")]
    public int Age { get; set; }
    public List<string> Languages { get; set; }
    public Company Company { get; set; }
}

public class Company
{
    public string Title { get; set; }
    public string Country { get; set; }
}
```

在该实例中添加了 RangeAttribute 特性，并且通过该特性限制最小年龄为 20 岁，设置错误提示为"信息填写错误，年龄必须大于 19 岁。"

上面已经对 Age 属性进行了设置，下面需要使用验证器检查这些值是否可以通过验证。

如代码 5-33 所示，首先使用 AddOptions 扩展方法注册选项模式的核心服务，并通过 Bind 方法进行配置的绑定操作，然后调用 ValidateDataAnnotations 扩展方法来确保可以触发验证机制。

代码5-33

```
IConfiguration configuration = new ConfigurationBuilder()
        .AddJsonFile("person.json")
        .Build();
var services = new ServiceCollection();
services.AddOptions<Person>()
    .Bind(configuration)
    .ValidateDataAnnotations();
var person = services.BuildServiceProvider()
    .GetRequiredService<IOptions<Person>>()
    .Value;
```

调用 IServiceCollection 对象的 BuildServiceProvider 扩展方法得到依赖注入容器的 IServiceProvider 对象后，可以先调用 GetRequiredService<T>扩展方法来获取 IOptions<Person> 服务，再通过调用属性 Value 返回 Person 对象，同时触发验证机制。

通过运行该程序，可以得到抛出的 OptionsValidationException 异常，如图 5-7 所示。

图 5-7　OptionsValidationException 异常

对于选项验证，还可以使用 delegate（委托）方式。开发人员可以通过调用 Validate 扩展方法实现配置的验证工作，如代码 5-34 所示。

代码5-34

```
var optionsBuilder = services.AddOptions<Person>()
    .Bind(configuration)
    .ValidateDataAnnotations()
    .Validate(config =>
    {
        if(config.Age < 19)
```

```
        return false;
    return true;
});
```

可以通过修改 Person 对象删除 RangeAttribute 属性，这样也可以对数据进行验证。当再次执行程序时，会抛出默认的异常信息，如图 5-8 所示。

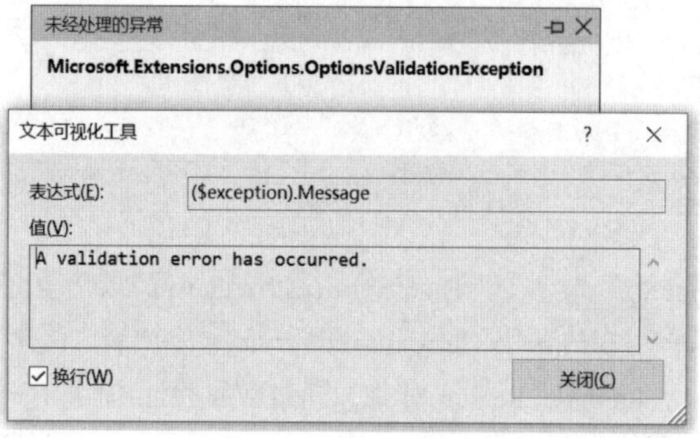

图 5-8　默认的异常信息

抛出的异常信息其实具有通用性，这是一个预期的效果。当开发人员要执行自己的验证逻辑时，可以自定义一条错误的消息内容，使它更友好一些，如代码 5-35 所示，自定义错误的消息内容为"填写信息有误，年龄必须大于 19 岁。"。

代码5-35

```
var optionsBuilder = services.AddOptions<Person>()
    .Bind(configuration)
    .ValidateDataAnnotations()
    .Validate(config =>
    {
        if (config.Age < 19)
            return false;
        return true;
    }, "填写信息有误，年龄必须大于19岁。");
```

当再次运行程序时，现有内容就变成自定义的异常信息，如图 5-9 所示。

对于验证的使用，还有另一种实现方式，就是通过 IValidateOptions 接口实现自定义的验证逻辑，开发人员可以将异常逻辑抽象到一个类中。为了做到这一点，可以新建一个 AgeConfigurationValidation 类并实现 IValidateOptions 接口，如代码 5-36 所示。

图 5-9　自定义的异常信息

代码5-36

```
public class AgeConfigurationValidation : IValidateOptions<Person>
{
    public ValidateOptionsResult Validate(string? name, Person options)
    {
        if (options.Age < 19)
        {
            return ValidateOptionsResult.Fail("填写信息有误,年龄必须大于19岁。");
        }
        return ValidateOptionsResult.Success;
    }
}
```

在该实例中,将 IValidateOptions 接口通过 AgeConfigurationValidation 类来实现,并且实现 Validate 方法。Validate 方法接收两个参数,即名称和选项(Options),可以看到此方法返回一个 ValidateOptionsResult 对象,这是返回验证结果的便捷方式。

如代码 5-37 所示,修改注册方式,注册 IValidateOptions 对象为 Singleton 模式。

代码5-37

```
IConfiguration configuration = new ConfigurationBuilder()
    .AddJsonFile("person.json")
    .Build();
var services = new ServiceCollection();
services.Configure<Person>(configuration);
services.AddSingleton<IValidateOptions<Person>, AgeConfigurationValidation>();
var person = services
```

```
.BuildServiceProvider()
.GetRequiredService<IOptions<Person>>()
.Value;
```

先通过 Configure 注册配置项，再注册 AgeConfigurationValidation 对象。需要注意的是，要先注册配置项，再注册配置验证对象。

如图 5-10 所示，使用 IValidateOptions 验证方式会得到同一个错误结果。

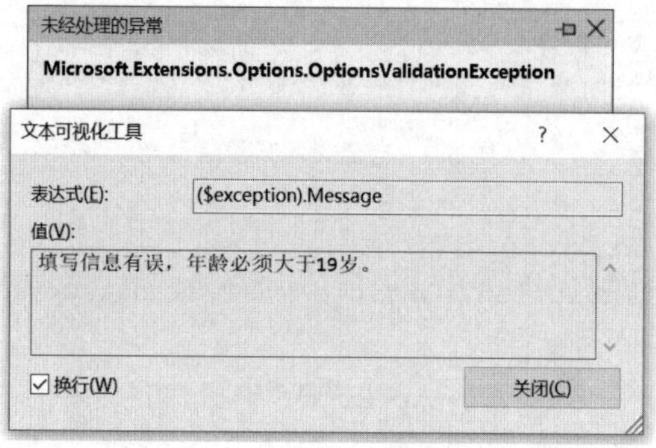

图 5-10　使用 IValidateOptions 验证方式

当然，对于 IValidateOptions 对象来说，开发人员也可以采用动态化的验证方式，如可以通过注入 DbContext 对象，或者通过数据库读取验证规则，从而实现验证逻辑的动态化，这些都是可行的。

代码 5-38 展示了 IValidateOptions 对象的实现，ValidateOptions<TOptions>类是对 IValidateOptions<TOptions>接口的实现。

代码5-38

```csharp
public class ValidateOptions<TOptions> : IValidateOptions<TOptions>
                                            where TOptions : class
{
    public ValidateOptions(string? name,
            Func<TOptions, bool> validation!!, string failureMessage)
    {
        Name = name;
        Validation = validation;
        FailureMessage = failureMessage;
    }

    public string? Name { get; }
```

```
public Func<TOptions, bool> Validation { get; }
public string FailureMessage { get; }

public ValidateOptionsResult Validate(string? name, TOptions options)
{
    if(Name == null || name == Name)
    {
        if(Validation.Invoke(options))
        {
            return ValidateOptionsResult.Success;
        }
        return ValidateOptionsResult.Fail(FailureMessage);
    }
    return ValidateOptionsResult.Skip;
}
```

从代码 5-38 中可以看到 ValidateOptions<TOptions>对象需要的参数，如 Options 的名称和 FailureMessage 属性，以及验证 Options 的 Func<TOptions, bool>对象和实现的 Validate 方法，在内部通过 Func<TOptions, bool>对象进行验证，并返回 ValidateOptionsResult 对象。

5.2.4　Options 后期配置

PostConfigure 扩展方法用于后期配置，主要是在 Options 配置完成之后再对其做出一些修改，覆盖前面的配置。下面介绍几种使用方式。

代码 5-39 展示了后期配置的使用方式。

代码5-39

```
services.PostConfigure<Person>(person =>
{
    person.Age = 10;
});
```

如代码 5-40 所示，可以用于命名选项的后期配置。

代码5-40

```
services.PostConfigure<Person>("person1", person =>
{
```

```
    person.Age = 10;
});
```

如代码 5-41 所示，可以使用 PostConfigureAll 对所有配置实例进行后期配置。

代码5-41

```
services.PostConfigureAll<Person>(person =>
{
    person.Age = 10;
});
```

如代码 5-42 所示，通过控制台实例来演示。首先创建 ServiceCollection 对象，通过 Configure 方法进行配置的装载，并且在调用 IServiceCollection 对象的 BuildServiceProvider 扩展方法得到依赖注入容器的 IServiceProvider 对象后，通过调用 GetRequiredService<T>扩展方法来获取 IOptions<Person>对象；然后调用 PostConfigure<TOptions>扩展方法，用于变更之前写入的配置。

代码5-42

```
IConfiguration configuration = new ConfigurationBuilder()
    .AddJsonFile("person.json")
    .Build();
var services = new ServiceCollection();
var person1 = services
    .Configure<Person>(configuration)
    .BuildServiceProvider()
    .GetRequiredService<IOptions<Person>>();

Console.WriteLine($"person1: {person1.Value.Age}");
services.PostConfigure<Person>(person =>
{
    person.Age = 10;
});
var person2 = services.BuildServiceProvider()
                      .GetRequiredService<IOptions<Person>>();
Console.WriteLine($"person2: {person2.Value.Age}");
```

该实例打印输出了两次配置信息，一次是装载配置后的原始信息，另一次是通过 PostConfigure 扩展方法修改后的数据。需要注意的是，此处存在一个顺序性，首先装载 Configure 配置，PostConfigure 扩展方法可以用于修改前后通过 Configure 的配置，也就是无论 PostConfigure 扩展方法在前或在后，配置都可以起到修改作用，然后将结果打印出来。通过使用 PostConfigure 扩展方法进行输出，如图 5-11 所示。

图 5-11　通过使用 PostConfigure 扩展方法进行输出

Options 对象的创建过程如代码 5-43 所示，通过 Create 方法创建指定的 Options 对象。可以很清晰地看出，选项模式会抽象出一个 OptionsFactory 来处理 Options 对象的初始化与创建工作。

代码5-43

```
public class OptionsFactory<TOptions> :
    IOptionsFactory<TOptions>
    where TOptions : class
{
    private readonly IConfigureOptions<TOptions>[] _setups;
    private readonly IPostConfigureOptions<TOptions>[] _postConfigures;
    private readonly IValidateOptions<TOptions>[] _validations;
    public OptionsFactory(IEnumerable<IConfigureOptions<TOptions>> setups,
            IEnumerable<IPostConfigureOptions<TOptions>> postConfigures)
        : this(setups,postConfigures,
            validations: Array.Empty<IValidateOptions<TOptions>>()){ }
    public OptionsFactory(IEnumerable<IConfigureOptions<TOptions>> setups,
            IEnumerable<IPostConfigureOptions<TOptions>> postConfigures,
            IEnumerable<IValidateOptions<TOptions>> validations)
    {
        _setups = setups as IConfigureOptions<TOptions>[] ?? new
                List<IConfigureOptions<TOptions>>(setups).ToArray();
        _postConfigures = postConfigures as
                IPostConfigureOptions<TOptions>[] ?? new
        List<IPostConfigureOptions<TOptions>>(postConfigures).ToArray();
        _validations = validations as IValidateOptions<TOptions>[] ?? new
                List<IValidateOptions<TOptions>>(validations).ToArray();
    }

    public TOptions Create(string name)
```

```csharp
{
    TOptions options = CreateInstance(name);
    foreach(IConfigureOptions<TOptions> setup in _setups)
    {
        if(setup is IConfigureNamedOptions<TOptions> namedSetup)
        {
            namedSetup.Configure(name, options);
        }
        else if(name == Options.DefaultName)
        {
            setup.Configure(options);
        }
    }
    foreach(IPostConfigureOptions<TOptions> post in _postConfigures)
    {
        post.PostConfigure(name, options);
    }

    if(_validations.Length > 0)
    {
        var failures = new List<string>();
        foreach(IValidateOptions<TOptions> validate in _validations)
        {
            ValidateOptionsResult result = validate.Validate(name, options);
            if(result is not null && result.Failed)
            {
                failures.AddRange(result.Failures);
            }
        }
        if(failures.Count > 0)
        {
            throw new OptionsValidationException(
                        name, typeof(TOptions), failures);
        }
    }
    return options;
}
```

```
protected virtual TOptions CreateInstance(string name)
{
    return Activator.CreateInstance<TOptions>();
}
}
```

Create 方法包含 3 个阶段，分别为 Configure 阶段、PostConfigure 阶段和 Validate 阶段，如图 5-12 所示，第一个阶段通过 Configure 方法对 Options 对象进行初始化。另外，在 IConfigureOptions<TOptions>的处理过程中，Options 对象如果存在名称，则以该名称进行初始化，如果不存在则采取默认的名称。随后，对 IPostConfigureOptions<TOptions>的 Options 对象进行初始化，初始化之后会对其进行验证，但仅在有验证设置的情况下。关于 Options 对象的验证可以参考 5.2.3 节。

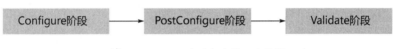

图 5-12　Create 方法包含的 3 个阶段

5.2.5　IOptionsSnapshot<TOptions>

IOptionsSnapshot<TOptions>可以用来自动重载配置，而不需要重新启动应用程序。IOptionsSnapshot 对象默认采用的生命周期模式为 Scoped，每次注入都会得到新的对象，而 IOptions 对象采用的生命周期模式是 Singleton。因此，Singleton 在应用程序的整个生命周期中都有效。由于 IOptionsSnapshot<TOptions>对象被注册为 Scoped，因此该对象获取的 Options 对象只会在当前请求上下文中保持一致。

接下来通过一个实例来演示 IOptions<TOptions>对象和 IOptionsSnapshot<TOptions>对象之间的差异。

如代码 5-44 所示，在应用程序中创建一个 ServiceCollection 对象，调用 Options 扩展方法注册 Options 对象的核心服务后，首先调用 Configure<Person>方法，然后注册 CountIncrement 实例，而 CountIncrement 类继承自 IConfigureOptions<T>接口，并且实现了 Configure 方法，用于处理对配置项更改的模拟操作，最后定义了 SetupInvokeCount 静态属性，用于保存每次累加的数值。使用本地函数 Print 可以模拟调用情况。

代码5-44

```
IConfiguration configuration = new ConfigurationBuilder()
    .AddJsonFile("person.json")
    .Build();
```

```
var services = new ServiceCollection();
var serviceProvider = services
    .AddOptions()
    .Configure<Person>(configuration)
    .AddSingleton<IConfigureOptions<Person>>(new CountIncrement())
    .BuildServiceProvider();
Print(serviceProvider);
Print(serviceProvider);

static void Print(IServiceProvider provider)
{
    var scopedProvider = provider
        .GetRequiredService<IServiceScopeFactory>()
        .CreateScope()
        .ServiceProvider;
    var options = scopedProvider.GetRequiredService<IOptions<Person>>()
                                                                    .Value;
    var optionsSnapshot1 =
        scopedProvider.GetRequiredService<IOptionsSnapshot<Person>>().Value;
    var optionsSnapshot2 =
        scopedProvider.GetRequiredService<IOptionsSnapshot<Person>>().Value;

    Console.WriteLine($"options:{options.Age}");
    Console.WriteLine($"optionsSnapshot1:{optionsSnapshot1.Age}");
    Console.WriteLine($"optionsSnapshot2:{optionsSnapshot2.Age}");
}

public class CountIncrement : IConfigureOptions<Person>
{
    static int SetupInvokeCount { get; set; }
    public void Configure(Person options)
    {
        SetupInvokeCount++;
        options.Age += SetupInvokeCount;
    }
}
```

可以通过 ServiceCollection 对象创建 IServiceProvider 对象来提供服务，但是通过 ServiceCollection 对象创建的 IServiceProvider 对象表示子容器的 IServiceProvider 对象，子容器的 IServiceProvider 对象相当于 ASP.NET Core 应用程序中针对当前请求创建的 IServiceProvider 对象（RequestServices）。随后分别针对 IOptions<TOptions>对象和 IOptionsSnapshot<TOptions>对象得到对应的 Person 对象之后，将 Age 属性值打印到控制台中，由此可知上述操作先后执行了两次，相当于 ASP.NET Core 应用程序处理了两次请求。

图 5-13 所示为使用 IOptionsSnapshot<TOptions>对象的输出结果，说明只有从同一个 IServiceProvider 对象获取的 IOptionsSnapshot<TOptions>对象才能提供一致的 Options 对象。但是对于所有来自同一个 IServiceProvider 对象来说，获取的 IOptions<TOptions>对象始终可以提供一致的 Options 对象。

图 5-13　使用 IOptionsSnapshot<TOptions>对象的输出结果

5.2.6　IOptionsMonitor<TOptions>

IOptionsMonitor 既可以做到热载，也可以做到在配置发生变动时触发事件。5.2.5 节提到，IOptionsSnapshot 对象可以做到热载，并且可以作为依赖项注册到容器中。但是需要注意的是，IOptionsSnapshot 对象采用的生命周期模式是 Scoped。Scoped 服务不能在 Singleton 服务中使用，因为这是依赖注入容器中的一项安全机制，被称为 ValidateScopes。IOptions 对象采用的生命周期模式为 Singleton，但是它并不能做到热载，因此要使用像 IOptionsSnapshot 对象的这种热载功能，必须处理验证问题，开发人员可以使用 IOptionsMonitor 对象。与 IOptions 对象一样，IOptionsMonitor 对象采用的生命周期模式也是 Singleton，意味着可以安全地将其注册到其他 Singleton 服务中。

如代码 5-45 所示，创建 IServiceCollection 对象，调用 BuildServiceProvider 扩展方法，由于该方法支持设置 validateScopes，因此将 validateScopes 设置为 true，以用于启动服务的范围验证。

代码5-45

```
IConfiguration configuration = new ConfigurationBuilder()
    .AddJsonFile("person.json", true, true)
```

```
        .Build();
var services = new ServiceCollection();
var optionsSnapshot = services
    .AddOptions()
    .Configure<Person>(configuration)
    .AddSingleton<ConfigureReader>()
    .BuildServiceProvider(validateScopes: true)
.GetRequiredService<ConfigureReader>();

public class ConfigureReader
{
    private readonly Person _person;
    public ConfigureReader(IOptionsSnapshot<Person> optionsSnapshot)
    {
        _person = optionsSnapshot.Value;
    }
    public int GetAge()
    {
        return _person.Age;
    }
}
```

随后创建 ConfigureReader 类，并以 Singleton 模式进行注册，在 ConfigureReader 类中获取 IOptionsSnapshot<TOptions>对象，应用程序运行后会触发验证错误。范围验证的错误信息如图 5-14 所示。

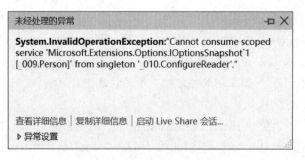

图 5-14　范围验证的错误信息

如代码 5-46 所示，先创建 ServiceCollection 对象，同时添加 Options 的核心服务，再装载配置，通过 GetRequiredService<T>方法获取服务对象。在 IOptionsMonitor<TOptions>对象中，OnChange 事件用于接收文件的变更通知，在收到变更通知时打印输出信息。

代码5-46

```
IConfiguration configuration = new ConfigurationBuilder()
        .AddJsonFile("person.json", true, true)
        .Build();
var services = new ServiceCollection();
var optionsMonitor = services
    .AddOptions()
    .Configure<Person>(configuration)
    .BuildServiceProvider().GetRequiredService<IOptionsMonitor<Person>>();

optionsMonitor.OnChange(o =>
{
    Console.WriteLine($"时间：{DateTime.Now}, Age：{o.Age}");
});
Console.ReadLine();
```

AddJsonFile 扩展方法具有重载方法。AddJsonFile 扩展方法中的 optional 参数表示文件是否可选，reloadOnChange 参数表示如果文件更改是否重载配置。因为此处需要重载，所以将 reloadOnChange 参数设置为 true。图 5-15 所示为文件更改时打印的信息，可以通过多次修改配置文件来达到热加载的状态。

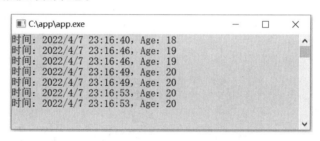

图 5-15　文件更改时打印的信息

5.3　设计一个简单的配置中心

上面介绍了配置文件的使用和运行机制，本节主要介绍配置中心。配置中心是许多分布式应用程序的基础，本节将创建一个配置中心的客户端，以便开发人员初步了解配置中心的设计。事实上，配置中心有许多开源组件可供挑选，虽然有时忌讳重新造轮子，但是探究一个组件的原理有助于开发人员构建更加扎实的技术体系。

5.3.1 什么是配置中心

配置中心，顾名思义，是一个统一管理项目配置的系统。在常规项目中，配置文件方便在应用程序中进行集中配置。随着微服务的流行，应用程序的粒度越来越小，配置文件的数量呈现几何级数增长的趋势，采用传统的本地配置文件，难免有点不合时宜，并且维护成本和同步成本都很高，进而造成产品运维成本的攀升，因此，许多复杂系统都需要引入配置中心。

如图 5-16 所示，当项目以单机服务运行时，配置通常存储在文件中，代码发布的时候，将配置文件和应用程序发布到服务器上。

随着业务和用户数量的增加，单机服务无法满足业务需求，这就需要对单体应用进行拆分，应用从而走向微服务，通常来说可能需要把服务进行多机器（集群）或多副本部署，在应用负载均衡配置的情况下，配置的发布就会变成如图 5-17 所示的形式。

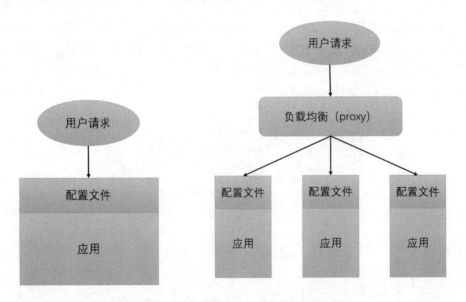

图 5-16　项目以单机服务运行　　　　图 5-17　应用负载均衡

由于应用分布在不同的服务器上，同时管理多份配置文件显然不现实，而利用配置中心可以做到在运行时在线修改配置信息，集中化管理应用配置，还有望支持配置的环境隔离（如果开发人员搭建了开发、测试、预发布、灰度/线上等多套环境，则可以通过为配置建立环境标签来进行物理隔离）。

5.3.2 ConfigurationSource

上面介绍了 .NET Core 配置及常规的配置方式，如文件、命令行参数等。由此可知，要扩展一个其他类型的配置源，需要重写 ConfigurationProvider 类，如代码 5-47 所示，首先定义一个名为 SimpleConfigurationProvider 的类，继承 ConfigurationProvider 类，然后重写 Load 方法。

代码5-47

```
public class SimpleConfigurationProvider : ConfigurationProvider
{
    public override void Load()
    {
        using(var httpClient = new HttpClient
        {
            BaseAddress = new Uri("https://localhost:44360")
        })
        {
            var response =
                httpClient.GetStringAsync("/configuration/ test.json")
                .ConfigureAwait(false).GetAwaiter().GetResult();
            if(!string.IsNullOrEmpty(response))
            {
                Data = JsonConvert
                    .DeserializeObject<IDictionary<string, string>>(response);
            }
        }
    }
}
```

在 Load 方法中，可以利用 HttpClient 对象发起 HTTP 请求，调用服务端配置，在接收到结果之后对其进行解析，同时绑定到数据字典的 Data 属性中。

5.3.3 ConfigurationProvider

如代码 5-48 所示，创建 SimpleConfigurationSource 类，实现 IConfigurationSource 接口，在 Build 方法中返回 SimpleConfigurationProvider 实例。

代码5-48

```
public class SimpleConfigurationSource : IConfigurationSource
{
    public IConfigurationProvider Build(IConfigurationBuilder builder)
    {
        return new SimpleConfigurationProvider();
    }
}
```

5.3.4 配置扩展方法

添加扩展方法，如代码 5-49 所示，创建一个名为 SimpleConfigurationExtensions 的静态类，同时添加一个名为 AddSimpleSource 的扩展方法，将 SimpleConfigurationSource 加入配置源集合中。

代码5-49

```
public static class SimpleConfigurationExtensions
{
    public static IConfigurationBuilder AddSimpleSource(this
                                       IConfigurationBuilder builder)
    {
        return builder.Add(new SimpleConfigurationSource());
    }
}
```

服务端向客户端应用返回的配置信息如代码 5-50 所示。

代码5-50

```
{"key":"1","key1":"Key1"}
```

如代码 5-51 所示，通过控制台实例进行演示，先调用 AddSimpleSource 扩展方法注册自定义的配置源，并创建 IConfiguration 对象，再利用 IConfiguration 对象读取配置项。

代码5-51

```
class Program
{
    static void Main(string[] args)
    {
        IConfiguration configuration = new ConfigurationBuilder()
```

```
        .AddSimpleSource()
        .Build();
    var key = configuration.GetValue<string>("key");
    Console.WriteLine(key);
    }
}
```

先通过 IConfiguration 对象调用 GetValue 方法,根据指定的键名读取指定的配置项,再将其打印输出到控制台中。GetValue 方法的输出结果如图 5-18 所示。

图 5-18　GetValue 方法的输出结果

5.3.5　配置动态更新(HTTP Long Polling 方案)

虽然 5.3.2~5.3.4 节中的实例能做到将配置进行统一化管理,但是如果配置发生变更,就无法保证应用程序随之改变配置信息,所以需要进行配置的拉取操作。对于拉取操作来说,可以选择轮询(Polling)、长轮询(Long Polling)、长连接或 WebSocket,本实例采用 Long Polling 方案实现。下面修改本节编写的应用程序。

如代码 5-52 所示,修改 SimpleConfigurationSource 类,增加的 ConfigurationServiceUri 属性用于设置配置中心服务端的地址,同时通过 RequestTimeout 属性设置 HTTPClient 的请求超时时间,添加 ConfigurationName 属性用于请求配置名称。细心的读者可能会发现,配置中心服务端的地址为/configuration/test.json,该地址以.json 的形式返回一组配置。当然,如果读者感兴趣也可以进行改造和扩展。

代码5-52

```
public class SimpleConfigurationSource : IConfigurationSource
{
    public string ConfigurationServiceUri { get; set; }
    public TimeSpan RequestTimeout { get; set; }
                              = TimeSpan.FromSeconds(60);

    public string ConfigurationName { get; set; }
    public IConfigurationProvider Build(IConfigurationBuilder builder)
```

```
        {
            return new SimpleConfigurationProvider(this);
        }
}
```

如代码 5-53 所示，修改 SimpleConfigurationProvider 类，继承并实现 IDisposable 接口用于 HttpClient 的释放操作。在 RequestConfigurationAsync 方法中，如果请求成功则调用 Notify 方法，在字典信息赋值后调用 OnReload 方法，进行配置的刷新操作。

代码5-53

```
public class SimpleConfigurationProvider : ConfigurationProvider,
 IDisposable
{
    private readonly SimpleConfigurationSource _source;
    private readonly Lazy<HttpClient> _httpClient;
    private bool _isDisposed;
    private CancellationTokenSource? _cts;
    private HttpClient HttpClient => _httpClient.Value;
    public SimpleConfigurationProvider(SimpleConfigurationSource source)
    {
        _source = source ??
                    throw new ArgumentException (nameof(source));
        _httpClient = new Lazy<HttpClient>(CreateHttpClient);
    }
    private HttpClient CreateHttpClient()
    {
        var handler = new HttpClientHandler();
        var client = new HttpClient(handler, true)
        {
            BaseAddress = new Uri(_source.ConfigurationServiceUri),
            Timeout = _source.RequestTimeout
        };
        return client;
    }
    private async Task RequestConfigurationAsync(
                        CancellationToken cancellationToken)
    {
```

```csharp
        var encodedConfigurationName =
            WebUtility.HtmlEncode(_source.ConfigurationName);
        while(!cancellationToken.IsCancellationRequested)
        {
            try
            {
                using(var response = await HttpClient.GetAsync(
                    encodedConfigurationName, cancellationToken)
                                            .ConfigureAwait(false))
                {
                    if(response.IsSuccessStatusCode)
                    {
                        using (var stream = await
                    response.Content.ReadAsStreamAsync(cancellationToken)
                                            .ConfigureAwait (false))
                        {
                            stream.Position = 0;
                            var data = new JsonConfigurationFileParser()
                                    .Parse(stream);
                            if(response.StatusCode == HttpStatusCode.OK)
                            {
                                Notify(data);
                            }
                        }
                    }
                }
            }
            finally
            {
                await Task.Delay(50, cancellationToken)
                                    . ConfigureAwait(false);
            }
        }
    }

private void Notify(IDictionary<string, string> data)
```

```csharp
    {
        Data = data;
        OnReload();
    }

    public override void Load()
    {
        LoadAsync();
    }

    private async Task LoadAsync()
    {
        CancellationTokenSource cancellationToken =
        Interlocked.Exchange(ref _cts,new CancellationTokenSource());
        if(cancellationToken != null)
        {
            return;
        }
        await RequestConfigurationAsync(_cts.Token);
    }

    public void Dispose()
    {
        if(_isDisposed)
        {
            return;
        }
        if(_httpClient?.IsValueCreated == true)
        {
            _httpClient.Value.Dispose();
        }
        _isDisposed = true;
    }
}
```

上述代码继承了 ConfigurationProvider 类，重写了 Load 方法，在首次加载时会调用 Load 方法，在 Load 方法中调用自定义的 LoadAsync 方法，随后利用 RequestConfigurationAsync

方法通过 while 语句拉取配置信息，当获取到信息后会调用 Notify 方法并赋值给 Data 属性，填充配置信息。

本实例采用的 Long Polling 方案需要等待 50 毫秒，如上述代码片段中的数据解析器提供的是解析的 JSON 格式，可以继续抽象，让它允许更多类型的支持，这样可以实现一个简易的配置中心客户端。

如代码 5-54 所示，笔者通过对配置源进行设置，同时循环打印并输出配置的 key 信息，因为服务端配置可能会在不固定的时间发生变更，所以可以通过随机数进行模拟变更。

代码5-54

```
class Program
{
    static async Task Main(string[] args)
    {
        var source = new SimpleConfigurationSource
        {
            ConfigurationName = "test.json",
            ConfigurationServiceUri = "https://localhost:44360/configuration/"
        };
        IConfiguration configuration = new ConfigurationBuilder()
            .AddSimpleSource(source)
            .Build();
        while (true)
        {
            var key = configuration.GetValue<string>("key");
            Console.WriteLine($"时间：{DateTime.Now}, Key：{key}");
            await Task.Delay(TimeSpan.FromSeconds(10));
        }
        Console.ReadLine();
    }
}
```

循环获取最新配置的输出结果，如图 5-19 所示。

图 5-19　循环获取最新配置的输出结果

5.4 小结

本章首先介绍了配置与选项的常规用法，包括如何使用 JSON 配置文件和环境变量，以及对配置进行动态更新等操作。配置是应用开发最不可或缺的一部分，基于 ASP.NET Core 提供的强大的配置能力，开发人员可以轻松地完成对单体应用配置的管理。随后，本章进一步介绍了在分布式应用场景下如何实现一个简单的配置中心，并基于配置中心实现了配置的动态刷新，限于篇幅，未对服务端配置管理展开介绍，如果读者感兴趣，可以考虑自行实现相关代码，获得更多收获。

第 6 章

使用 IHostedService 和 BackgroundService 实现后台任务

在应用程序开发过程中，开发人员可能常常需要创建用于后台执行的任务。例如，在应用程序内部需要定时调用第三方接口获得某些响应数据来为业务提供支持，或者需要定时对历史数据进行某些计算操作，这些都可能会用到定时任务或后台任务。

.NET 生态体系中与定时任务或后台任务有关的技术实现有多种方式，其中有 4 种比较常见，分别是 IHostedService、后台任务（BackgroundService）、Quartz.NET 和 Hangfire。其中，前两者都是.NET 原生提供的方式，IHostedService 是一种接口，BackgroundService 则是基于主机托管服务的一种实现，使用这两种方式可以轻松、快速地创建后台托管服务。

6.1 IHostedService

IHostedService 允许开发人员在后台创建多个任务（托管服务），并支持在 Web 应用程序后台运行。IHostedService 的生命周期采用的是 Singleton 模式，在应用程序启动时注册实例化，应用程序将通过一个独立的线程来维护该实例化任务，在应用程序退出时，注册的任务也会随着应用程序的退出而退出。

如代码 6-1 所示，IHostedService 接口定义了 StartAsync 方法和 StopAsync 方法，.NET 应用程序会在启动和关闭时分别调用 IHostedService 接口的 StartAsync 方法和 StopAsync 方法。

代码6-1

```csharp
public interface IHostedService
{
    Task StartAsync(CancellationToken cancellationToken);
    Task StopAsync(CancellationToken cancellationToken);
}
```

从 IHostedService 接口的命名来看，它需要与通用主机绑定，本节将使用控制台应用程序进行探讨。如代码 6-2 所示，以控制台应用程序为例，在创建了通用主机之后，调用 ConfigureServices 方法用于配置和注册服务项，在该方法中通过调用 AddHostedService 扩展方法来注册后台任务 MyHostedService 对象，最终创建到 HostBuilder 对象上。为了能够更好地查看效果，本实例还调用了 ConfigureLogging 扩展方法，用于添加配置日志依赖项。由于这是一个控制台应用程序，该对象实例的 RunConsoleAsync 方法会在应用程序启动后等待 Ctrl + C 组合键或 SIGTERM（终止信号）退出，表明若没有明确告诉应用程序退出，则它不会退出。

代码6-2

```csharp
var host = new HostBuilder()
    .ConfigureLogging(logging =>
    {
        logging.AddConsole();
    })
    .ConfigureServices((hostContext, services) =>
    {
        services.AddHostedService<MyHostedService>();
    });
await host.RunConsoleAsync();
```

如代码 6-3 所示，创建一个名为 MyHostedService 的类，继承 IHostedService 接口，实现 StartAsync 方法和 StopAsync 方法，在 StartAsync 方法中每隔 5 秒触发并执行一次 DoWork 方法，该方法主要用于输出时间，并测试后台任务的执行情况。

代码6-3

```csharp
public class MyHostedService : IHostedService
{
    private readonly ILogger<MyHostedService> _logger;
    private const int WAITTIME = 5000;//定义等待时间
    public MyHostedService(ILogger<MyHostedService> logger)
```

第 6 章　使用 IHostedService 和 BackgroundService 实现后台任务

```
        {
            _logger = logger;
        }
        public async Task StartAsync(CancellationToken cancellationToken)
        {
            logger.LogInformation(
                        "Starting IHostedService registered in Startup");
            while(true)
            {
                await Task.Delay(WAITTIME);
                DoWork();
            }
        }
        private void DoWork()
        {
            _logger.LogInformation($"Hello World! - {DateTime.Now}");
        }

        public Task StopAsync(CancellationToken cancellationToken)
        {
            _logger.LogInformation(
                        "Stopping IHostedService registered in Startup");
            return Task.CompletedTask;
        }
}
```

图 6-1 所示为 IHostedService 后台任务的输出结果，在应用程序运行后，IHostedService 会在后台将消息输出到控制台中。

图 6-1　IHostedService 后台任务的输出结果

> **注意**：可以创建多个 IHostedService 对象，并且每个实现 IHostedService 的对象都会在应用程序启动时调用它们的 StartAsync 方法和 StopAsync 方法。

6.2 BackgroundService

6.1 节对 IHostedService 对象进行了详细的探讨，本节主要介绍 BackgroundService 的抽象类，使用 BackgroundService 类来搭建一个后台任务应用程序。

如代码 6-4 所示，以控制台项目为例，创建 HostBuilder 对象，先调用 ConfigureLogging 扩展方法，添加配置日志依赖项，再调用 ConfigureServices 扩展方法用于配置服务项，并在该方法中调用 AddHostedService 扩展方法，注册后台任务 MyBackgroundService 对象，最后调用 HostBuilder 实例对象的 RunConsoleAsync 扩展方法，将它注册为控制台生命周期服务。

代码6-4

```
var host = new HostBuilder()
    .ConfigureLogging(logging =>
    {
        logging.AddConsole();
    })
    .ConfigureServices((hostContext, services) =>
    {
        services.AddHostedService<MyBackgroundService>();
    });
await host.RunConsoleAsync();
```

通过 NuGet 包安装 Microsoft.Extensions.Hosting，如代码 6-5 所示，创建 MyBackgroundService 类继承 Microsoft.Extensions.Hosting 命名空间下的 BackgroundService 类。

代码6-5

```
public class MyBackgroundService : BackgroundService
{
    private readonly ILogger<MyBackgroundService> _logger;
    public MyBackgroundService(ILogger<MyBackgroundService> logger)
    {
        _logger = logger;
    }
    protected override async Task ExecuteAsync(
```

第 6 章　使用 IHostedService 和 BackgroundService 实现后台任务

```
                             CancellationToken stoppingToken)
{
    _logger.LogInformation(
               "Starting IHostedService registered in Startup");
    while(true)
    {
        DoWork();
        await Task.Delay(5000);
    }
}
private void DoWork()
{
    _logger.LogInformation($"Hello World! - {DateTime.Now}");
}
}
```

后台服务需要实现 ExecuteAsync 方法，重写该方法，随后添加 while 循环语句，在循环体内部调用 Task.Delay 方法，使其每隔 5 秒触发并执行一次 DoWork 方法。DoWork 方法主要用于输出时间和测试后台任务的执行情况。图 6-2 所示为 BackgroundService 类的输出结果。

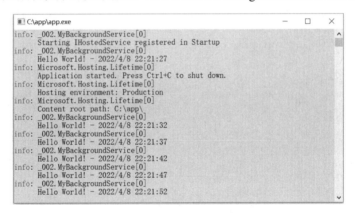

图 6-2　BackgroundService 类的输出结果

⌛ 扩展延伸

上面已经提到 BackgroundService 类是基于 IHostedService 的一种实现，下面介绍 BackgroundService 类的本质。在 BackgroundService 类中定义一个名为 ExecuteTask 的 Task 类型的属性，在启动主机托管服务后，BackgroundService 类会创建一个任务，用于执行 ExecuteAsync 方法，如代码 6-6 所示。

135

代码6-6

```csharp
public abstract class BackgroundService : IHostedService, IDisposable
{
    private Task? _executeTask;
    private CancellationTokenSource? _stoppingCts;
    public virtual Task? ExecuteTask => _executeTask;
    protected abstract Task ExecuteAsync(CancellationToken stoppingToken);

    public virtual Task StartAsync(CancellationToken cancellationToken)
    {
        _stoppingCts = CancellationTokenSource
                        .CreateLinkedTokenSource(cancellationToken);
        _executeTask = ExecuteAsync(_stoppingCts.Token);
        if(_executeTask.IsCompleted)
        {
            return _executeTask;
        }
        return Task.CompletedTask;
    }

    public virtual async Task StopAsync(CancellationToken cancellationToken)
    {
        if(_executeTask == null)
        {
            return;
        }
        try
        {
            _stoppingCts!.Cancel();
        }
        finally
        {
            Await Task.WhenAny(_executeTask, Task.Delay(
                Timeout.Infinite, cancellationToken)).ConfigureAwait(false);
        }
    }

    public virtual void Dispose()
```

```
{
    _stoppingCts?.Cancel();
}
}
```

BackgroundService 类简化了 IHostedService 对象的使用方式，并对它进行了抽象。图 6-3 所示为 IHostedService 关系图。

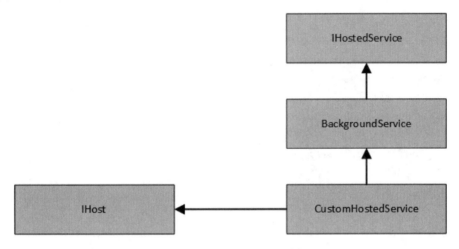

图 6-3　IHostedService 关系图

6.3　任务调度

Quartz 是任务库，在其他技术语言框架中都有应用。Quartz.NET 是 Quartz 在 .NET 中的实现，主要使用 Cron 语法来定义任务的执行方式，从 .NET Framework 时代开始就受到开发人员的欢迎。Hangfire 最大的优点是提供了 Hangfire Dashboard，可以提供可视化界面来管理定时任务。Hangfire 包括商业版和社区版两种版本，无论是大型商业项目，还是中小型企业级应用或互联网产品，都有适用的版本。

6.3.1　Hangfire

Hangfire 是在 .NET 平台下开源的任务调度框架，可以很方便地集成到项目中，并且功能强大，提供了可集成的可视化界面，支持中文界面。Hangfire 任务调度库位于"Hangfire" NuGet 包中。

如代码 6-7 所示，笔者以 WebAPI 项目为例，引入的 NuGet 包为 Hangfire，先调用 AddHangfire 扩展方法，再调用 UseSqlServerStorage 扩展方法，配置 Hangfire 以 SqlServer 数据库为持久化存储服务，最后调用 AddHangfireServer 扩展方法注册 Hangfire 核心服务。

代码6-7

```
var builder = WebApplication.CreateBuilder(args);

//注册 Hangfire 数据库持久服务
builder.Services.AddHangfire(x =>
    x.UseSqlServerStorage
(@"Server=.\;Database=Hangfire.Sample;Trusted_Connection=True;"));
//注册 Hangfire 核心服务
builder.Services.AddHangfireServer();

var app = builder.Build();
//启动 Hangfire 面板
app.UseHangfireDashboard();

app.Run();
```

调用 UseHangfireDashboard 扩展方法用于注册 Hangfire 的可视化面板界面。

在 Hangfire 配置完成后，便可以注册任务，周期性任务是日常开发中常用的任务模式之一，RecurringJob.AddOrUpdate 方法用于创建周期性任务，如代码 6-8 所示。

代码6-8

```
RecurringJob.AddOrUpdate("周期性任务", () =>
    Console.WriteLine($"周期性任务：{DateTime.Now.ToLongTimeString()}"),
Cron.Minutely);
```

接下来启动程序，先对配置的数据库进行检查，判断是否包含 Hangfire 的数据库结构，如果不包含则动态生成如图 6-4 所示的 Hangfire 的数据库结构。

在默认情况下，可视化界面的路径为/hangfire。图 6-5 所示为 Hangfire 仪表盘。

从仪表盘首页中可以清晰地看到实时图表走势和历史图表走势，由此可以对任务执行情况进行综合概述。另外，仪表盘首页的导航栏中提供了"作业"菜单、"重试"菜单、"周期性作业"菜单和"服务器"菜单。

作业：所有作业都将在此处展示，并且以不同的状态呈现，如计划、执行中、完成、失败、等待中等状态。

重试：处于失败状态的任务会进入"重试"菜单中。

周期性作业：显示定义的周期性作业列表。

服务器：显示注册的 Hangfire 服务器，这些服务器用于处理作业。

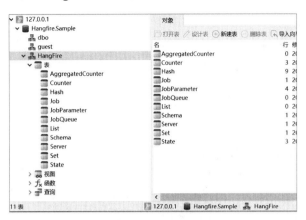

图 6-4　Hangfire 的数据库结构

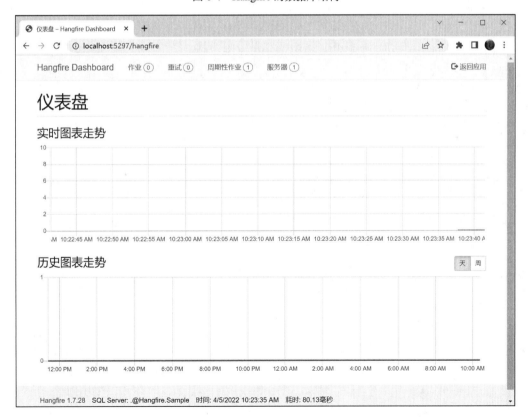

图 6-5　Hangfire 仪表盘

6.3.2 Quartz.NET

Quartz.NET 是在.NET 平台下开源的任务调度框架，如代码 6-9 所示，笔者以 WebAPI 项目为例，引入的 NuGet 包为 Quartz.AspNetCore，先调用 AddQuartz 扩展方法注册核心服务，再调用 AddQuartzServer 扩展方法注册 QuartzHostedService 服务。

代码6-9

```
var builder = WebApplication.CreateBuilder(args);

builder.Services.AddQuartz(q =>
{
    q.UseMicrosoftDependencyInjectionJobFactory();
    var jobKey = new JobKey("CustomJob");

    q.AddJob<CustomJob>(o => o.WithIdentity(jobKey));

    q.AddTrigger(opts => opts
        .ForJob(jobKey)
        .WithIdentity(jobKey + "trigger")
        //从第 0 秒开始，每 5 秒执行一次
        .WithCronSchedule("0/5 * * * * ?"));
});

builder.Services.AddQuartzServer(q =>
{
    q.WaitForJobsToComplete = true;
});

var app = builder.Build();

app.MapGet("/", () =>
{
    return "Hello World!";
});

app.Run();
```

在 AddQuartz 扩展方法中，首先调用 UseMicrosoftDependencyInjectionJobFactory 扩展方法，通过 DI（Dependency Injection，依赖注入）容器来管理这些作业，然后创建 JobKey 实例，接收一个字符串参数表示任务的名称，接着调用 AddJob 扩展方法注册 CustomJob 任务，最后调用 AddTrigger 扩展方法为作业创建一个触发器，其中，ForJob 扩展方法用于指定某个任务，WithIdentity 扩展方法用于设置触发器的名称，同时通过 WithCronSchedule 扩展方法设置任务每 5 秒执行一次。

代码 6-10 定义了 CustomJob 类，它继承了 IJob 接口，实现 Execute 方法，在该方法中输出日志。

代码6-10

```
public class CustomJob : IJob
{
    private readonly ILogger<CustomJob> _logger;
    public CustomJob(ILogger<CustomJob> logger)
    {
        _logger = logger;
    }
    public Task Execute(IJobExecutionContext context)
    {
        _logger.LogInformation("Job {dateTime}", DateTime.UtcNow);
        return Task.CompletedTask;
    }
}
```

图 6-6 所示为运行 Quartz.NET 控制台实施后台任务的输出结果。

图 6-6　运行 Quartz.NET 控制台实施后台任务的输出结果

6.4 小结

在.NET 平台下可以简化创建定时任务，使开发人员在编码时有更多的时间关注业务层。本章先介绍 IHostedService 和 BackgroundService，再介绍第三方任务调度的使用。通过学习本章，希望读者可以从本质上来了解这些内容，从思维上来掌握。

第 7 章

中间件

ASP.NET Core 引入了中间件（Middleware）的概念。由官方文档可知，中间件是一种从装配到应用管道以处理请求和响应的软件。ASP.NET Core 内置了多种中间件，开发人员也可以定义自己的中间件。使用中间件可以对 ASP.NET Core 应用程序的每个请求和响应进行处理，从而实现某些特定的功能。例如，在某些场景下，开发人员需要对请求携带的某个参数进行校验，这可以通过中间件来实现。

管道在 ASP.NET 中较为常见，主要包括 HttpHandlers 和 HttpModules 两部分。中间件是对管道的进一步包装，极大地减轻了开发人员的工作量，使开发人员能够以更加便捷的方式实现相关需求。本章将对 ASP.NET Core 的中间件展开介绍。

7.1 中间件的作用

中间件是构建在管道之上的一层组件，允许开发人员以更优雅的形式对请求管道进行某些特定的处理。多个中间件以责任链模式形成一种串行的顺序结构，每个中间件都会在其传递时进行处理并传递到下一个中间件，而响应内容则以相反的顺序返回。另外，利用中间件可以对应用程序的代码进行解耦。

图 7-1 所示为 Logger 请求中间件的执行过程。Logger 是一个负责记录所有请求和响应的中间件，会在请求时记录所有的请求和响应。

例如，在授权中间件中（如图 7-2 所示的授权响应阻止后），若中间件验证失败，则不会继续调用下一个中间件，并且在 Web 应用程序响应之前拒绝请求。

```
                request              > log request           > process request
SERVER ================ LOGGER ================ WEB APPLICATION
               response             < log response          < return response
```

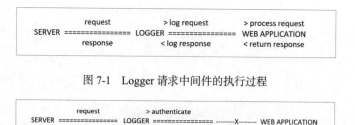

图 7-1　Logger 请求中间件的执行过程

```
               request              > authenticate
SERVER =============== LOGGER =============== --------X-------- WEB APPLICATION
              response              < deny
```

图 7-2　授权响应阻止后

通常，ASP.NET Core 应用程序具有多个中间件，但按照顺序依次调用。图 7-3 所示为中间件请求响应的过程。

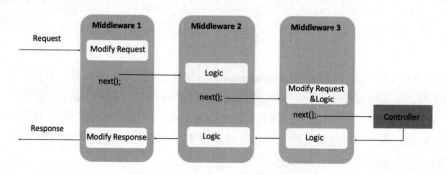

图 7-3　中间件请求响应的过程

对请求而言，可以确定每个中间件是否满足执行规则，若不满足执行规则，则直接忽略，并使请求直接进入下一个中间件。

7.2　中间件的调用过程

目前，开发人员普遍追求简单明了的设计理念，偏向于“复杂问题简单化”，中间件正好体现了这种设计理念。本章开头提到了管道，所谓请求管道，从本质上来说，就是对 HTTP 上下文（HttpContext 对象）的一系列操作，HttpContext 对象承载了单次请求的上下文信息，主要包括请求（Request）属性和响应（Response）属性两个主要成员，使用 Request 属性可以获取到当前请求的所有信息，使用 Response 属性可以获取和设置响应内容。对于请求处理，ASP.NET Core 应用程序中定义了 RequestDelegate 委托类型，主要表示请求管道中的一个步骤，如代码 7-1 所示。

代码7-1

```
public delegate Task RequestDelegate(HttpContext context);
```

处理者接收 HttpContext 对象，并且会通过 Request 属性获取所有关于该请求的信息，而处理的结果则在 Response 属性中表示，应用程序可以通过修改 Response 属性来对当前的响应内容进行更改。

而对请求处理管道的注册是通过 Func<RequestDelegate,RequestDelegate>类型委托对象的，相信读者会发现一个问题，为什么中间件的输入和输出都是 RequestDelegate 对象。对于管道中的中间件来说，前一个中间件的输出会成为下一个中间件的输入，所以，中间件的输入和输出都是 RequestDelegate 对象。

中间件的 Func<RequestDelegate,RequestDelegate>对象表示请求处理器，该对象的转换是通过 IApplicationBuilder 接口来完成的，如代码 7-2 所示。

代码7-2

```
public interface IApplicationBuilder
{
    IServiceProvider ApplicationServices { get; set; }
    IFeatureCollection ServerFeatures { get; }
    IDictionary<string, object?> Properties { get; }

    ///<summary>
    ///将中间件的委托方法添加到应用程序的请求管道中
    ///</summary>
    IApplicationBuilder Use(Func<RequestDelegate, RequestDelegate> middleware);
    ///<summary>
    ///创建一个新的 IApplicationBuilder 对象
    ///</summary>
    ///<returns></returns>
    IApplicationBuilder New();
    RequestDelegate Build();
}
```

为了方便读者理解，笔者将 ApplicationBuilder 类做了精简处理，对于中间件处理对象，无论上层进行了什么样的调用，底部的设计都会调用 Use 方法，从而将 Func<RequestDelegate, RequestDelegate>对象添加到_components 集合中，如代码 7-3 所示。

代码7-3

```
public class ApplicationBuilder : IApplicationBuilder
```

```csharp
{
    Private readonly IList<Func<RequestDelegate,RequestDelegate>>
     _components = new List<Func<RequestDelegate, RequestDelegate>>();

    public IApplicationBuilder Use(
                Func<RequestDelegate, RequestDelegate> middleware)
    {
        _components.Add(middleware);
        return this;
    }
    public RequestDelegate Build()
    {
        RequestDelegate app = context =>
        {
            var endpoint = context.GetEndpoint();
            var endpointRequestDelegate = endpoint?.RequestDelegate;
            if(endpointRequestDelegate != null)
            {
                var message =
     $"The request reached the end of the pipeline without executing
            the endpoint: '{endpoint!.DisplayName}'. Please register
            the EndpointMiddleware using' {nameof(IApplicationBuilder)}
                            .UseEndpoints(...)' if using routing.";
                throw new InvalidOperationException(message);
            }
            context.Response.StatusCode = StatusCodes.Status404NotFound;
            return Task.CompletedTask;
        };

        for(var c = _components.Count - 1; c >= 0; c--)
        {
            app = _components[c](app);
        }
        return app;
```

        }
    }

Build 方法创建一个 RequestDelegate 类型的委托，中间件的请求链也是通过它完成的。Build 方法先定义 404 的委托对象，再调用_components 集合的 Count 属性，并以 c-- 的形式从大到小循环，将注册的中间件反转构成一个链式结构。所以，先执行最后一个中间件，再依次执行。其实，构建过程类似俄罗斯套娃，按照注册顺序从里到外，一层套一层。在没有自定义处理时，默认使用 404 的委托对象作为响应返回。

7.3 编写自定义中间件

自定义一个中间件，开发人员需要创建一个带有 RequestDelegate 作为参数的构造函数类，以及一个返回 Task 类型和名为 Invoke 或 InvokeAsync 的方法，并且参数为 HttpContext。如代码 7-4 所示，创建一个名为 CustomMiddleware 的中间件，该中间件主要用于记录请求路径。

代码7-4

```
public class CustomMiddleware
{
    private readonly RequestDelegate _next;
    private readonly Ilogger<CustomMiddleware> _logger;

    public CustomMiddleware(RequestDelegate next,
                            ILogger <CustomMiddleware> logger)
    {
        this._next = next;
        this._logger = logger;
    }

    public async Task InvokeAsync(HttpContext context)
    {
        _logger.LogWarning($"Before Invoke {context.Request.Path}");
        await _next(context);
        _logger.LogWarning($"After Invoke {context.Request.Path}");
    }
}
```

在 InvokeAsync 方法中，通过打印输出 Before Invoke {context.Request.Path}日志模拟执行该中间件逻辑，并且在执行完成之后，调用_next 方法来执行后续的中间件操作。

接下来为这个中间件创建扩展方法，用于简化调用。要注册中间件，就需要通过 IApplicationBuilder 对象调用 UseMiddleware 扩展方法，如代码 7-5 所示。

代码7-5

```
public static class MiddlewareExtension
{
    public static void UseCustomExtension(this IApplicationBuilder app)
    {
        app.UseMiddleware<CustomMiddleware>();
    }
}
```

如代码 7-6 所示，调用 app.UseCustomExtension 扩展方法。

代码7-6

```
var builder = WebApplication.CreateBuilder(args);

var app = builder.Build();
app.UseCustomExtension();

app.MapGet("/", () =>
{
    return "Hello World!";
});

app.Run();
```

注册中间件需要调用 UseMiddleware 扩展方法注册，如代码 7-7 展示了 UseMiddleware 的扩展方法的实现，笔者简化了扩展方法。

代码7-7

```
public static class UseMiddlewareExtensions
{
    internal const string InvokeMethodName = "Invoke";
    internal const string InvokeAsyncMethodName = "InvokeAsync";
```

```csharp
public static IApplicationBuilder UseMiddleware(
    this IApplicationBuilder app, Type middleware, params object?[] args)
{
    //如果middleware对象实现了IMiddleware接口，则执行UseMiddlewareInterface方法
    if(typeof(IMiddleware).IsAssignableFrom(middleware))
    {
        return UseMiddlewareInterface(app, middleware);
    }
    var applicationServices = app.ApplicationServices;
    return app.Use(next =>
    {
        var methods = middleware.GetMethods(
                BindingFlags.Instance | BindingFlags.Public);
        //查看middleware中实现的Invoke方法和InvokeAsync方法
        var invokeMethods = methods.Where(m =>
                string.Equals(m.Name, InvokeMethodName,
            StringComparison.Ordinal)|| string.Equals(m.Name,
        InvokeAsyncMethodName,StringComparison.Ordinal)).ToArray();

        if(invokeMethods.Length > 1)
        {
            //不允许Invoke方法和InvokeAsync方法同时存在
            throw new InvalidOperationException(
             Resources.FormatException_UseMiddleMutlipleInvokes(
                    InvokeMethodName, InvokeAsyncMethodName));
        }

        if(invokeMethods.Length == 0)
        {
            //必须定义Invoke方法或InvokeAsync方法
            throw new InvalidOperationException(
             Resources.FormatException_UseMiddlewareNoInvokeMethod(
                InvokeMethodName, InvokeAsyncMethodName, middleware));
        }

        var methodInfo = invokeMethods[0];
        if(!typeof(Task).IsAssignableFrom(methodInfo.ReturnType))
```

```csharp
{
    //如果方法的返回值不是 Task 类型，则抛出异常
    throw new InvalidOperationException(
        Resources.FormatException_UseMiddlewareNonTaskReturnType(
            InvokeMethodName, InvokeAsyncMethodName, nameof(Task)));
}

var parameters = methodInfo.GetParameters();
if (parameters.Length == 0 || parameters[0].ParameterType
                                != typeof(HttpContext))
{
    //方法的参数必须是 HttpContext
    throw new InvalidOperationException(
        Resources.FormatException_UseMiddlewareNoParameters(
            InvokeMethodName, InvokeAsyncMethodName, nameof(HttpContext)));
}

var ctorArgs = new object[args.Length + 1];
ctorArgs[0] = next;
Array.Copy(args, 0, ctorArgs, 1, args.Length);
var instance = ActivatorUtilities.CreateInstance(
        app.ApplicationServices, middleware, ctorArgs);
if (parameters.Length == 1)
{
    return (RequestDelegate)methodInfo.CreateDelegate(
                    typeof (RequestDelegate), instance);
}
var factory = Compile<object>(methodInfo, parameters);
return context =>
{
    var serviceProvider =
            context.RequestServices ?? applicationServices;
    if (serviceProvider == null)
    {
        throw new InvalidOperationException(
            Resources.FormatException_UseMiddlewareIServiceProviderNotAvailable(
                        nameof(IServiceProvider)));
```

```
            }
            return factory(instance, context, serviceProvider);
        };
    });
}
```

在 UseMiddleware 方法中，如果 middleware 对象实现了 IMiddleware 接口，就执行 UseMiddlewareInterface 方法，否则表示是非 IMiddleware 接口约束的对象，并检查中间件的合法性，如是否定义了 Invoke 方法和 HttpContext 参数等，最终通过 ActivatorUtilities 类型的 CreateInstance 方法创建中间件实例。

7.4 在过滤器中应用中间件

中间件还可以以过滤器（Filter）的形式注册。ASP.NET Core 应用程序中提供了 MiddlewareFilterAttribute 对象，中间件可以在过滤器管道中被调用。要使用 MiddlewareFilterAttribute 对象，需要创建一个对象，如代码 7-8 所示，该对象需要满足以下条件：

- 无参的构造函数。
- 具有名称为 Configure 的公共方法，同时接收 IApplicationBuilder 对象参数。

代码7-8

```
public class CustomPipeline
{
    public void Configure(IApplicationBuilder app)
    {
        app.UseMiddleware<CustomMiddleware>();
    }
}
```

接下来可以使用自定义的 CustomPipeline 对象。代码 7-9 使用的是 MiddlewareFilterAttribute 对象。

代码7-9

```
[ApiController]
[Route("[controller]")]
[MiddlewareFilter(typeof(CustomPipeline))]
```

```
public class ValuesController : ControllerBase
{
    public OkResult Get()
    {
        return Ok();
    }
}
```

如代码 7-9 所示，每个访问 ValuesController 的请求都会触发使用 CustomPipeline 对象定义的中间件。

如代码 7-10 所示，MiddlewareFilterAttribute 类实现了 IFilterFactory 接口，用于将服务注册到 MVC 过滤器的接口，这个接口只定义了一个 CreateInstance 方法。

代码7-10

```
[AttributeUsage(AttributeTargets.Class | AttributeTargets.Method,
 AllowMultiple = true, Inherited = true)]
public class MiddlewareFilterAttribute : Attribute,
                                          IFilterFactory, IOrderedFilter
{
    public MiddlewareFilterAttribute(Type configurationType)
    {
        if (configurationType == null)
        {
            throw new ArgumentNullException(nameof(configurationType));
        }
        ConfigurationType = configurationType;
    }

    public Type ConfigurationType { get; }

    public int Order { get; set; }

    public bool IsReusable => true;

    public IFilterMetadata CreateInstance(IServiceProvider serviceProvider)
    {
        if (serviceProvider == null)
        {
```

```
            throw new ArgumentNullException(nameof(serviceProvider));
        }

        var middlewarePipelineService =
            serviceProvider.GetRequiredService<MiddlewareFilterBuilder>();
        var pipeline = middlewarePipelineService
                            .GetPipeline (ConfigurationType);
        return new MiddlewareFilter(pipeline);
    }
}
```

首先根据 MiddlewareFilterBuilder 从依赖注入容器中获取一个实例对象，然后调用 GetPipeline 方法传入定义的 ConfigurationType 属性（上面定义的 CustomPipeline 对象）。GetPipeline 方法返回的是 RequestDelegate 委托对象，如代码 7-11 所示，代表一个中间件管道，它接收一个 HttpContext 参数并返回一个 Task 对象。

代码7-11

```
public delegate Task RequestDelegate(HttpContext context);
```

将 RequestDelegate 以参数形式在构建 MiddlewareFilter 对象时传递。如代码 7-12 所示，可以看到MiddlewareFilter构造方法接收RequestDelegate类型的委托，并且在MiddlewareFilter 方法中实现了 IAsyncResourceFilter 接口。另外，OnResourceExecutionAsync 方法记录了 ResourceExecutingContext 对象，以及下一个要执行的过滤器 ResourceExecutionDelegate 委托对象，作为一个新的 MiddlewareFilterFeature 对象存储在 HttpContext 上下文中，因此可以在其他地方使用并访问。

代码7-12

```
internal class MiddlewareFilter : IAsyncResourceFilter
{
    private readonly RequestDelegate _middlewarePipeline;

    public MiddlewareFilter(RequestDelegate middlewarePipeline)
    {
        if (middlewarePipeline == null)
        {
            throw new ArgumentNullException(nameof(middlewarePipeline));
        }
```

```
        _middlewarePipeline = middlewarePipeline;
    }

    public Task OnResourceExecutionAsync(ResourceExecutingContext context
                                      ,ResourceExecutionDelegate next)
    {
        var httpContext = context.HttpContext;
        var feature = new MiddlewareFilterFeature()
        {
            ResourceExecutionDelegate = next,
            ResourceExecutingContext = context
        };
        httpContext.Features.Set<IMiddlewareFilterFeature>(feature);
        return _middlewarePipeline(httpContext);
    }
}
```

通常，开发人员会通过 await next() 在管道中执行下一个中间件，但是目前只是从 RequestDelegate 的调用中返回 Task，那么这个管道应该如何继续呢？

如代码 7-13 所示，该对象的 BuildPipeline 方法用于执行过滤器管道，然后通过 HttpContext 对象加载 IMiddlewareFilterFeature 对象，并通过它获取 ResourceExecutionDelegate 来访问下一个过滤器。

代码7-13

```
internal class MiddlewareFilterBuilder
{
    private readonly ConcurrentDictionary<Type,
            Lazy<RequestDelegate>> _pipelinesCache
        = new ConcurrentDictionary<Type, Lazy<RequestDelegate>>();
    private readonly MiddlewareFilterConfigurationProvider _
                                        configurationProvider;

    public IApplicationBuilder? ApplicationBuilder { get; set; }

    public MiddlewareFilterBuilder(
        MiddlewareFilterConfigurationProvider configurationProvider)
    {
```

```csharp
        _configurationProvider = configurationProvider;
    }

    public RequestDelegate GetPipeline(Type configurationType)
    {
        var requestDelegate = _pipelinesCache.GetOrAdd(
            configurationType,
            key => new Lazy<RequestDelegate>(() => BuildPipeline(key)));

        return requestDelegate.Value;
    }

    private RequestDelegate BuildPipeline(Type middlewarePipelineProviderType)
    {
        if(ApplicationBuilder == null)
        {
            throw new InvalidOperationException(
                Resources.FormatMiddlewareFilterBuilder_NullApplicationBuilder
(nameof(ApplicationBuilder)));
        }

        var nestedAppBuilder = ApplicationBuilder.New();

        var configureDelegate = MiddlewareFilterConfigurationProvider
                .CreateConfigureDelegate(middlewarePipelineProviderType);
        configureDelegate(nestedAppBuilder);

        nestedAppBuilder.Run(async (httpContext) =>
        {
            var feature = httpContext.Features.GetRequiredFeature
                                        <IMiddlewareFilterFeature>();

            var resourceExecutionDelegate = feature. ResourceExecutionDelegate!;

            var resourceExecutedContext = await resourceExecutionDelegate();
            if(resourceExecutedContext.ExceptionHandled)
            {
```

```
            return;
        }
        resourceExecutedContext.ExceptionDispatchInfo?.Throw();
        if(resourceExecutedContext.Exception != null)
        {
            throw resourceExecutedContext.Exception;
        }
    });

    return nestedAppBuilder.Build();
  }
}
```

7.5 制作简单的 API 统一响应格式与自动包装

在实际的项目中，可能需要对外提供统一的响应内容，这样开发人员就需要定义统一的结构化内容。但是如果每个方法都按照一个响应类进行包装，那么响应类的格式一旦发生变化，需要修改的地方可能就会比较多，这时开发人员就需要考虑有没有一种便捷的方式：无须关注每个方法返回的内容，只需要在 HTTP 请求响应时将内容进行包装，生成统一的响应格式，并返回给客户端。这时就可以使用中间件来实现这个功能。

7.5.1 创建一个中间件

如代码 7-14 所示，首先定义一个 CustomApiResponse 类，用于统一响应内容的格式，Status 属性表示状态码，RequestId 属性表示请求的唯一编码，Result 属性表示一个结果，结果可能是数组也可能是单个对象，所以定义为 object。

代码 7-14

```
public class CustomApiResponse
{
    public static CustomApiResponse Create(int statusCode,
                                    object result, string requestId)
    {
        return new CustomApiResponse(statusCode, result, requestId);
    }
```

```
    internal CustomApiResponse(int statusCode, object result,
                                              string requestId)
    {
        Status = statusCode;
        RequestId = requestId;
        Result = result;
    }

    public int Status { get; set; }
    public string RequestId { get; set; }
    public object Result { get; set; }
}
```

如代码 7-15 所示，按照约束定义的一个结果响应包装中间件，笔者在构造函数中注入了一个 RequestDelegate 对象，并且在用于请求处理的 InvokeAsync 方法中定义了表示当前 HttpContext 上下文的参数。

代码7-15

```
public class CustomMiddleware
{
    private readonly RequestDelegate _next;

    public CustomMiddleware(RequestDelegate next)
    {
        _next = next;
    }

    public async Task InvokeAsync(HttpContext context)
    {
        var originalResponseBodyStream = context.Response.Body;

        using var memoryStream = new MemoryStream();
        context.Response.Body = memoryStream;

        await _next(context);

        context.Response.Body = originalResponseBodyStream;
```

```
        memoryStream.Seek(0, SeekOrigin.Begin);

        var readToEnd = await new StreamReader(memoryStream).ReadToEndAsync();
        var objResult = JsonConvert.DeserializeObject(readToEnd);
        var result = CustomApiResponse.Create(
           context.Response.StatusCode, objResult, context.TraceIdentifier);
        await context.Response.WriteAsync(JsonConvert.SerializeObject(result));
    }
}
```

该中间件在完成自身的请求操作后，通过构造函数中注入的 RequestDelegate 对象可以将请求分发给后续的中间件，在后续的内容响应完成后会执行_next 后面的代码逻辑。

使用上述代码可以获取控制器方法输出的 Response 响应内容，从而读取响应内容。创建一个 CustomApiResponse 类，并调用 Create 扩展方法，将返回的状态码和当前请求的唯一标识及返回的内容进行包装，最终序列化为 JSON 格式（包装后的响应如图 7-4 所示），返回给客户端。

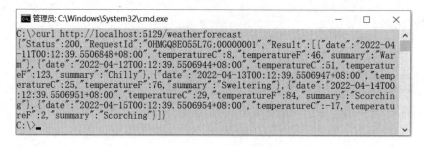

图 7-4　包装后的响应

7.5.2　配置扩展方法

在中间件创建完成后，创建一个扩展类，在使用时只需要调用扩展方法进行注册即可，如代码 7-16 所示。

代码7-16

```
public static class MiddlewareExtension
{
    public static void UseCustomExtension(this IApplicationBuilder app)
    {
        app.UseMiddleware<CustomMiddleware>();
```

}
}

7.6 延伸阅读：责任链模式

责任链模式（Chain of Responsibility Pattern）是一种行为设计模式，这种模式允许将调用以链式的形式沿着处理者程序按顺序执行和传递，在请求到达后，每个处理者均可对请求进行处理，或者将其直接传递给下一个处理者程序。使用责任链模式可以避免请求发送者与接收者之间的耦合关系，将它们形成一个链，对象与对象之间无须知道具体的处理者是谁，该链上的每个处理者都有机会参与处理请求。如果调用链上的某个处理者无法处理当前的请求，那么它会把相同的请求传递给下一个对象。

下面通过一个控制台实例来介绍责任链模式。如代码 7-17 所示，创建一个名为 IHandler 的接口，每个处理者都需要继承统一的 IHandler 接口。定义 Process 方法用于约束其执行的对象，随后新建一个名为 Processor 的对象，该对象继承 IHandler 接口并实现 Process 方法。在 Processor 对象中定义私有 IHandler 属性，并定义 Next 方法用于设置当前实例的下一个要执行的实例对象，实现 Process 方法用于执行 DoProcess 方法。另外，还需要判断 IHandler 对象是否为空，若不为空，则执行下一个实例对象的 Process 方法，这样依次向下继续执行。

代码7-17

```
public interface IHandler
{
    void Process(string msg);
}
public abstract class Processor : IHandler
{
    private IHandler _next;

    public Processor Next(IHandler next)
    {
        this._next = next;
        return this;
    }
```

```
public void Process(string msg)
{
    DoProcess(msg);
    if (_next != null)
    {
        _next.Process(msg);
    }
}
public abstract void DoProcess(string msg);
}
```

在控制台实例中定义 3 个 Handler 实例对象,在第二个实例对象中通过调用 Next 方法传递上一个实例对象,并依次向下传递依赖顺序(责任链模式的输出结果如图 7-5 所示),这样方便对依赖关系进行调用,但是会导致关系变成强依赖,如果出现中间一个环节没有传递的情况,那么这个链接就会在此处终止,如代码 7-18 所示,调用最后一个实例对象的 Process 方法,用于标识该链接程序的起始之处。

代码7-18

```
IHandler handler1 = new Handler1();
IHandler handler2 = new Handler2().Next(handler1);
IHandler handler3 = new Handler3().Next(handler2);
handler3.Process("hello");

public class Handler1 : Processor
{
    public override void DoProcess(string msg)
    {
        Console.WriteLine($"{nameof(Handler1)}: {msg}");
    }
}

public class Handler2 : Processor
{
    public override void DoProcess(string msg)
    {
        Console.WriteLine($"{nameof(Handler2)}: {msg}");
    }
```

```
}
public class Handler3 : Processor
{
    public override void DoProcess(string msg)
    {
        Console.WriteLine($"{nameof(Handler3)}: {msg}");
    }
}
```

图 7-5 责任链模式的输出结果

在实例中对象以递归执行的方式实现了相关代码，代码层次不清晰，类与类的耦合性较高，不便于对代码进行维护，因此必须打破这种关系。如代码 7-19 所示，删除了 Processor 对象，增加了 IHandler 的集合属性成员（用于存放每个 IHandler 的实例对象），并循环 IHandler 对象集合，每个实例对象调用 Process 方法，最终的输出结果如图 7-6 所示。

代码7-19

```
List<IHandler> handlers = new List<IHandler>();
handlers.Add(new Handler1());
handlers.Add(new Handler2());
handlers.Add(new Handler3());
foreach (var handler in handlers)
{
    handler.Process("hello");
}

public class Handler1 : IHandler
{
    public void Process(string msg)
    {
        Console.WriteLine($"{nameof(Handler1)}: {msg}");
    }
}
public class Handler2 : IHandler
{
```

```
    public void Process(string msg)
    {
        Console.WriteLine($"{nameof(Handler2)}: {msg}");
    }
}
public class Handler3 : IHandler
{
    public void Process(string msg)
    {
        Console.WriteLine($"{nameof(Handler3)}: {msg}");
    }
}
```

图 7-6　最终的输出结果

将代码结构由递归型改为顺序型，使代码层次分明，以便于后期的开发和扩展。

7.7　延伸阅读：中间件常见的扩展方法

与以前的版本相比，ASP.NET Core 应用程序中的中间件更加优雅、简单，开发人员可以使用 Use 方法、Run 方法和 Map 方法来定义中间件。

7.7.1　Run 方法和 Use 方法

Run 是 IApplicationBuilder 对象的扩展方法。使用该方法可以将中间件添加到管道中。如代码 7-20 所示，调用 app.Run 方法，在程序运行后，通过访问该程序得到的输出结果如图 7-7 所示。

代码7-20

```
var builder = WebApplication.CreateBuilder(args);

var app = builder.Build();
```

```
app.Run(async context =>
{
    await context.Response.WriteAsync("Hello, World!");
});

app.Run();
```

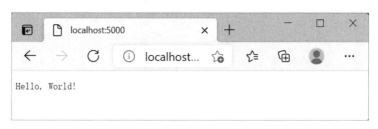

图 7-7　输出结果

由此可知，即使在后面再添加其他中间件，在执行过程中也不会调用，这是因为使用了 app.Run 中间件，它会在此处返回相应的结果从而终止请求，不再调用后续的中间件。

如代码 7-21 所示，笔者精简了 Run 方法的定义。该方法先接收 RequestDelegate 委托对象，再调用 app.Use 方法将 RequestDelegate 对象传入 Use 方法中，因为 Use 方法接收的是一个 Func 委托对象，这是一个带有返回值的类型的委托。

代码7-21

```
public static class RunExtensions
{
    public static void Run(
                this IApplicationBuilder app, RequestDelegate handler)
    {
        app.Use(_ => handler);
    }
}
```

需要注意的是，代码 7-21 中使用的 _=>（**弃元，Discards**）表示忽略使用的变量。当然，还可以使用 **app.Use(r=>handler)** 调用，r 对象是没有意义的，所以没有必要声明它，可以直接忽略，就是将 RequestDelegate 对象添加到中间件管道中。

Use 方法不会终止请求。无论上层调用任何方法将中间件注册到管道中，最终在内部都会调用 Use 方法进行注册。如代码 7-22 所示，笔者调用了 app.Use 方法和 app.Run 方法，先定义第一个中间件 app.Use 将内容写入响应对象，再通过 next 方法调用下一个中间件，并再次将内容写入响应对象，定义 app.Run 中间件写入响应对象并返回。

代码7-22

```
var builder = WebApplication.CreateBuilder(args);

var app = builder.Build();

app.Use(next =>
{
    return async (context) =>
    {
        await context.Response.WriteAsync(
                            "Before Invoke from 1st app.Use()\n");
        await next(context);
        await context.Response.WriteAsync(
                            "After Invoke from 1st app.Use()\n");
    };
});

app.Run(async (context) =>
{
    await context.Response.WriteAsync(
                            "Hello from 1st app.Run()\n");
});

app.Run(async context =>
{
    await context.Response.WriteAsync("Hello, World!");
});

app.Run();
```

如图 7-8 所示，定义两个 app.Run 中间件来验证本节开头的结论，即"使用 app.Run 中间件，它会在此处返回相应的结果从而终止请求，不再调用后续的中间件。"，app.Run 方法前面的 app.Use 方法执行了，但是第二个 app.Run 方法并没有执行，所以 app.Run 相当于一个终结中间件，它后面的中间件不会被执行。

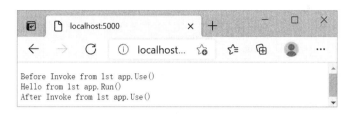

图 7-8 调用 Use 方法和 Run 方法

在请求管道执行过程中（见图 7-9），当执行完第一个 app.Run 方法后，请求就会终止，不继续往下执行，而是输出响应内容。

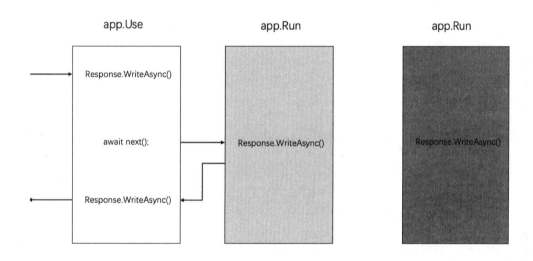

图 7-9 请求管道的执行过程

7.7.2 Map 方法

顾名思义，Map，是指映射到路径上，并为该路径指定独立的执行管道。在日常开发中，开发人员需要自定义路径执行不同的操作，而使用 Map 方法可以实现对指定路由前缀的映射。

如代码 7-23 所示，创建一个 Web 实例。先定义两个 app.Map 方法，再定义路由/user 和/test 的中间件。

代码7-23

```
var builder = WebApplication.CreateBuilder(args);
```

```
var app = builder.Build();

app.Map("/user", HandleMapUser);
app.Map("/test", appMap =>
{
    appMap.Run(async context =>
    {
        await context.Response.WriteAsync("Hello from 2nd app.Map()");
    });
});
app.Run(async context =>
{
    await context.Response.WriteAsync("Hello, World!");
});

app.Run();

static void HandleMapUser(IApplicationBuilder app)
{
    app.Run(async context =>
    {
        await context.Response.WriteAsync("Hello from 1st app.Map()");
    });
}
```

请求和响应的参照表如表 7-1 所示，其中 Map 方法用于匹配路径前缀的规则。

表 7-1　请求和响应的参照表

请求	响应
http://localhost:5000/	Hello, World!
http://localhost:5000/user	Hello from 1st app.Map()
http://localhost:5000/user/1	Hello from 1st app.Map()
http://localhost:5000/test	Hello from 2st app.Map()
http://localhost:5000/test/1	Hello from 2st app.Map()

代码 7-24 展示了 MapExtensions 类中的 Map 方法的定义。在 IApplicationBuilder 对象中还有一个常用的 New 方法，通常用来创建分支。可以先根据实际情况使用 New 方法创建一

个新的 ApplicationBuilder 对象，然后根据当前匹配的路径和 Branch 分支管道构建一个 MapOptions 对象，最后调用 app.Use 方法注册中间件。

代码7-24

```csharp
public static class MapExtensions
{
    public static IApplicationBuilder Map(this IApplicationBuilder app,
            string pathMatch, Action<IApplicationBuilder> configuration)
    {
        return Map(
         app, pathMatch, preserveMatchedPathSegment: false, configuration);
    }

    public static IApplicationBuilder Map(this IApplicationBuilder app,
         PathString pathMatch, Action<IApplicationBuilder> configuration)
    {
        return Map(
         app, pathMatch, preserveMatchedPathSegment: false, configuration);
    }

    public static IApplicationBuilder Map(this IApplicationBuilder app,
            PathString pathMatch, bool preserveMatchedPathSegment,
            Action<IApplicationBuilder> configuration)
    {
        var branchBuilder = app.New();
        configuration(branchBuilder);
        var branch = branchBuilder.Build();

        var options = new MapOptions
        {
          Branch = branch,
          PathMatch = pathMatch,
          PreserveMatchedPathSegment = preserveMatchedPathSegment
        };
        return app.Use(next => new MapMiddleware(next, options).Invoke);
    }
}
```

可以发现，代码 7-24 中新建了一个 MapMiddleware 方法，该方法的定义如代码 7-25 所示。笔者精简了 MapMiddleware 方法的定义的内容，先构建 MapMiddleware 对象，并且将其 Invoke 方法包装为 RequestDelegate 对象，再判断要匹配的内容是否匹配，若匹配，则进入分支管道，否则继续在当前管道中。通过判断 _options 对象的 PreserveMatchedPathSegment 属性，先重新设置 PathBase 属性和 Path 属性，再调用 _options 对象的 Branch 方法进入分支管道，最后还原 PathBase 路径和 Path 路径，因为可能有其他中间件继续处理 HttpContext。

代码7-25

```csharp
public class MapMiddleware
{
    private readonly RequestDelegate _next;
    private readonly MapOptions _options;

    public MapMiddleware(RequestDelegate next, MapOptions options)
    {
        _next = next;
        _options = options;
    }

    public Task Invoke(HttpContext context)
    {
        if(context == null)
        {
            throw new ArgumentNullException(nameof(context));
        }

        if(context.Request.Path.StartsWithSegments(_options.PathMatch,
                        out var matchedPath, out var remainingPath))
        {
            if(!_options.PreserveMatchedPathSegment)
            {
                return InvokeCore(context, matchedPath, remainingPath);
            }
            return _options.Branch!(context);
        }
        return _next(context);
    }
```

```
private async Task InvokeCore(HttpContext context,
                    string matchedPath, string remainingPath)
{
    var path = context.Request.Path;
    var pathBase = context.Request.PathBase;

    //更新 Path
    context.Request.PathBase = pathBase.Add(matchedPath);
    context.Request.Path = remainingPath;

    try
    {
        await _options.Branch!(context);
    }
    finally
    {
        context.Request.PathBase = pathBase;
        context.Request.Path = path;
    }
}
```

7.7.3 MapWhen 方法

Map 方法主要用于通过路由规则来判断是否进入分支管道，若开发人员有其他的需求，如想对 Method 为 Delete 的请求用特殊的管道处理，则可以使用 MapWhen 方法。MapWhen 是一种通用的 Map 方法，可以由使用者来决定什么时候进入分支管道，可以自定义相关的条件判断。通常来说，Map 方法定义了一种通用情况，而 MapWhen 方法提供了更强大的判断策略。

如代码 7-26 所示，app.MapWhen 方法定义一个条件，用于判断是否进入执行分支管道，若满足条件，则进入分支管道，否则继续在当前管道中，如判断请求参数中是否存在 q 键（Key）。

代码7-26

```
var builder = WebApplication.CreateBuilder(args);
```

```
var app = builder.Build();

app.MapWhen(context => context.Request.Query.Keys.Contains("q"),
            appMap =>
            {
                appMap.Run(async context =>
                {
                    await context.Response.WriteAsync(
                    $"Hello Test:{context.Request.Query["q"]}");
                });
            });
app.Run(async context =>
{
    await context.Response.WriteAsync("Hello, World!");
});

app.Run();
```

代码7-27展示了MapWhen方法的定义。MapWhen方法的第一个参数是IApplicationBuilder对象；第二个参数是Func<HttpContext,bool>委托对象，输入参数为HttpContext对象，输出参数是bool类型；第三个参数是Action<IApplicationBuilder>委托对象，表示也可以将实现Action委托的方法添加到中间件管道中执行。

代码7-27

```
using Predicate = Func<HttpContext, bool>;

public static class MapWhenExtensions
{
    public static IApplicationBuilder MapWhen(this IApplicationBuilder app,
        Predicate predicate, Action<IApplicationBuilder> configuration)
    {
        //create branch
        var branchBuilder = app.New();
        configuration(branchBuilder);
        var branch = branchBuilder.Build();

        //put middleware in pipeline
        var options = new MapWhenOptions
        {
```

```
            Predicate = predicate,
            Branch = branch,
        };
        return app.Use(next => new MapWhenMiddleware(next, options).Invoke);
    }
}
```

和 Map 方法一样,先创建 ApplicationBuilder 对象,再创建 MapWhen 对象用于保存相关条件和分支,最后将 MapWhenMiddleware 中间件注册到管道中。

如代码 7-28 所示,在 MapWhenMiddleware 中间件中实现还是比较简单的,若委托对象返回 true,则进入分支管道,否则继续向下执行下一个中间件。

代码7-28

```
public class MapWhenMiddleware
{
    private readonly RequestDelegate _next;
    private readonly MapWhenOptions _options;

    public MapWhenMiddleware(RequestDelegate next, MapWhenOptions options)
    {
        _next = next;
        _options = options;
    }

    public Task Invoke(HttpContext context)
    {
        if (context == null)
        {
            throw new ArgumentNullException(nameof(context));
        }

        if (_options.Predicate!(context))
        {
            return _options.Branch!(context);
        }
        return _next(context);
    }
}
```

7.7.4 UseWhen 方法

如代码 7-29 所示，UseWhen 方法也是根据指定的条件判断是否执行当前的方法，核心是 UseWhen 方法基于指定的条件执行创建的请求管道分支。如果这个分支在执行时不中断或不使用 app.Run 中间件，就重新加入主管道中执行，而 MapWhen 方法只在满足条件时执行管道分支（通过 app.UseWhen 方法判断后的输出结果如图 7-10 所示）。

代码7-29

```
var builder = WebApplication.CreateBuilder(args);

var app = builder.Build();

app.UseWhen(
    context =>
    context.Request.Query.Keys.Contains("q"),
            appMap =>
            {
                appMap.Use(next =>
                {
                    return async (context) =>
                    {
                        await context.Response.WriteAsync(
                            "Before Invoke from 1st app.Use()\n");
                        await next(context);
                        await context.Response.WriteAsync(
                            "After Invoke from 1st app.Use()\n ");
                    };
                });
            });
app.Run(async context =>
{
    await context.Response.WriteAsync("Hello, World!\n");
});

app.Run();
```

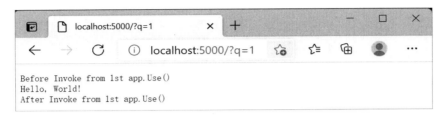

图 7-10　通过 app.UseWhen 方法判断后的输出结果

UseWhen 方法如代码 7-30 所示，先调用 IApplicationBuilder 对象的 New 方法创建一个管道分支，将 configuration 委托注册到该分支上，再将 main（也就是后续要执行的中间件）也注册到该分支上，最后根据 predicate 来判断，执行新分支还是继续在之前的管道中执行。

代码7-30

```
using Predicate = Func<HttpContext, bool>;

public static class UseWhenExtensions
{
    public static IApplicationBuilder UseWhen(this IApplicationBuilder app,
            Predicate predicate, Action<IApplicationBuilder> configuration)
    {
        var branchBuilder = app.New();
        configuration(branchBuilder);

        return app.Use(main =>
        {
            branchBuilder.Run(main);
            var branch = branchBuilder.Build();

            return context =>
            {
                if(predicate(context))
                {
                    return branch(context);
                }
                else
                {
                    return main(context);
                }
```

173

```
            };
        });
    }
}
```

7.8 小结

通过学习本章，读者可以深入了解 ASP.NET Core 的中间件。中间件技术可以为开发人员对 HTTP 请求上下文的某些操作带来极大的便利，因此，它也是 ASP.NET Core 一项非常重要的核心技术。

第 8 章

缓存

互联网的普及、用户和访问量的增加可以带来更多的并发量，这对于应用程序来说是一个考验。由于应用程序的资源有限，数据库能承载的请求次数也有限，使用缓存不仅能减少计算还可以提升响应速度，因此应该尽可能利用有限的资源提供更大的吞吐量。

本章先介绍内存缓存和分布式缓存。内存缓存将缓存内容序列化存储在应用程序进程中，分布式缓存可以借助一个第三方存储中间件进行保存，如通过 Redis 和 SQL Server 来存储。除了这些方式，ASP.NET Core 还提供了一个响应缓存，用来实现基于 HTTP 缓存的设计规范。

8.1 内存缓存

通俗来讲，内存缓存就是将任意类型的对象内容存储在缓存中。内存缓存的使用比较简单，只需要通过 IMemoryCache 对象进行对象的设置和读取即可。另外，本节也会涉及存储结构的相关内容。

8.1.1 IMemoryCache

在 .NET 中，对象是内存缓存的核心内容之一。下面以控制台项目为例，创建 ServiceCollection 对象，调用 AddMemoryCache 扩展方法注册内存缓存核心服务，通过调用 IServiceCollection 对象的 BuildServiceProvider 扩展方法得到依赖注入容器的 IServiceProvider

对象之后，可以直接调用 GetRequiredService<T>扩展方法来获取 IMemoryCache 对象。

如代码 8-1 所示，要设置缓存可以调用 Set 扩展方法。该方法有两个参数，Key 用来表示缓存的对象的名称，Value 表示要存储的值。如果要从缓存中检索对象，并且开发人员不确定缓存中是否存在特定的键，就可以调用 TryGetValue 方法。TryGetValue 方法返回一个 bool 值（布尔值），指示当前的键是否存在，最后一个参数使用 out 关键字，按照引用传递。

代码8-1

```
var cache = new ServiceCollection()
    .AddMemoryCache()
    .BuildServiceProvider()
    .GetRequiredService<IMemoryCache>();

cache.Set("m1", "dotnet");
cache.TryGetValue("m1", out var value);
Console.WriteLine(value);
```

调用 TryGetValue 方法可以读取缓存值，并将结果输出到控制台中，如图 8-1 所示。

图 8-1 输出结果

如代码 8-2 所示，调用 Get<T>方法可以在缓存中检索对象。

代码8-2

```
cache.Get<string>("m1");
```

如代码 8-3 所示，调用 GetOrCreate<T>方法可以根据对应的键检索对象。如果对应的键不存在，那么 GetOrCreate<T>方法会选择创建指定的键。

代码8-3

```
cache.GetOrCreate<string>("m2", cacheEntry =>
{
    return "dotnet";
});
```

IMemoryCache 接口承载了基于内存的读操作和写操作。IMemoryCache 接口可以针对缓存数据进行设置、获取和移除，如代码 8-4 所示。

代码8-4

```
public interface IMemoryCache : IDisposable
```

```
{
    bool TryGetValue(object key, out object value);
    ICacheEntry CreateEntry(object key);
    void Remove(object key);
}
```

IMemoryCache 接口定义了 TryGetValue 方法和 Remove 方法，Set 方法的实现会调用一个名为 CacheExtensions 的静态类，该类为开发人员提供了缓存设置的方法，可以设置缓存。CreateEntry 方法可以根据指定的键创建一个 ICacheEntry 对象。

8.1.2 ICacheEntry

内存缓存在.NET 的内部以一个标准对象存储，该对象的名称为 CacheEntry。CacheEntry 对象承载了缓存的特征。

如代码 8-5 所示，先调用 CreateEntry 方法创建一个缓存对象，并通过 using 代码块包裹起来，再调用 Get<T>方法获取存储的缓存值。

代码8-5

```
using (var entry = cache.CreateEntry("m3"))
{
    entry.Value = "dotnet";
}
cache.Get<string>("m3");
```

如代码 8-6 所示，ICacheEntry 接口定义了如下属性：Key 属性作为存储对象的键；Value 属性表示获取或设置的值；AbsoluteExpiration 属性表示获取或设置对象的绝对过期日期，也就是设置一个特定时间点过期；AbsoluteExpirationRelativeToNow 属性相当于设置一个有效期，从当前时间开始计算，如 20 分钟失效；SlidingExpiration 属性提供了一种在指定时间内不活动或不常访问的缓存失效策略；ExpirationTokens 属性提供导致缓存内容过期的 IChangeToken 实例；PostEvictionCallbacks 属性表示获取或设置在缓存中失效后触发的回调；Priority 属性表示获取或设置缓存项优先级，默认为 CacheItemPriority.Normal，还包括 High、Low 和 NeverRemove，在缓存满载的情况下，对缓存设置保留优先级。

代码8-6

```
public interface ICacheEntry : IDisposable
{
    object Key { get; }
    object Value { get; set; }
```

```
    DateTimeOffset? AbsoluteExpiration { get; set; }
    TimeSpan? AbsoluteExpirationRelativeToNow { get; set; }
    TimeSpan? SlidingExpiration { get; set; }
    IList<IChangeToken> ExpirationTokens { get; set; }
    IList<PostEvictionCallbackRegistration> PostEvictionCallbacks { get; }
    CacheItemPriority Priority { get; set; }
    long? Size { get; set; }
}
```

CreateEntry 方法并没有显式地执行 Add 操作，而是使用 using 语句执行隐式操作。因为 C#中的 using 语句会在语句块结束时自动调用 Dispose 方法，所以 Add 操作被定义在 Dispose 方法中。接下来介绍 Dispose 方法的实现。如代码 8-7 所示，笔者简化了 CacheEntry 的实现，只保留了 Dispose 方法的释放操作。调用_state 对象的 IsDisposed 属性，其标识了内容是否已经被释放，若没有被释放，则执行 if 语句块内的代码片段，将 IsDisposed 属性设置为 true。另外，调用 IsValueSet 属性判断是否设置了 Value 值，如果设置了 Value 值，就调用 MemoryCache 对象的 SetEntry 方法保存 CacheEntry 对象。

代码8-7

```
internal sealed partial class CacheEntry : ICacheEntry
{
    private readonly MemoryCache _cache;
    public void Dispose()
    {
        if(!_state.IsDisposed)
        {
            _state.IsDisposed = true;

            CacheEntryHelper.ExitScope(this, _previous);
            if(_state.IsValueSet)
            {
                _cache.SetEntry(this);

                if(_previous != null && CanPropagateOptions())
                {
                    PropagateOptions(_previous);
                }
            }
```

```
            _previous = null;
        }
    }
}
```

如代码 8-8 所示,MemoryCache 承载了缓存的操作,而 MemoryCache 对象会直接利用一个 ConcurrentDictionary<object,CacheEntry>字典对象来维护缓存内容,所以获取、创建和删除都是基于这个字典对象来操作的。ConcurrentDictionary 是一个线程安全的集合类,正是因为 MemoryCache 对象使用了这样的存储结构,所以它默认支持多线程并发。

代码8-8
```
public class MemoryCache : IMemoryCache
{
    private readonly ConcurrentDictionary<object, CacheEntry> _entries;
    bool TryGetValue(object key, out object value);
    ICacheEntry CreateEntry(object key);
    void Remove(object key);
}
```

MemoryCache 对象提供了 CacheEntry 方法,用于将 CacheEntry 对象添加到缓存字典中,而 CacheEntry 对象正是通过 CacheEntry 方法中的 Dispose 方法来执行调用的。

8.1.3 缓存大小的限制

MemoryCache 对象可以通过设置 MemoryCacheOptions 对象的 SizeLimit 属性来限制缓存的大小。缓存大小并没有定义一个明确的度量单位,因为缓存没有度量条目大小的机制,如果启用了缓存大小限制,那么所有的条目都必须指定 Size 属性。.NET 中没有提供根据内存压力限制内存大小的方式,需要开发人员自行限制缓存大小。

如代码 8-9 所示,以控制台实例为例,创建 ServiceCollection 对象,调用 AddMemoryCache 扩展方法注册内存缓存核心服务。AddMemoryCache 扩展方法将 Action<MemoryCacheOptions> 委托对象作为参数,设置 MemoryCacheOptions 对象的 SizeLimit 属性的值为 1,表示最大的缓存,随后调用 IServiceCollection 对象的 BuildServiceProvider 扩展方法得到依赖注入容器的 IServiceProvider 对象,这样就可以直接调用 GetRequiredService<T>扩展方法来提供 IMemoryCache 对象。

代码8-9
```
var cache = new ServiceCollection()
```

```
    .AddMemoryCache(options =>
    {
        options.SizeLimit = 1;
    })
    .BuildServiceProvider()
    .GetRequiredService<IMemoryCache>();

var options = new MemoryCacheEntryOptions().SetSize(1);
cache.Set("key1", "value1", options);
cache.Set("key2", "value2");

Console.WriteLine(cache.Get("key1"));
Console.WriteLine(cache.Get("key2"));
```

创建一个 MemoryCacheEntryOptions 对象，调用 SetSize 扩展方法，设置 Size 属性的值。如果设置了 SizeLimit 属性的最大值，就需要在添加每个缓存时显示设置该属性，该属性的值必须小于或等于最大值，若不设置，则抛出异常（当设置了 SizeLimit 属性时，缓存项必须指定 Size 属性的值，如图 8-2 所示）。如果设置的数值累计超出 SizeLimit 属性的大小，那么缓存不生效。

如果超出限制的容量，那么超出部分的缓存不会被添加进去（缓存大小受限，未存储超出的内容，如图 8-3 所示）。

图 8-2　当设置了 SizeLimit 属性时，缓存项必须指定 Size 属性的值

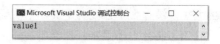

图 8-3　缓存大小受限，未存储超出的内容

8.2　分布式缓存

集群会导致每个节点服务器中的缓存出现不一致的情况，而采用分布式缓存可以有效地搭建"中心化"的存储，解决缓存共享的问题，从而实现应用水平扩展自动伸缩的目标。

8.2.1 IDistributedCache

分布式缓存的核心对象是 IDistributedCache 接口（见代码 8-10），分布式缓存的管理除了基础的设置、获取和删除，还包括缓存的刷新。在 IDistributedCache 接口中还定义了同步和异步的操作，缓存的添加过程采用了字节数组，因此需要对缓存的对象进行序列化和反序列化操作。

代码8-10

```
public interface IDistributedCache
{
    byte[] Get(string key);
    Task<byte[]> GetAsync(string key, CancellationToken token
                                    = default(CancellationToken));

    void Set(string key, byte[] value,
                        DistributedCacheEntryOptions options);
    Task SetAsync(string key, byte[] value,
                        DistributedCacheEntryOptions options,
            CancellationToken token = default(CancellationToken));

    void Refresh(string key);
    Task RefreshAsync(string key, CancellationToken token =
                                    default(CancellationToken));

    void Remove(string key);
    Task RemoveAsync(string key, CancellationToken token =
                                    default(CancellationToken));
}
```

Get 方法/GetAsync 方法可以用来获取缓存，并返回一个 byte 类型的数组；Set 方法/SetAsync 方法用于设置或添加缓存；Refresh 方法/RefreshAsync 方法用于刷新缓存条目中的最后访问时间，并将它设置为当前时间；Remove 方法/RemoveAsync 方法用于根据指定键删除缓存。

分布式缓存可以通过 DistributedCacheEntryOptions 对象来指定相关的过期策略，绝对时间和滑动时间都可以通过该对象的属性来指定，如代码 8-11 所示。

代码8-11

```csharp
public class DistributedCacheEntryOptions
{
    private DateTimeOffset? _absoluteExpiration;
    private TimeSpan? _absoluteExpirationRelativeToNow;
    private TimeSpan? _slidingExpiration;

    public DateTimeOffset? AbsoluteExpiration { get;set; }
    public TimeSpan? AbsoluteExpirationRelativeToNow { get;set; }
    public TimeSpan? SlidingExpiration { get; set; }
}
```

如代码 8-12 所示，分布式缓存 DistributedCacheEntryOptions 对象可以通过调用扩展方法来操作对象属性。

代码8-12

```csharp
public static class DistributedCacheEntryExtensions
{
    public static DistributedCacheEntryOptions SetAbsoluteExpiration(
        this DistributedCacheEntryOptions options,
        TimeSpan relative)
    {
        options.AbsoluteExpirationRelativeToNow = relative;
        return options;
    }

    public static DistributedCacheEntryOptions SetAbsoluteExpiration(
        this DistributedCacheEntryOptions options,
        DateTimeOffset absolute)
    {
        options.AbsoluteExpiration = absolute;
        return options;
    }

    public static DistributedCacheEntryOptions SetSlidingExpiration(
        this DistributedCacheEntryOptions options,
        TimeSpan offset)
    {
```

```
        options.SlidingExpiration = offset;
        return options;
    }
}
```

如代码 8-13 所示，IDistributedCache 对象可以通过调用扩展方法 Set 和 SetAsync 来设置缓存的 Key 与 Value，但需要自行将缓存内容序列化成字节数组。当然，也可以直接调用扩展方法 SetString 和 SetStringAsync，从而节省序列化的操作，开发人员不需要关注字符串序列化，只需要传递相应的 String 类型值即可。开发人员可以采用这种方式获取对象，分布式缓存的存储内容为序列化的字节数组，而取出的内容也是序列化字节数组，但 DistributedCacheExtensions 类提供了扩展方法 GetString 和 GetStringAsync，因此，开发人员可以调用这两个方法来省略反序列化操作。值得注意的是，缓存的设置和读取默认使用 UTF-8 进行字符串的编码与解码。

代码8-13

```
public static class DistributedCacheExtensions
{
    public static void Set(this IDistributedCache cache, string key!!,
                                                          byte[] value!!)
    {
        cache.Set(key, value, new DistributedCacheEntryOptions());
    }

    public static Task SetAsync(this IDistributedCache cache, string key!!,
      byte[] value!!, CancellationToken token = default(CancellationToken))
    {
        return cache.SetAsync(key, value, new DistributedCacheEntryOptions(),
                                                                    token);
    }

    public static void SetString(this IDistributedCache cache, string key,
                                                          string value)
    {
        cache.SetString(key, value, new DistributedCacheEntryOptions());
    }

    public static void SetString(this IDistributedCache cache, string key!!,
                    string value!!,DistributedCacheEntryOptions options)
```

```csharp
    {
        cache.Set(key, Encoding.UTF8.GetBytes(value), options);
    }

    public static Task SetStringAsync(this IDistributedCache cache,
                string key, string value,CancellationToken token
                                        = default(CancellationToken))
    {
        return cache.SetStringAsync(key, value,
                        new DistributedCacheEntryOptions(), token);
    }

public static Task SetStringAsync(this IDistributedCache cache,
        string key!!,string value!!,DistributedCacheEntryOptions options,
         string key!!,string value!!,DistributedCacheEntryOptions options,
                CancellationToken token = default (CancellationToken))
    {
        Return cache.SetAsync(key, Encoding.UTF8.GetBytes(value),
                                                    options, token);
    }

    public static string? GetString(this IDistributedCache cache, string key)
    {
        byte[]? data = cache.Get(key);
        if (data == null)
        {
            return null;
        }
        return Encoding.UTF8.GetString(data, 0, data.Length);
    }

public static async Task<string?> GetStringAsync(
                    this IDistributedCache cache, string key,
            CancellationToken token = default(CancellationToken))
    {
        byte[]? data = await cache.GetAsync(key, token)
                                        .ConfigureAwait (false);
        if (data == null)
```

```
        {
            return null;
        }
        return Encoding.UTF8.GetString(data, 0, data.Length);
    }
}
```

8.2.2 基于 Redis 的分布式缓存

Redis 是一个开源的内存数据库，还是一个基于 C 语言编写的键值数据库，也是一个 NoSQL 数据库，支持相当多的数据结构，如 String（字符串）、Hash（哈希）、List（列表）等，通常用作分布式缓存。.NET 提供了针对 Redis 数据库的分布式缓存的支持，该操作定义在 "Microsoft.Extensions.Caching.StackExchangeRedis" NuGet 包中。

如代码 8-14 所示，先调用 AddStackExchangeRedisCache 扩展方法注册 Redis 服务，再获取 IDistributedCache 对象，调用 SetString 扩展方法设置缓存值。指定的缓存内容保存在 Redis 数据库中，而与其相关的还有绝对过期时间和滑动过期时间。AbsoluteExpiration 属性表示该缓存的绝对过期时间，可以将其设置为当前系统时间 30 秒后过期；SlidingExpiration 属性表示滑动过期时间，设置为 10 秒。

代码8-14

```
var cache = new ServiceCollection()
    .AddStackExchangeRedisCache(options =>
    {
        options.Configuration = "localhost";
    }).BuildServiceProvider()
    .GetRequiredService<IDistributedCache>();
cache.SetString("key1", "value", new DistributedCacheEntryOptions
{
    AbsoluteExpiration = DateTimeOffset.UtcNow.AddSeconds(30),
    SlidingExpiration = TimeSpan.FromSeconds(10)
});
var value = cache.GetString("key1");
```

在上述实例执行完之后，缓存数据会被保存，并且通过 Redis 命令行的方式查看 Redis 数据库中的缓存内容。接下来按照如图 8-4 所示的方式执行 hgetall key1 命令，并输出结果，由输出结果可以看出，设置的内容都被保存到 Redis 数据库中，而 absexp 和 sldexp

是以纳秒量级存储的，DateTimeOffset 对象和 TimeSpan 对象都对应 Ticks 属性的值（一个 Tick 代表 1 纳秒，DateTimeOffset 对象的 Ticks 返回距离"0001 年 1 月 1 日午夜 12:00:00"的纳秒数）。

图 8-4　查看 Redis 数据库中的数据

8.2.3　基于 SQL Server 的分布式缓存

对于分布式缓存来说，数据除了可以放在 Redis 数据库中，还可以放在关系型数据库 SQL Server 中。与 Redis 缓存相比，SQL Server 可能不是最流行的缓存存储方式，但是它可以将应用程序状态保持在进程本身之外，从而实现基于数据库的分布式副本缓存的共享。

如代码 8-15 所示，以控制台实例为例，调用 AddDistributedSqlServerCache 扩展方法注册 SQL Server 分布式缓存相应的服务。因为需要将缓存数据存储到数据库中，所以通过 SqlServerCacheOptions 对象设置数据库和数据表的信息，ConnectionString 属性表示目标数据库的连接字符串，SchemaName 属性和 TableName 属性分别表示缓存数据库的 Schema 和表。接下来获取 IDistributedCache 实例对象，对缓存进行设置、删除等操作。

代码8-15

```
var cache = new ServiceCollection()
    .AddDistributedSqlServerCache(options =>
    {
        options.ConnectionString = "data source=.;integrated security=True;
                    User ID=sa;initial catalog=TestDb;Password=sa;";
        options.SchemaName = "dbo";
        options.TableName = "AspNetCache";
    }).BuildServiceProvider()
    .GetRequiredService<IDistributedCache>();
cache.SetString("key", "value", new DistributedCacheEntryOptions
{
```

```
    AbsoluteExpiration = DateTimeOffset.Now.AddSeconds(20)
});
```

如代码 8-16 所示，安装 CLI 工具，该命令行工具用于 SQL Server 数据库中分布式缓存的表结构。

代码8-16

```
dotnet tool install --global dotnet-sql-cache
```

如代码 8-17 所示，执行 dotnet sql-cache create 命令，创建 SQL Server 数据库缓存表。dotnet sql-cache create 命令的格式为 dotnet sql-cache create <connection string> <schema> <table name>。

代码8-17

```
dotnet sql-cache create "data source=.;integrated security=True;
User ID=sa;Password=sa;initial catalog=TestDb" dbo AspNetCache
```

图 8-5 所示为 SQL Server 数据库缓存表结构，该表结构中的 Id 属性和 Value 属性分别表示缓存的 Key 和 Value，SlidingExpirationInSeconds 属性和 AbsoluteExpiration 属性表示滑动过期时间和绝对过期时间，ExpiresAtTime 属性则表示存储到期时间。需要注意的是，bigint 数据类型意味着将 SlidingExpirationInSeconds 属性的总秒数添加到当前时间日期中可以创建该值。

图 8-5 SQL Server 数据库缓存表结构

SQL Server 数据库缓存中的 AddDistributedSqlServerCache 扩展方法的配置由 SqlServerCacheOptions 对象定义（见代码 8-18）。SystemClock 属性返回一个本地时间，提供同步时间的系统时钟；ExpiredItemsDeletionInterval 属性表示缓存清除的时间，默认时间间隔为 30 分钟，与 Redis 数据库做分布式缓存差异最大的一点是删除过期的缓存数据，Redis 支持内存淘汰策略，当设置的数据过期时间过期后，Redis 会自动清除过期的缓存记录。而 SQL Server 清除缓存的策略较为简单，在每次调用 Get 方法/GetAsync 方法时通过

ScanForExpiredItemsIfRequired 方法扫描过期的缓存，同时执行清空过期缓存的 SQL 语句 DELETE FROM {0} WHERE @UtcNow > ExpiresAtTime。

代码8-18

```
public class SqlServerCacheOptions : IOptions<SqlServerCacheOptions>
{
    public ISystemClock SystemClock { get; set; }
    public TimeSpan? ExpiredItemsDeletionInterval { get; set; }

    public string ConnectionString { get; set; }
    public string SchemaName { get; set; }
    public string TableName { get; set; }

    public TimeSpan DefaultSlidingExpiration { get; set; }
                                             = TimeSpan.FromMinutes(20);

    SqlServerCacheOptions IOptions<SqlServerCacheOptions>.Value
    {
        get
        {
            return this;
        }
    }
}
```

ConnectionString 属性表示目标数据库的连接字符串，TableName 属性和 SchemaName 属性表示对应的缓存表名称和 Schema，DefaultSlidingExpiration 属性表示缓存的滑动过期时间。

8.3 HTTP 缓存

HTTP 缓存分为两种，即强缓存（本地缓存）和协商缓存（弱缓存），主要用于提高资源的获取速度，减少网络传输的开销，缓解服务端的压力。在 HTTP 规范下，一般可以缓存 GET 请求和 HEAD 请求，而 PUT 请求和 DELETE 请求不可缓存，POST 请求在大多数情况下是不可以缓存的。本节从 HTTP 的角度开始介绍响应缓存，然后介绍 ASP.NET Core 中的响应缓存核心 ResponseCachingMiddleware 中间件的基本机制。

8.3.1 响应缓存

缓存会根据相应的规则以副本的形式将响应内容存储到缓存数据库中，并且可以匹配已经存在的资源副本。另外，缓存并不是永久保存的，缓存资源是有一定的时效性的，即使缓存过期，也不会直接被清除或忽略，而是采用替换的形式，将旧的资源替换为新的资源。图 8-6 所示为第一次请求资源和写入缓存的过程。

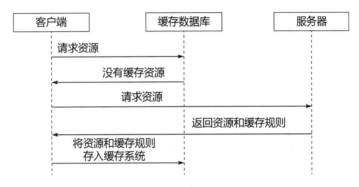

图 8-6　第一次请求资源和写入缓存的过程

HTTP 缓存的类型

HTTP 缓存策略通常分为两种，即强缓存和协商缓存。强缓存用于强制使用缓存，协商缓存需要和服务器协商确认这个缓存是否可以使用。

浏览器在加载资源时，根据请求头中的信息（Expires 属性和 Cache-Control 属性）来判断是否命中强缓存。如果命中强缓存，那么直接从缓存中读取资源，不会将请求发到服务器；如果没有命中强缓存，那么浏览器会将请求发送到服务器，并通过 Last-Modified 属性和 ETag 属性验证资源是否命中协商缓存。如果命中，服务器就将这个请求返回，但是不会返回这个资源的数据，依然从缓存中读取资源。如果两者都没有命中，那么直接从服务器中读取资源。

HTTP 缓存控制

在 HTTP 中，可以通过设置响应报文及请求头中对应的 Header 信息来控制缓存策略。缓存主要由几个 Header 属性共同控制，强缓存由 Expires 属性、Cache-Control 属性和 Pragma 属性共同控制，协商缓存由 Last-Modified 属性或 ETag 属性控制。

Expires

Expires 响应头是 HTTP/1.0 的属性，表示缓存资源的过期时间，并且是一个绝对过期时间。在响应报文中，Expires 响应头会告诉浏览器在该时间过期之前可以直接从浏览器中获取缓存的资源，由于该属性是绝对过期时间，因此客户端与服务端的时间存在时间时差或误差这就会出现时间不一致的情况，最终缓存会存在误差。

Fri, 24 Sep 2021 07:42:34 GMT

Cache-Control

Cache-Control 响应头是 HTTP/1.1 的属性，表示缓存资源的最大有效时间，在有效时间内，客户端不需要向服务器发送请求。Cache-Control 属性的优先级高于 Expires 属性的优先级，两者的区别就是 Expires 属性表示的是绝对时间，而 Cache-Control 属性允许设置相对时间。

可缓存性

- Cache-Control: public：共享缓存，又称为公共缓存，资源可以被客户端和代理服务器缓存。
- Cache-Control: private：私有缓存，资源只可以被客户端缓存，代理服务器不能缓存，不能作为共享缓存。
- Cache-Control: no-store：缓存不以任何形式存储，该指令用于阻止响应内容被缓存，通常用于阻止带有敏感性信息或随时都会变化的数据响应。
- Cache-Control: no-cache：在每次请求时，缓存都会将请求发送到服务器，表示不使用缓存的资源。

过期

- Cache-Control: max-age={seconds}：设置缓存存储的最大周期，超过这个时间缓存则被认为过期。
- Cache-Control: s-maxage={seconds}：设置共享缓存可以应用该指令，覆盖 max-age 指令或 Expires 属性。该指令仅适用于共享缓存，私有缓存不会应用，而是选择忽略。
- Cache-Control: max-stale 或 Cache-Control: max-stale={seconds}：表示缓存可以获取已过期的资源，客户端可以设置一个以秒为单位的时间（可选），以指定超时后有效的时间，表示该响应即使过期，只要不超过该指定时间就可以使用。
- Cache-Control: min-fresh={seconds}：表示服务器返回 min-fresh 时间内的资源。

重新验证和重新加载

- Cache-Control: must-revalidate：缓存可以在本地存储当前资源，但是在资源过期后必须经过验证，确定一致性之后才能返回给客户端。

- Cache-Control: proxy-revalidate：和 must-revalidate 指令一样，但是该指令仅用于共享缓存。

其他

- Cache-Control: only-if-cached：不进行网络请求，仅使用缓存资源。
- Cache-Control: no-transform：中间代理有时会改变图片及文件的格式，从而节省缓存提高性能，而 no-transform 指令告诉中间代理不要改变和修改任何响应。

Last-Modified 属性和 If-Modified-Since 属性

Last-Modified 属性和 If-Modified-Since 属性用来表示资源的最后修改时间，在客户端第一次请求时，服务端把资源的最后修改时间添加到 Last-Modified 响应头中，在第二次发起请求时，请求头会带上第一次响应头中的 Last-Modified 属性的时间，并设置在 If-Modified-Since 属性中，当服务端接收到这个请求后，会利用这些信息判断资源是否发生过变化。如果资源一直没有发生变化，那么返回 304 Not Modified 的响应，浏览器从缓存中获取信息；如果资源发生变化，那么开始传输一个最新的资源，同时服务器返回的状态码为 200 OK。

在请求时会出现一种特殊的情况，即资源被修改，但内容没有发生过任何变化，Last-Modified 属性的时间匹配不上，导致返回整个内容给客户端。为了解决这个问题，HTTP/1.1 增加了 ETag 属性。ETag 属性的优先级高于 Last-Modified 属性的优先级。

ETag 属性和 If-None-Match 属性

ETag 是对资源生成的唯一标识，像一个指纹，资源只要发生变化就会导致 ETag 属性发生变化，和最后的修改时间没有关系。ETag 属性可以保证每个资源都是唯一的。

浏览器发起请求后，请求报文中会包含 If-None-Match 属性，该属性的值就是上次返回的 ETag 属性的值，当服务器发现 If-None-Match 属性后，会与请求资源的唯一标识进行比较。若相同，则说明资源没有修改，返回 304 Not Modified 的响应，浏览器直接从缓存中获取数据信息；若不同，则说明资源发生变化，响应资源内容，返回的状态码为 200 OK。

8.3.2 ResponseCachingMiddleware 中间件

ASP.NET Core 提供了服务端响应缓存。它的核心是一个名为 ResponseCachingMiddleware 的中间件，与 HTTP/1.1 缓存的设计规范兼容，可以为应用程序添加服务端的缓存功能。

如代码 8-19 所示，Startup 类先调用 AddResponseCaching 扩展方法注册响应缓存，再调用 UseResponseCaching 扩展方法将 ResponseCachingMiddleware 中间件添加到 ASP.NET Core 的请求管道中。

代码8-19

```
var builder = WebApplication.CreateBuilder(args);

builder.Services.AddResponseCaching(options => {
    options.UseCaseSensitivePaths = true;
    options.MaximumBodySize = 1024;
});
var app = builder.Build();

app.UseResponseCaching();
app.Run();
```

如代码 8-20 所示,在注册响应缓存服务时,可以通过 ResponseCachingOptions 对象设置缓存响应中间件的相关配置。

代码8-20

```
public class ResponseCachingOptions
{
    public long SizeLimit { get; set; } = 100 * 1024 * 1024;
    public long MaximumBodySize { get; set; } = 64 * 1024 * 1024;
    public bool UseCaseSensitivePaths { get; set; }
    [EditorBrowsable(EditorBrowsableState.Never)]
    internal ISystemClock SystemClock { get; set; } = new SystemClock();
}
```

SizeLimit 属性用于设置缓存响应中间件缓存的大小,默认为 100MB;MaximumBodySize 属性用于设置缓存响应正文的最大值,默认为 64MB;UseCaseSensitivePaths 属性用于标记缓存是否区分请求路径的大小写,默认为 false。

缓存操作通过一个名为 IResponseCache 的接口表示,在接口中定义了两个方法,Set 方法和 Get 方法分别实现了缓存的读和写,如代码 8-21 所示。

代码8-21

```
internal interface IResponseCache
{
    IResponseCacheEntry? Get(string key);
    void Set(string key, IResponseCacheEntry entry, TimeSpan validFor);
}
```

ResponseCachingMiddleware 中间件会利用 IMemoryCache 对象设置缓存响应内容,并且通过 CachedResponse 类定义结构化的缓存内容,如代码 8-22 所示。CachedResponse 类定义

了创建时间、状态码、报头集合和主体内容。

代码8-22

```
internal class CachedResponse : IResponseCacheEntry
{
    public DateTimeOffset Created { get; set; }

    public int StatusCode { get; set; }

    public IHeaderDictionary Headers { get; set; } = default;

    public CachedResponseBody Body { get; set; } = default;
}
```

8.4 小结

通过学习本章，读者可以掌握多样化缓存的存储方式（如常见的内存缓存和分布式缓存）和 ResponseCachingMiddleware 中间件。

第 9 章

本地化

本地化是指在应用交付到国际环境下时，需要根据特定国家的语言或市场的特点对软件的某些功能进行组织和调整。本地化（Localization）又简称为 L10n，10 用来表示 L 和 n 中间省略的 10 个字母。本地化的基础是提供多语言的支持，如查阅.NET 的官方文档也可以看到其提供了多语言的支持。

本章将对.NET 中的本地化功能进行解析，主要包括两个方面：一是了解请求携带语言文化信息参数的传递形式，以及如何自定义属于自己的语言文化信息解析器；二是了解资源文件中字符串文本的读取，以多样化的数据源响应结果。

9.1 内容本地化

.NET 中提供了开箱即用的本地化功能，开发人员可以通过几行代码快速实现一个支持多语言的应用程序。

9.1.1 提供文本多语言支持

为了快速构建一个实例，下面以.NET Minimal APIs 项目为例展开介绍，项目的文件结构如图 9-1 所示。先调用 WebApplication 对象的 CreateBuilder 扩展方法获取 WebApplicationBuilder 对象；再调用它的 Services 属性返回一个 IServiceCollection 对象，并调用 AddLocalization 扩展方法注册本地化相关的服务；最后调用 WebApplicationBuilder 对象的 Build 方法创建

WebApplication 对象，并利用 WebApplication 对象调用 RunAsync 方法启动该服务，如代码 9-1 所示。

代码9-1
```
var builder = WebApplication.CreateBuilder(args);

builder.Services.AddLocalization();

var app = builder.Build();

await app.RunAsync();
```

通常，开发人员会创建资源文件存储多语言资源，笔者分别创建名为 SharedResource.resx 和 SharedResource.zh.resx 的资源文件（为对应的资源文件设置不同的语言，如图 9-2 所示）。在这两个资源文件中定义一个名为 Message 的字符串资源，并且为这两个资源文件分别定义中文"你好，世界！"和英文"Hello World!"。

图 9-1　项目的文件结构

图 9-2　为对应的资源文件设置不同的语言

如代码 9-2 所示，通过 WebApplication 对象调用 MapGet 扩展方法，创建一个支持 GET 请求的方法，通过一个名为 culture 的参数获得请求传递的语种信息，在 MapGet 方法中注入一个 IStringLocalizerFactory 服务，并调用它的 Create 方法创建一个用于提供本地化字符串的 IStringLocalizer 对象。先利用 culture 参数创建出对应语种的 CultureInfo 对象，再设置当前线程的 CurrentCulture 属性和 CurrentUICulture 属性，最后调用 IStringLocalizer 对象的 GetString 扩展方法获取指定的资源字符串内容。

代码9-2

```
app.MapGet("/", (string? culture,
 IStringLocalizerFactory localizerFactory) =>
{
    var localizer = localizerFactory.Create("SharedResource", "App");
    if (!string.IsNullOrEmpty(culture))
    {
        CultureInfo.CurrentCulture = CultureInfo.CurrentUICulture =
                                        new CultureInfo(culture);
    }
    return Results.Content(localizer.GetString("Message").Value);
});
```

接下来通过 curl 命令行工具发起 GET 请求。图 9-3 所示为针对不同语言文化输出的响应结果，通过 culture 参数在请求中携带语言文化代码，若参数设置为 zh-cn，则返回中文内容，若请求时没有设置参数或参数不匹配默认情况，则采用 SharedResource.resx 资源文件中的内容。

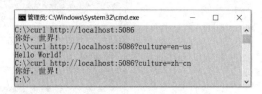

图 9-3　针对不同语言文化输出的响应结果

9.1.2　RequestLocalizationMiddleware 中间件

通过学习 9.1.1 节的实例，读者可以了解多语言文化的使用，但是对于语言文化来说，笔者一直都是在方法内部设置的，如果使用多个方法，那么通常来说要做很多重复的工作。.NET 中已经为开发人员考虑到了这一点，所以关于语言文化属性的设置可以利用 RequestLocalizationMiddleware 中间件来完成。下面修改实例，如代码 9-3 所示。

代码9-3

```
var builder = WebApplication.CreateBuilder(args);
builder.Services.AddLocalization();

var app = builder.Build();
```

```
app.UseRequestLocalization(options => options
    .AddSupportedCultures("en", "zh")
.AddSupportedUICultures("en", "zh"));

await app.RunAsync();
```

开发人员只需要通过 IApplicationBuilder 对象调用 UseRequestLocalization 扩展方法即可完成对中间件 RequestLocalizationMiddleware 的注册。接下来调用 AddSupportedCultures 方法和 AddSupportedUICultures 方法设置 en 和 zh，将这两种语言文化添加到配置选项的 Culture 集合与 UICulture 集合中。修改 MapGet 方法中的代码，删除对语言文化设置的代码内容，因为这些操作中间件已经帮助开发人员自动处理了，如代码 9-4 所示。

代码9-4

```
app.MapGet("/", (IStringLocalizerFactory localizerFactory) =>
{
    var localizer = localizerFactory.Create("SharedResource", "App");
    return Results.Content(localizer.GetString("Message").Value);
});
```

中间件默认提供了 3 种语言文化参数的设置方式，如 URL 参数、Cookie 和指定的 Accept-Language 请求头。可以通过任意一种方式传递语言文化信息，最终通过 RequestLocalizationMiddleware 中间件解析该请求传递的语言文化信息，如图 9-4 所示，设置 Accept-Language 请求头的值，通过 curl 命令行工具并携带 Accept-Language 请求头作为语言文化信息传递的方式，最终将请求信息和响应结果输出到控制台中。

图 9-4　设置 Accept-Language 请求头的值

如图 9-5 所示，通过 Cookie 设置语言文化，以 Cookie 的方式传递当前的语言文化信息，这是一种常用的方式。默认使用.AspNetCore.Culture 作为 Cookie 的名称，Cookie 的值以 c={Culture}|uic={UICulture}的格式来传递，以 curl 命令行工具发起 HTTP 请求，使用 Cookie 的方式设置语言文化。

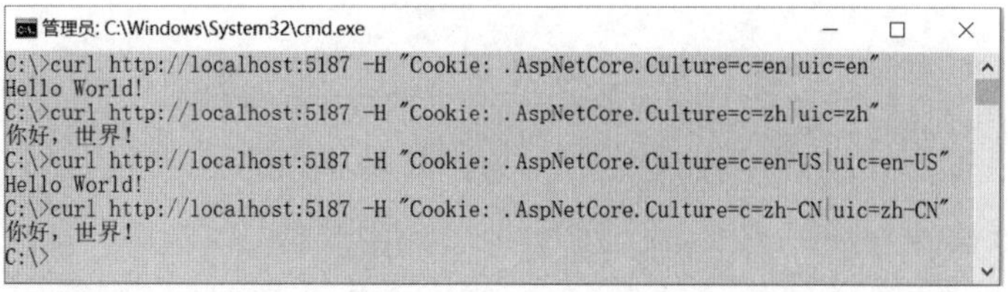

图 9-5 通过 Cookie 设置语言文化

至此，读者已经基本了解了 RequestLocalizationMiddleware 中间件的使用，接下来详细介绍在中间件中是如何利用多语言文化解析器的。如代码 9-5 所示，简化 RequestLocalizationMiddleware 中间件的实现，在其内部定义 RequestLocalizationOptions 配置选项对象，在中间件中判断该对象的 RequestCultureProviders 属性是否为空。如果 RequestCultureProviders 属性不为空，那么执行 RequestLocalizationMiddleware 中间件。另外，RequestCultureProviders 属性用于存储语言文化的解析器对象。在 RequestCultureProviders 属性不为空的情况下循环该解析器，获取 IRequestCultureProvider 对象并调用它的 DetermineProviderCultureResult 方法获取当前解析器返回的结果。如果 DetermineProviderCultureResult 方法返回的结果为空，那么继续循环；如果 DetermineProviderCultureResult 方法返回的结果不为空，那么调用 SetCurrentThreadCulture 方法设置当前线程的语言文化。

代码9-5

```
public class RequestLocalizationMiddleware
{
    private readonly RequestLocalizationOptions _options;
    public async Task Invoke(HttpContext context)
    {
        if(_options.RequestCultureProviders != null)
        {
            foreach(var provider in _options.RequestCultureProviders)
            {
                var providerResultCulture =
                  await provider.DetermineProviderCultureResult(context);
                if(providerResultCulture == null){
                    continue;
                }

                var cultures = providerResultCulture.Cultures;
                var uiCultures = providerResultCulture.UICultures;
```

```
            if(_options.SupportedCultures != null)
            {
                cultureInfo = GetCultureInfo(
                    cultures,
                    _options.SupportedCultures,
                    _options.FallBackToParen、tCultures);
            }

            if(_options.SupportedUICultures != null)
            {
                uiCultureInfo = GetCultureInfo(
                    uiCultures,
                    _options.SupportedUICultures,
                    _options.FallBackToParentUICultures);
            }

            cultureInfo ??= _options.DefaultRequestCulture.Culture;
            uiCultureInfo ??= _options.DefaultRequestCulture.UICulture;

            var result = new RequestCulture(cultureInfo, uiCultureInfo);
            requestCulture = result;
            break;
        }
    }
    SetCurrentThreadCulture(requestCulture);
    await _next(context);
}
```

RequestLocalizationMiddleware 中间件的配置选项定义在 RequestLocalizationOptions 对象中，如代码 9-6 所示，DefaultRequestCulture 属性用于中间件获取不到有效的语言文化或语言文化没有在受支持的列表中，此时使用 DefaultRequestCulture 属性的设置作为默认语言文化，当然，默认语言文化也可以通过调用 SetDefaultCulture 方法来指定。FallBackToParentCultures 属性和 FallBackToParentUICultures 属性表示如果 IRequestCultureProvider 对象提供的语言文化（如 en-US、en-GB）并不在当前受支持的列表中，但它的 Parent（en）是受支持的语言之一，则 RequestLocalizationMiddleware 中间件会使用它的 Parent 作为当前语言的文化。当然，FallBackToParentCultures 属性和 FallBackToParentUICultures 属性默认为 true，开发人员可以修改基于 Parent 的语言文化策略，可以将其属性设置为 false。ApplyCurrentCultureToResponseHeaders 属性表示是否设置

Content-Language 响应头，若设置为 true，则告知客户端当前使用的语言是 zh-CN 或 en-US 等。SupportedCultures 属性和 SupportedUICultures 属性表示当前应用程序支持的语言文化列表。RequestCultureProviders 属性表示当前支持的语言文化解析器集合。AddSupportedCultures 方法和 AddSupportedUICultures 方法可以用来将指定的语言文化添加到 SupportedCultures 列表和 SupportedUICultures 列表中。SetDefaultCulture 方法可以用来设置当前默认的语言文化。

代码9-6

```
public class RequestLocalizationOptions
{
    public RequestCulture DefaultRequestCulture {get; set; }
    public bool FallBackToParentCultures { get; set; } = true;
    public bool FallBackToParentUICultures { get; set; } = true;
    public bool ApplyCurrentCultureToResponseHeaders { get; set; }
    public IList<CultureInfo>? SupportedCultures { get; set; } =
                new List<CultureInfo> { CultureInfo.CurrentCulture };
    public IList<CultureInfo>? SupportedUICultures { get; set; } =
                new List<CultureInfo> { CultureInfo.CurrentUICulture };
    public IList<IRequestCultureProvider> RequestCultureProviders
                                                    { get; set; }

    public RequestLocalizationOptions AddSupportedCultures(
                                        params string[] cultures);
    public RequestLocalizationOptions AddSupportedUICultures(
                                        params string[] uiCultures);
    public RequestLocalizationOptions SetDefaultCulture(
                                        string defaultCulture);
}
```

9.1.3 使用 RouteDataRequestCultureProvider

使用 RequestLocalizationMiddleware 中间件，开发人员可以快速创建基于本地化的应用程序。RequestLocalizationMiddleware 中间件提供了多种方式可以对语言文化进行设置，如 Cookie、请求参数等。

本节主要讨论基于 URL 的方式传递语言文化，通过 RouteDataRequestCultureProvider 对象即可以 URL 的形式传递语言文化。

下面创建一个项目实例，如代码 9-7 所示，先调用 RequestLocalizationOptions 对象的 RequestCultureProviders 属性，再调用 Insert 方法注册 RouteDataRequestCultureProvider 对象。

代码9-7

```
var builder = WebApplication.CreateBuilder(args);

builder.Services.AddLocalization();

var app = builder.Build();

app.UseRequestLocalization(options =>
{
    options
        .AddSupportedCultures("en", "zh")
        .AddSupportedUICultures("en", "zh")
        .SetDefaultCulture("en")
        .RequestCultureProviders.Insert(0,new
                            RouteDataRequestCultureProvider());
});

await app.RunAsync();
```

上述代码实现了对 URL 提供者的注册。如代码 9-8 所示，修改 MapGet 方法，将路径修改为{culture}/docs。culture 是必须项，如它可以是一个 docs 页面，主要用于接收语言文化参数。

代码9-8

```
app.MapGet("{culture}/docs", (IStringLocalizerFactory localizerFactory) =>
{
    var localizer = localizerFactory.Create("SharedResource", "App1");
    return Results.Content(localizer.GetString("Message").Value);
});
```

如图 9-6 所示，可以通过 URL 设置语言文化，通过 curl 命令行工具发起 HTTP 请求，以 URL 的形式设置请求语言文化。

图 9-6 通过 URL 设置语言文化

9.1.4 自定义 RequestCultureProvider 类

.NET 中默认提供了 3 个语言文化解析器，除此之外，开发人员还可以创建自定义语言文化解析器。如代码 9-9 所示，先调用 RequestLocalizationOptions 对象的 RequestCultureProviders 属性，再调用 Clear 方法清除所有的解析器对象，最后调用 RequestCultureProviders 属性的 Add 方法注册自定义语言文化解析器。

代码9-9

```
var builder = WebApplication.CreateBuilder(args);

builder.Services.AddLocalization();

var app = builder.Build();

app.UseRequestLocalization(options =>
{
    options
        .AddSupportedCultures("en", "zh")
        .AddSupportedUICultures("en", "zh")
        .RequestCultureProviders.Clear();
    options.RequestCultureProviders.Add(new
CustomRequestCultureProvider());
});
await app.RunAsync();
```

如代码 9-10 所示，先创建一个名为 CustomRequestCultureProvider 的类，该类继承 RequestCultureProvider 类，再实现一个名为 DetermineProviderCultureResult 的方法，该方法返回 ProviderCultureResult 对象。

代码9-10

```
public class CustomRequestCultureProvider : RequestCultureProvider
{
    public override Task<ProviderCultureResult?>
            DetermineProviderCultureResult (HttpContext httpContext)
    {
        if(httpContext == null)
        {
```

```
            throw new ArgumentNullException(nameof(httpContext));
        }

        if(!httpContext.User.Identity.IsAuthenticated)
        {
            return Task.FromResult((ProviderCultureResult)null);
        }

        var culture = httpContext.User.GetCulture();
        if(culture == null)
        {
            return Task.FromResult((ProviderCultureResult)null);
        }

        return Task.FromResult(new ProviderCultureResult(culture));
    }
}
```

9.2 多样化的数据源

在 .NET 中,已知的是以 .resx 为后缀的资源文件,用于存储多语言资源,用户还可以自定义数据源类型,如可以定义自己的一套方案(存储到 JSON、TXT 甚至数据库中都是可以的)。本节主要介绍自定义资源存储。

9.2.1 基于 JSON 格式的本地化

为了使读者可以更好地理解,笔者以 JSON 格式的文件作为数据源,存储资源内容。如代码 9-11 所示,创建一个名为 SharedResource.json 的文件,定义一个 Key 为 Message,方便定位指定的值,同时该 Key 对应两个值,分别为 zh-CN 和 en-US 两种语言文化内容。

代码9-11
```
//SharedResource.json
{
  "Message": {
    "LocalizedValue": {
```

```
    "zh-CN": "你好,世界!",
    "en-US": "Hello World!"
  },
  "Key": "Message"
}
```

如代码 9-12 所示,先调用 WebApplicationBuilder 对象的 Services 属性,该属性返回一个 IServiceCollection 接口,再调用 AddJsonLocalizer 扩展方法注册自定义 JSON 数据源实现。

代码9-12

```
var builder = WebApplication.CreateBuilder(args);

builder.Services.AddLocalization();
builder.Services.AddJsonLocalizer(
                        builder.Environment.ContentRootFileProvider);

var app = builder.Build();

app.UseRequestLocalization(options =>
{
    options
        .AddSupportedCultures("en-US", "zh-CN")
        .AddSupportedUICultures("en-US", "zh-CN");
});

app.MapGet("/", (IStringLocalizer<SharedResource> localizer) =>
{
    return Results.Content(localizer.GetString("Message").Value);
});

app.Run();
```

如代码 9-13 所示,创建一个名为 LocalizationStringEntry 的类,用于对应 JSON 文件中的内容。LocalizationStringEntry 类的 Key 属性表示内容的键;LocalizedValue 属性通过一个字典类型表示,该字典的 Key 和 Value 分别为语种和对应的本地化内容。

代码9-13

```
public class LocalizationStringEntry
```

```
{
    public string Key { get; set; }
    public Dictionary<string, string> LocalizedValue =
                                    new Dictionary <string, string>();
}
```

如代码9-14所示，实现IStringLocalizer接口，创建一个名为JsonStringLocalizer的类，在内部通过Dictionary<string,LocalizationStringEntry>属性承载本地化内容。

代码9-14

```
public class JsonStringLocalizer : IStringLocalizer
{
    private readonly Dictionary<string, LocalizationStringEntry> _localizaion;
    public JsonStringLocalizer(
                Dictionary<string, LocalizationStringEntry> directory)
    {
        _localizaion =
            new Dictionary<string, LocalizationStringEntry> (directory);
    }

    public LocalizedString this[string name]
    {
        get
        {
            var value = GetString(name);
            return new LocalizedString(
                    name, value ?? name, resourceNotFound: value == null);
        }
    }

    public LocalizedString this[string name, params object[] arguments]
    {
        get
        {
            var format = GetString(name);
            var value = string.Format(format ?? name, arguments);
            return new LocalizedString(
                    name, value, resourceNotFound: format == null);
        }
```

```csharp
    }

    public IEnumerable<LocalizedString> GetAllStrings(bool includeParentCultures)
    {
        return _localizaion.Where(
          l => l.Value.LocalizedValue.Keys.Any(lv => lv ==
                        CultureInfo.CurrentCulture.Name))
                    .Select(l => new
                        LocalizedString(l.Key,
          l.Value.LocalizedValue[CultureInfo.CurrentCulture.Name], true));
    }

    public IStringLocalizer WithCulture(CultureInfo culture)
    {
        throw new NotImplementedException();
    }

    private string GetString(string name)
    {
        var query = _localizaion.Where(
            l => l.Value.LocalizedValue. Keys.Any(lv => lv ==
                                CultureInfo.CurrentCulture.Name));
        var value = query.FirstOrDefault(l => l.Value.Key == name).Value;
        return value.LocalizedValue[CultureInfo.CurrentCulture.Name];
    }
}
```

如代码9-15所示，通过IStringLocalizerFactory接口实现并创建一个JsonStringLocalizerFactory类，在内部定义一个构造函数将 IFileProvider 对象作为参数，用于按照指定的路径读取 JSON 文件，最终读取出来的内容被反序列化为一个 Dictionary<string,LocalizedStringEntry>对象。通过 Dictionary<String,LocalizedStringEntry>对象创建 JsonStringLocalizer 对象，并将该实例对象缓存起来。

代码9-15

```csharp
public class JsonStringLocalizerFactory : IStringLocalizerFactory
{
    private readonly ConcurrentDictionary
                        <string, IStringLocalizer> _localizers;
```

```csharp
private readonly IFileProvider _fileProvider;
public JsonStringLocalizerFactory(IFileProvider fileProvider)
{
    _localizers = new ConcurrentDictionary<string, IStringLocalizer>();
    _fileProvider = fileProvider;
}

public IStringLocalizer Create(Type resourceSource)
{
    var path = ParseFilePath(resourceSource);
    return _localizers.GetOrAdd(path, _ =>
    {
        return CreateStringLocalizer(_);
    });
}

public IStringLocalizer Create(string baseName, string location)
{
    var path = ParseFilePath(location, baseName);
    return _localizers.GetOrAdd(path, _ =>
    {
        return CreateStringLocalizer(_);
    });
}

private IStringLocalizer CreateStringLocalizer(string path)
{
    var file = _fileProvider.GetFileInfo(path);
    if (!file.Exists)
    {
        return new JsonStringLocalizer(
            new Dictionary<string, LocalizationStringEntry>());
    }
    var dictionary = JsonConvert.DeserializeObject<Dictionary<string,
            LocalizationStringEntry>>(File.ReadAllText(path));
    return new JsonStringLocalizer(dictionary);
}
```

```csharp
private string ParseFilePath(string location, string baseName)
{
    var path = location + "." + baseName;
    return path.Replace("..", ".")
        .Replace('.', Path.DirectorySeparatorChar) + ".json";
}

private string ParseFilePath(Type resourceSource)
{
    var rootNamespaceAttribute = resourceSource.Assembly
                .GetCustomAttribute<RootNamespaceAttribute>()
       ?.RootNamespace ?? new AssemblyName(
                    resourceSource.Assembly.FullName).Name;
    return TrimPrefix(resourceSource.FullName, rootNamespaceAttribute + ".")
                                                        + ".json";
}

private string TrimPrefix(string name, string prefix)
{
    if (name.StartsWith(prefix, StringComparison.Ordinal))
    {
        return name.Substring(prefix.Length);
    }
    return name;
}
}
```

如代码 9-16 所示，创建一个名为 AddJsonLocalizer 的扩展方法，针对 IServiceCollection 接口创建扩展方法，并且利用 IFileProvider 对象创建 JsonStringLocalizerFactory 对象，将该对象注册为 Singleton 服务实例，同时将 IStringLocalizer 对象注册为 Transient 服务实例。

代码9-16

```csharp
public static class ServiceCollectionExtensions
{
    public static IServiceCollection AddJsonLocalizer(
            this IServiceCollection services, IFileProvider fileProvider)
    {
        services.AddSingleton<IStringLocalizerFactory>
```

```
                            (new JsonStringLocalizerFactory(fileProvider));
    services.AddTransient(typeof(IStringLocalizer<>),
                                        typeof(StringLocalizer<>));
    return services;
    }
}
```

9.2.2 基于 Redis 存储的本地化

对于本地化内容来说，放在配置文件中也许不是最好的选择，也可以将其保存到数据库或 Redis 中进行持久化存储，以方便开发人员在运行时动态设置本地化资源。本节以 Redis 为例存储本地化内容。

对于 Redis 操作，笔者先引入 StackExchange.Redis 框架作为 Redis 操作的客户端，再对 9.2.1 节中创建的项目进行调整。为了可以更好地理解，下面的简化代码只保留有修改的方法。如代码 9-17 所示，先引入 Redis 的操作类，再通过 IDatabase 对象调用 HashGetAll 方法，根据指定的键查询相应的值，取值后创建 Dictionary<string,LocalizationStringEntry>对象。

代码9-17

```
public class JsonStringLocalizerFactory : IStringLocalizerFactory
{
    private readonly ConcurrentDictionary<
                                string, IStringLocalizer> _localizers;
    private readonly IFileProvider _fileProvider;
    private volatile IConnectionMultiplexer conn =
                            ConnectionMultiplexer.Connect("localhost");
    private IDatabase _cache;

    public JsonStringLocalizerFactory(IFileProvider fileProvider)
    {
        _localizers = new ConcurrentDictionary<string, IStringLocalizer>();
        _fileProvider = fileProvider;
        _cache = conn.GetDatabase();
    }
    …
    private IStringLocalizer CreateStringLocalizer(string path)
    {
        var file = _fileProvider.GetFileInfo(path);
```

```
    if (!file.Exists)
    {
        return new JsonStringLocalizer(
                new Dictionary<string, LocalizationStringEntry>());
    }
    var result = _cache.HashGetAll(path);
    Dictionary<string, LocalizationStringEntry> dictionary =
                new Dictionary<string, LocalizationStringEntry>();
    dictionary.Add(path, new LocalizationStringEntry
    {
        Key = result[0].Value,
        LocalizedValue = JsonConvert.DeserializeObject
                    <Dictionary<string, string>>(result[1].Value)
    });
    return new JsonStringLocalizer(dictionary);
}
…
}
```

在上述代码中，先读取 Redis 中的数据，再通过 Redis 命令行工具查看数据库中存储的内容。图 9-7 所示为执行 hgetall SharedResource.json 命令输出的结果。

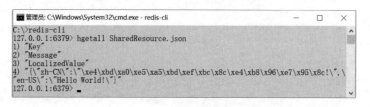

图 9-7　执行 hgetall SharedResource.json 命令输出的结果

9.3　小结

通过学习本章，读者可以了解如何构建一个支持多语言的应用程序，以及语言文化参数的传递形式和工作机制。关于多语言资源内容的存储，默认 .NET 中提供的是以 .resx 为后缀的资源文件，也可以自定义数据源，自行实现一个数据源提供器，如 JSON 文件或基于 Redis。虽然本章的实例比较详细，但是主要是为了使读者更好地理解，不建议用在项目生产中。另外，读者可以对本章的实例自行扩展。

第 10 章

健康检查

对于面向互联网的线上服务来说,通常有基础监控和健康检查,以判断服务状态是否可用。在 .NET 中可以轻松地构建出包含健康检查的应用程序,客户端会维护一个定时任务,每隔一个指定时间发送一次心跳请求,以确保自身处于活跃状态。

10.1 检查当前应用的健康状态

.NET 中提供了 HealthCheckMiddleware 中间件,通过该中间件可以对应用程序做可用性检查,利用 "Microsoft.Extensions.Diagnostics.HealthChecks" NuGet 包中的 IHealthCheck 接口,可以根据实际需求编写自定义检查逻辑。

10.1.1 查看当前应用的可用性

在 .NET 中可以很轻松地对应用程序的健康状态做出响应。通常,在集群部署中需要对节点服务的可用性进行检查,以确保服务可以正常做出正确的响应。当然,开发人员需要对外暴露出一个终节点,以确定服务的可用性,如代码 10-1 所示。

代码10-1

```
var builder = WebApplication.CreateBuilder(args);

builder.Services.AddHealthChecks();
```

```
var app = builder.Build();

app.UseHealthChecks("/health");

app.Run();
```

健康检查功能位于"Microsoft.Extensions.Diagnostics.HealthChecks"NuGet 包中，如代码 10-1 所示，先通过 IServiceCollection 接口的 AddHealthChecks 扩展方法注册健康检查所依赖的核心服务，再利用 WebApplication 对象调用 UseHealthChecks 扩展方法注册 HealthCheckMiddleware 中间件，并将参数/health 作为健康检查终节点指定的路径。图 10-1 所示为当前健康检查返回的状态。

图 10-1　当前健康检查返回的状态

10.1.2　自定义健康检查逻辑

开发人员可以自定义健康状态，以方便对应用程序的"健康度"进行监测，在 10.1.1 节的代码实例中，应用程序始终会返回 Healthy 状态，显然这不利于对应用程序健康程度的表示。假设存在一个更复杂的场景，需要检查数据库和其他一些服务的可用性，直接引用且不做自定义逻辑处理显然是行不通的。

如代码 10-2 所示，创建一个 Check 方法，用于返回健康检查的结果状态，使用随机数对应 3 种健康状态（Healthy、Unhealthy 和 Degraded）并返回。在调用 AddHealthChecks 扩展方法注册所需的依赖服务后，会返回一个 IHealthChecksBuilder 对象，通过该对象继续追加调用 AddCheck 扩展方法注册 IHealthCheck 对象，通过 Check 方法执行并返回健康状态。

代码10-2

```
var builder = WebApplication.CreateBuilder(args);

builder.Services.AddHealthChecks()
    .AddCheck("default", Check);
```

```
var app = builder.Build();

app.UseHealthChecks("/health");

app.Run();

HealthCheckResult Check()
{
    var rnd = new Random().Next(1, 4);
    return rnd switch
    {
        1 => HealthCheckResult.Healthy(),
        2 => HealthCheckResult.Unhealthy(),
        _ => HealthCheckResult.Degraded()
    };
}
```

如代码 10-3 所示，在实例运行后，通过 curl 命令行工具访问该应用程序，会随机返回 3 种健康状态的响应报文。

代码10-3

```
C:\Users\hueif>curl -i http://localhost:5000/health
HTTP/1.1 503 Service Unavailable
Content-Type: text/plain
Date: Thu, 30 Sep 2021 14:10:18 GMT
Server: Kestrel
Cache-Control: no-store, no-cache
Expires: Thu, 01 Jan 1970 00:00:00 GMT
Pragma: no-cache
Transfer-Encoding: chunked

Unhealthy
C:\Users\hueif>curl -i http://localhost:5000/health
HTTP/1.1 200 OK
Content-Type: text/plain
Date: Thu, 30 Sep 2021 14:10:19 GMT
Server: Kestrel
Cache-Control: no-store, no-cache
```

```
Expires: Thu, 01 Jan 1970 00:00:00 GMT
Pragma: no-cache
Transfer-Encoding: chunked

Degraded
C:\Users\hueif>curl -i http://localhost:5000/health
HTTP/1.1 200 OK
Content-Type: text/plain
Date: Thu, 30 Sep 2021 14:10:20 GMT
Server: Kestrel
Cache-Control: no-store, no-cache
Expires: Thu, 01 Jan 1970 00:00:00 GMT
Pragma: no-cache
Transfer-Encoding: chunked

Healthy
C:\Users\hueif>
```

如代码10-4所示，在IHealthCheck接口中定义CheckHealthAsync方法，健康检查逻辑也需要在该方法中实现，HealthCheckContext表示与当前执行相关的上下文对象，CancellationToken对象用于执行长时间健康检查任务时中止执行中的健康检查。CancellationAsync方法最终返回一个HealthCheckResult对象，表示健康检查的结果。

代码10-4

```
public interface IHealthCheck
{
    Task<HealthCheckResult> CheckHealthAsync(
                    HealthCheckContext context,
                    CancellationToken cancellationToken = default);
}
```

代码10-5展示了HealthCheckResult对象的定义。其中，Data属性表示可以附加的数据；Description属性表示该健康检查对象结果的状态描述；Exception属性表示检查过程中的异常信息；Status属性表示健康检查的状态，健康检查的状态对应Unhealthy、Degraded和Healthy。在HealthCheckResult中还提供了3个静态方法，分别对应3种状态，开发人员可以更方便地返回健康检查的结果。

代码10-5

```csharp
public struct HealthCheckResult
{
    public HealthCheckResult(HealthStatus status,
            string? description = null, Exception? Exception = null,
                IReadOnlyDictionary<string, object>? data = null);
    public IReadOnlyDictionary<string, object> Data { get; }
    public string? Description { get; }
    public Exception? Exception { get; }
    public HealthStatus Status { get; }

    public static HealthCheckResult Healthy(string? description = null,
        IReadOnlyDictionary<string,object>? Data = null) =>
            new HealthCheckResult(status: HealthStatus.Healthy,
                        description, exception: null, data);
    public static HealthCheckResult Degraded(string? description = null,
        Exception? exception = null,
          IReadOnlyDictionary<string, object>? data = null)
                            => new HealthCheckResult(
    status: HealthStatus.Degraded, description, exception: exception, data);
    public static HealthCheckResult Unhealthy(
        string? description = null, Exception? exception =
                null, IReadOnlyDictionary<string, object>? data = null)
            => new HealthCheckResult(
             status: HealthStatus.Unhealthy, description, exception, data);
}

public enum HealthStatus
{
    Unhealthy = 0,
    Degraded = 1,
    Healthy = 2,
}
```

DelegateHealthCheck 是 IHealthCheck 接口的实现类，简化了健康检查逻辑的创建，可以利用委托对象来实现健康检查逻辑，如代码 10-6 所示，DelegateHealthCheck 类提供了一个 Func<CancellationToken, Task<HealthCheckResult>>类型的委托对象。

代码10-6

```
internal sealed class DelegateHealthCheck : IHealthCheck
{
    private readonly Func<CancellationToken, Task<HealthCheckResult>> _check;
    public DelegateHealthCheck(Func<CancellationToken,
                                    Task <HealthCheckResult>> check)
    {
        _check = check ?? throw new ArgumentNullException(nameof(check));
    }
    public Task<HealthCheckResult> CheckHealthAsync(
    HealthCheckContext context,CancellationToken cancellationToken = default) =>
                                            _check (cancellationToken);
}
```

如代码 10-7 所示，IHealthChecksBuilder 接口用于注册 IHealthChecks 对象，对外提供 Add 方法，并将一个 HealthCheckRegistration 对象作为参数。Services 属性是辅助类型，并且是一个 IServiceCollection 对象，可以通过该属性注册服务。

代码10-7

```
public interface IHealthChecksBuilder
{
    IHealthChecksBuilder Add(HealthCheckRegistration registration);
    IServiceCollection Services { get; }
}
```

如代码 10-8 所示，HealthChecksBuilder 是 IHealthChecksBuilder 接口的实现类，Add 方法将添加的 HealthCheckRegistration 对象保存到 HealthCheckServiceOptions 配置选项中。

代码10-8

```
internal class HealthChecksBuilder : IHealthChecksBuilder
{
    public HealthChecksBuilder(IServiceCollection services)
    {
        Services = services;
    }
    public IServiceCollection Services { get; }

    public IHealthChecksBuilder Add(HealthCheckRegistration registration)
```

```
    {
        Services.Configure<HealthCheckServiceOptions>(options =>
        {
            options.Registrations.Add(registration);
        });
        return this;
    }
}
```

HealthCheckServiceOptions 配置选项用于存储 HealthCheckRegistration 对象,如代码 10-9 所示。

代码10-9

```
public sealed class HealthCheckServiceOptions
{
    public ICollection<HealthCheckRegistration> Registrations { get; } =
                                    new List<HealthCheckRegistration>();
}
```

代码 10-10 展示了 HealthCheckRegistration 类的定义,该类提供了多个构造方法重载。其中,Factory 属性用于获取或设置一个 IHealthCheck 对象的委托;FailureStatus 属性表示该健康检查对象在失败的情况下对应的健康检查状态;Timeout 属性表示健康检查的超时时限;Name 属性表示当前注册的 IHealthCheck 对象的名称;Tags 属性表示存储的标签集合,通常该属性用于对健康检查逻辑进行分类。

代码10-10

```
public sealed class HealthCheckRegistration
{
    public HealthCheckRegistration(
        string name, IHealthCheck instance, HealthStatus? failureStatus,
        IEnumerable<string>? tags) : this(name, instance, failureStatus,
                                                  tags, default) { }
    public HealthCheckRegistration(
        string name, IHealthCheck instance, HealthStatus? failureStatus,
                        IEnumerable<string>? tags, TimeSpan? timeout);
    public HealthCheckRegistration(
      string name, Func<IServiceProvider, IHealthCheck> factory,
      HealthStatus? failureStatus, IEnumerable<string>? Tags) : this(
                      name, factory, failureStatus, tags, default) { }
```

```csharp
public HealthCheckRegistration(
    string name, Func<IServiceProvider, IHealthCheck> factory,
        HealthStatus? failureStatus, IEnumerable<string>? tags,
                                            TimeSpan? timeout);

public Func<IServiceProvider, IHealthCheck> Factory { get; set; }
public HealthStatus FailureStatus { get; set; }
public TimeSpan Timeout { get; set; }
public string Name { get; set; }
public ISet<string> Tags { get; set; }
}
```

如代码 10-11 所示，HealthChecksBuilderAddCheckExtensions 为扩展类，该类中定义了多个 AddCheck 扩展方法、AddCheck<T>泛型方法及 AddTypeActivatedCheck<T>扩展方法，用于注册 IHealthCheck 对象。AddCheck<T>泛型方法和 AddTypeActivatedCheck<T>扩展方法最终都会调用 IHealthChecksBuilder 对象的 Add 方法添加 HealthCheckRegistration 对象，而这两个方法的不同之处在于：当实例化 IHealthCheck 对象时，一个调用的是 ActivatorUtilities 类型的 GetServiceOrCreateInstance<T>方法，该方法复用现有的服务实例；另一个调用的是 ActivatorUtilities 类型的 CreateInstance<T>方法，始终创建一个新的实例。

代码10-11

```csharp
public static class HealthChecksBuilderAddCheckExtensions
{
    public static IHealthChecksBuilder AddCheck(
        this IHealthChecksBuilder builder, string name, IHealthCheck instance,
        HealthStatus? failureStatus, IEnumerable<string> tags) =>
            AddCheck(builder, name, instance, failureStatus, tags, default);

    public static IHealthChecksBuilder AddCheck(
        this IHealthChecksBuilder builder, string name, IHealthCheck instance,
        HealthStatus? failureStatus = null, IEnumerable<string>? tags = null,
        TimeSpan? Timeout = null) =>
            builder.Add(new HealthCheckRegistration(
                    name, instance, failureStatus, tags, timeout));

    public static IHealthChecksBuilder AddCheck<T>(
        this IHealthChecksBuilder builder, string name,
```

```csharp
        HealthStatus? failureStatus, IEnumerable<string> tags) where T : class,
    IHealthCheck =>
            AddCheck<T>(builder, name, failureStatus, tags, default);

public static IHealthChecksBuilder AddCheck<T>(
    this IHealthChecksBuilder builder, string name,
        HealthStatus? failureStatus = null,
    IEnumerable<string>? tags = null, TimeSpan? timeout = null)
      where T : class, IHealthCheck =>
            builder.Add(new HealthCheckRegistration(name, s =>
                ActivatorUtilities.GetServiceOrCreateInstance<T>(s),
                                    failureStatus, tags, timeout));

public static IHealthChecksBuilder AddTypeActivatedCheck<T>(
      this IHealthChecksBuilder builder, string name,
        params object[] args) where T : class, IHealthCheck =>
            AddTypeActivatedCheck<T>(
                builder, name, failureStatus: null, tags: null, args);

public static IHealthChecksBuilder AddTypeActivatedCheck<T>(
    this IHealthChecksBuilder builder, string name,
    HealthStatus? failureStatus, params object[] args)
      where T : class, IHealthCheck
          => AddTypeActivatedCheck<T>(
                builder, name, failureStatus, tags: null, args);
public static IHealthChecksBuilder AddTypeActivatedCheck<T>(
    this IHealthChecksBuilder builder, string name,
     HealthStatus? failureStatus,IEnumerable<string>? tags,
      params object[] args) where T : class, IHealthCheck =>
          builder.Add(new HealthCheckRegistration(
            name, s =>
             ActivatorUtilities.CreateInstance<T>(s, args),
                                    failureStatus, tags));

public static IHealthChecksBuilder AddTypeActivatedCheck<T>(
    this IHealthChecksBuilder builder, string name,
    HealthStatus? failureStatus,
```

```
        IEnumerable<string> tags, TimeSpan timeout,
        params object[] args) where T : class, IHealthCheck =>
         builder.Add(new HealthCheckRegistration(name, s =>
                ActivatorUtilities.CreateInstance<T>(s, args),
                                failureStatus, tags, timeout));
}
```

如代码 10-12 所示，在 HealthChecksBuilderDelegateExtensions 扩展类下，每个扩展方法都需要传递一个委托对象，而委托对象都会通过 DelegateHealthCheck 对象进行实例化。

代码10-12

```
public static class HealthChecksBuilderDelegateExtensions
{
    public static IHealthChecksBuilder AddCheck(
        this IHealthChecksBuilder builder, string name,
        Func<HealthCheckResult> check, IEnumerable<string> tags) =>
                    AddCheck(builder, name, check, tags, default);

    public static IHealthChecksBuilder AddCheck(
       this IHealthChecksBuilder builder, string name,
       Func <HealthCheckResult> check, IEnumerable<string>? tags = null,
                                    TimeSpan? timeout = default)
    {
        var instance = new DelegateHealthCheck((ct) =>
                                    Task.FromResult(check()));
        return builder.Add(
            new HealthCheckRegistration(name,
                    instance, failureStatus: null, tags, timeout));
    }

    public static IHealthChecksBuilder AddCheck(
        this IHealthChecksBuilder builder, string name,
          Func<CancellationToken, HealthCheckResult> check,
            IEnumerable<string>? tags) =>
                    AddCheck(builder, name, check, tags, default);
    public static IHealthChecksBuilder AddCheck(
        this IHealthChecksBuilder builder, string name,
```

```csharp
            Func<CancellationToken, HealthCheckResult> check,
        IEnumerable<string>? tags = null, TimeSpan? timeout = default)
{
    var instance = new DelegateHealthCheck((ct) =>
                                    Task.FromResult(check(ct)));
    return builder.Add(
            new HealthCheckRegistration(
                name, instance, failureStatus: null, tags, timeout));
}

public static IHealthChecksBuilder AddAsyncCheck(
    this IHealthChecksBuilder builder, string name,
     Func<Task<HealthCheckResult>> check, IEnumerable<string> tags) =>
        AddAsyncCheck(builder, name, check, tags, default);

public static IHealthChecksBuilder AddAsyncCheck(
    this IHealthChecksBuilder builder, string name, Func<Task<HealthCheckResult>> check,
 IEnumerable<string>? tags = null, TimeSpan? timeout = default)
{
    var instance = new DelegateHealthCheck((ct) => check());
    return builder.Add(new HealthCheckRegistration(name, instance,
failureStatus: null, tags,
 timeout));
}

public static IHealthChecksBuilder AddAsyncCheck(
    this IHealthChecksBuilder builder, string name,
        Func<CancellationToken, Task<HealthCheckResult>> check,
            IEnumerable<string> tags) => AddAsyncCheck(
                                builder, name, check, tags, default);
public static IHealthChecksBuilder AddAsyncCheck(
    this IHealthChecksBuilder builder, string name,
        Func<CancellationToken, Task<HealthCheckResult>> check,
            IEnumerable<string>? tags = null, TimeSpan? timeout = default)
{
    var instance = new DelegateHealthCheck((ct) => check(ct));
```

```
        return builder.Add(
            new HealthCheckRegistration(
                    name, instance, failureStatus: null, tags, timeout));
    }
}
```

10.1.3 改变响应状态码

开发人员可以分别对健康状态设置自定义的状态码，通过状态码做出业务的逻辑处理和判断，如代码 10-13 所示。

代码10-13
```
var builder = WebApplication.CreateBuilder(args);

builder.Services.AddHealthChecks()
    .AddCheck("default", Check);

var app = builder.Build();
var options = new HealthCheckOptions
{
    ResultStatusCodes =
    {
        [HealthStatus.Unhealthy] = 420,
        [HealthStatus.Healthy] = 298,
        [HealthStatus.Degraded] = 299
    }
};

app.UseHealthChecks("/health", options);

app.Run();

HealthCheckResult Check()
{
    var rnd = new Random().Next(1, 4);
    return rnd switch
    {
```

```
        1 => HealthCheckResult.Healthy(),
        2 => HealthCheckResult.Unhealthy(),
        _ => HealthCheckResult.Degraded()
    };
}
```

如上所示，调用 UseHealthChecks 扩展方法注册 HealthCheckMiddleware 中间件，并且创建一个 HealthCheckOptions 对象作为该中间件的配置选项。HealthCheckOptions 对象通过 ResultStatusCodes 属性设置对应的健康状态和状态码之间的关系，该属性为 IDictionary<HealthStatus,int>类型，笔者将 Unhealthy 的状态码设置为 420，将 Healthy 和 Degraded 的状态码设置为 298 和 299。

运行该应用程序，通过 curl 命令行工具请求该应用程序，输出结果如代码 10-14 所示。

代码10-14

```
C:\Users\hueif>curl -i http://localhost:5000/health
HTTP/1.1 299
Content-Type: text/plain
Date: Fri, 01 Oct 2021 00:23:24 GMT
Server: Kestrel
Cache-Control: no-store, no-cache
Expires: Thu, 01 Jan 1970 00:00:00 GMT
Pragma: no-cache
Transfer-Encoding: chunked

Degraded
C:\Users\hueif>curl -i http://localhost:5000/health
HTTP/1.1 298
Content-Type: text/plain
Date: Fri, 01 Oct 2021 00:23:25 GMT
Server: Kestrel
Cache-Control: no-store, no-cache
Expires: Thu, 01 Jan 1970 00:00:00 GMT
Pragma: no-cache
Transfer-Encoding: chunked

Healthy
C:\Users\hueif>
```

10.1.4 定制响应内容

至此，读者已经基本了解自定义逻辑及状态码。如代码 10-15 所示，通过一个 Web 实例自定义健康报告，并将自定义响应内容返回给客户端。

代码10-15

```
var builder = WebApplication.CreateBuilder(args);

builder.Services.AddHealthChecks()
    .AddCheck("Foo", Check)
    .AddCheck("Bar", Check)
    .AddCheck("Baz", Check);

var app = builder.Build();
var options = new HealthCheckOptions
{
    ResponseWriter = ReportAsync
};

app.UseHealthChecks("/health", options);

app.Run();

HealthCheckResult Check()
{
    var rnd = new Random().Next(1, 4);
    return rnd switch
    {
        1 => HealthCheckResult.Healthy(),
        2 => HealthCheckResult.Unhealthy(),
        _ => HealthCheckResult.Degraded()
    };
}

static Task ReportAsync(HttpContext context, HealthReport report)
{
    var result = JsonConvert.SerializeObject(
```

```
    new
    {
        status = report.Status.ToString(),
        responseTimeStamp =
            new DateTimeOffset(DateTime.UtcNow).ToUnixTimeSeconds(),
        errors = report.Entries.Select(e => new { key = e.Key, value=
                Enum.GetName(typeof(HealthStatus), e.Value.Status) })
    });
    context.Response.ContentType = MediaTypeNames.Application.Json;
    return context.Response.WriteAsync(result);
}
```

首先通过 AddCheck 扩展方法注册多个 IHealthCheck 对象，分别为 Foo、Bar 和 Baz 的服务；然后调用 UseHealthChecks 扩展方法注册 HealthCheckMiddleware 中间件，指定 HealthCheckOptions 对象作为配置选项，并将其 ResponseWriter 属性作为健康报告的处理方法。创建一个名为 ReportAsync 的方法作为委托对象对应的响应方法，ResponseWriter 属性接收一个 Func<HttpContext,HealthReport,Task>类型。使用 ReportAsync 方法可以序列化响应内容并将其返回。图 10-2 所示为健康报告。

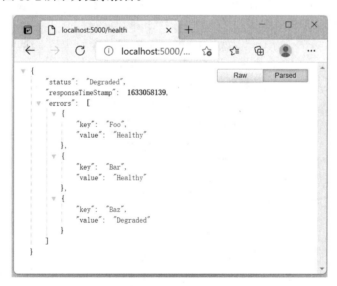

图 10-2　健康报告

HealthCheckMiddleware 中间件将一个 HealthCheckOptions 配置选项作为搭配。其中，Predicate 属性表示获取或设置 Func<HealthCheckRegistration,bool>类型的委托对象，该属性用于过滤已注册的 IHealthCheck 对象，也就是说，即使已经注册过 IHealthCheck 对象，如果

不包含在筛选的条件内，也不会执行；ResultStatusCodes 属性表示健康状态和状态码之间的映射关系；ResponseWriter 属性表示设置或获取响应内容的格式，该属性是一个 Func<HttpContext, HealthReport,Task>类型的委托；AllowCachingResponses 属性表示健康检查的响应是否需要被客户端缓存，如代码 10-16 所示。

代码10-16

```
public class HealthCheckOptions
{
    public Func<HealthCheckRegistration, bool>? Predicate { get; set; }
    public IDictionary<HealthStatus, int> ResultStatusCodes { get; set; }
    public Func<HttpContext, HealthReport, Task> ResponseWriter { get; set; }
                = HealthCheckResponseWriters.WriteMinimalPlaintext;
    public bool AllowCachingResponses { get; set; }
}
internal static class HealthCheckResponseWriters
{
    private static readonly byte[] DegradedBytes =
            Encoding.UTF8.GetBytes(HealthStatus.Degraded.ToString());
    private static readonly byte[] HealthyBytes =
            Encoding.UTF8.GetBytes(HealthStatus.Healthy.ToString());
    private static readonly byte[] UnhealthyBytes =
            Encoding.UTF8.GetBytes(HealthStatus.Unhealthy.ToString());

    public static Task WriteMinimalPlaintext(
                    HttpContext httpContext, HealthReport result)
    {
        httpContext.Response.ContentType = "text/plain";
        return result.Status switch
        {
            HealthStatus.Degraded =>
              httpContext.Response.Body.WriteAsync(
                DegradedBytes.AsMemory()).AsTask(), HealthStatus.Healthy =>
                httpContext.Response.Body.WriteAsync(
                  HealthyBytes.AsMemory()).AsTask(),HealthStatus.Unhealthy =>
                  httpContext.Response.Body.WriteAsync(
                        UnhealthyBytes.AsMemory()).AsTask(),
            _ => httpContext.Response.WriteAsync(result.Status.ToString())
```

```
            };
    }
}
```

如代码 10-17 所示，HealthCheckMiddleware 中间件首先通过 HealthCheckService 服务获取 HealthReport 对象，然后通过 HealthCheckOptions 配置选项做出业务判断，如判断 AllowCachingResponses 属性是否启用客户端缓存，同时判断 ResponseWriter 属性是否为空，该属性是一个委托类型。如代码 10-16 所示，在默认情况下会通过 HealthCheckResponseWriters 类中的 WriteMinimalPlaintext 方法做内容的响应。

代码10-17

```
public class HealthCheckMiddleware
{
    private readonly RequestDelegate _next;
    private readonly HealthCheckOptions _healthCheckOptions;
    private readonly HealthCheckService _healthCheckService;

    public HealthCheckMiddleware(
        RequestDelegate next, IOptions<HealthCheckOptions> healthCheckOptions,
                              HealthCheckService healthCheckService)
    {
        _next = next;
        _healthCheckOptions = healthCheckOptions.Value;
        _healthCheckService = healthCheckService;
    }

    public async Task InvokeAsync(HttpContext httpContext)
    {
        var result =
                await _healthCheckService.CheckHealthAsync(
            _healthCheckOptions.Predicate, httpContext.RequestAborted);
        if(!_healthCheckOptions.ResultStatusCodes.TryGetValue(
                              result.Status, out var statusCode))
        {
            var message =
                $"No status code mapping found for {nameof(HealthStatus)} value:
            {result.Status}."+$"{nameof(HealthCheckOptions)}.
```

```
            {nameof(HealthCheckOptions.ResultStatusCodes)} must contain" +
        $"an entry for {result.Status}.";
        throw new InvalidOperationException(message);
    }

    httpContext.Response.StatusCode = statusCode;
    if(!_healthCheckOptions.AllowCachingResponses)
    {
        var headers = httpContext.Response.Headers;
        headers.CacheControl = "no-store, no-cache";
        headers.Pragma = "no-cache";
        headers.Expires = "Thu, 01 Jan 1970 00:00:00 GMT";
    }
    if(_healthCheckOptions.ResponseWriter != null)
    {
        await _healthCheckOptions.ResponseWriter(httpContext, result);
    }
}
```

如代码 10-18 所示，HealthCheckService 是一个抽象类，在该类中定义了两个 CheckHealthAsync 重载方法。HealthCheckService 抽象类相当于一个操作器，在进行健康检查时会调用 CheckHealthAsync 方法。

代码10-18

```
public abstract class HealthCheckService
{
    public Task<HealthReport> CheckHealthAsync(
            CancellationToken cancellationToken = default)
                => CheckHealthAsync(predicate: null, cancellationToken);

    public abstract Task<HealthReport> CheckHealthAsync(
            Func<HealthCheckRegistration, bool>? predicate,
                CancellationToken cancellationToken = default);
}
```

如代码 10-19 所示，DefaultHealthCheckService 类是 HealthCheckService 抽象类的实现，

并且会通过 HealthCheckServiceOptions 配置选项获取存储的 HealthCheckRegistration 对象集合。它是健康检查模块的核心部分，主要负责调度其他对象方法的使用，可以看到在 CheckHealthAsync 方法中，首先利用 Func<HealthCheckRegistration,bool>类型的参数筛选出符合条件的 HealthCheckRegistration 对象（Func<HealthCheckRegistration,bool>参数是在注册中间件时，通过 HealthCheckOptions 配置选项对象传递的），然后循环 HealthCheckRegistration 对象集合，通过 Task.Run 创建任务，将 RunCheckAsync 作为 Task.Run 执行的委托方法，在该方法中创建 HealthCheckContext 对象，并利用 HealthCheckRegistration 对象注册 IHealthCheck 对象，调用 IHealthCheck 对象的 CheckHealthAsync 方法，获取当前返回的 HealthCheckResult 对象。最终利用这些返回值创建并返回一个 HealthReport 对象。

代码10-19

```csharp
internal partial class DefaultHealthCheckService : HealthCheckService
{
    private readonly IServiceScopeFactory _scopeFactory;
    private readonly IOptions<HealthCheckServiceOptions> _options;
    private readonly ILogger<DefaultHealthCheckService> _logger;

    public DefaultHealthCheckService(
                    IServiceScopeFactory scopeFactory,
                    IOptions<HealthCheckServiceOptions> options,
                    ILogger<DefaultHealthCheckService> logger)
    {
        //…
    }
    public override async Task<HealthReport> CheckHealthAsync(
                Func<HealthCheckRegistration, bool>? predicate,
                CancellationToken cancellationToken = default)
    {
        var registrations = _options.Value.Registrations;
        if (predicate != null)
        {
            registrations = registrations.Where(predicate).ToArray();
        }
        var totalTime = ValueStopwatch.StartNew();
        var tasks = new Task<HealthReportEntry>[registrations.Count];
```

```csharp
    var index = 0;
    foreach (var registration in registrations)
    {
        tasks[index++] = Task.Run(() =>
     RunCheckAsync(registration, cancellationToken), cancellationToken);
    }
    await Task.WhenAll(tasks).ConfigureAwait(false);
    index = 0;
    var entries = new Dictionary<string, HealthReportEntry>(
                                StringComparer.OrdinalIgnoreCase);
    foreach (var registration in registrations)
    {
        entries[registration.Name] = tasks[index++].Result;
    }
    var totalElapsedTime = totalTime.GetElapsedTime();
    var report = new HealthReport(entries, totalElapsedTime);
    return report;
}

private async Task<HealthReportEntry> RunCheckAsync (
    HealthCheckRegistration registration, CancellationToken cancellationToken)
{
    cancellationToken.ThrowIfCancellationRequested();
    using(var scope = _scopeFactory.CreateScope())
    {
        var healthCheck = registration.Factory(scope.ServiceProvider);
        using(_logger.BeginScope(
                    new HealthCheckLogScope(registration. Name)))
        {
            var stopwatch = ValueStopwatch.StartNew();
            var context =
                    new HealthCheckContext { Registration = registration };
            HealthReportEntry entry;
            CancellationTokenSource? timeoutCancellationTokenSource
                                                            = null;
```

```csharp
try
{
    HealthCheckResult result;
    var checkCancellationToken = cancellationToken;
    if(registration.Timeout > TimeSpan.Zero)
    {
        timeoutCancellationTokenSource =
CancellationTokenSource.CreateLinkedTokenSource(cancellationToken);
timeoutCancellationTokenSource.CancelAfter(registration.Timeout);
checkCancellationToken = timeoutCancellationTokenSource.Token;
    }
    result = await
      healthCheck.CheckHealthAsync(
        context,checkCancellationToken).ConfigureAwait(false);
    var duration = stopwatch.GetElapsedTime();
    entry = new HealthReportEntry(
    status: result.Status, description: result.Description,
        duration: duration, exception: result.Exception,
            data: result.Data, tags: registration.Tags);
}
catch(OperationCanceledException ex)when(
                !cancellationToken.IsCancellationRequested)
{
    var duration = stopwatch.GetElapsedTime();
    entry = new HealthReportEntry(
        status: registration. FailureStatus,
        description: "A timeout occurred while running check.",
        duration: duration, exception: ex, data: null,
                                tags: registration.Tags);
}
catch(Exception ex)when(
                ex as OperationCanceledException == null)
{
    var duration = stopwatch.GetElapsedTime();
    entry = new HealthReportEntry(
```

```csharp
                            status: registration.FailureStatus,
                            description: ex.Message,duration: duration,
                            exception: ex, data: null, tags: registration.Tags);
            }
            finally
            {
                timeoutCancellationTokenSource?.Dispose();
            }
            return entry;
        }
    }
}

private static void ValidateRegistrations(
                    IEnumerable<HealthCheckRegistration> registrations)
{
    StringBuilder? builder = null;
    var distinctRegistrations =
            new HashSet<string>(StringComparer.OrdinalIgnoreCase);
    foreach(var registration in registrations)
    {
        if(!distinctRegistrations.Add(registration.Name))
        {
            builder ??= new StringBuilder(
            "Duplicate health checks were registered with the name(s): ");
            builder.Append(registration.Name).Append(", ");
        }
    }
    if(builder is not null)
    {
        throw new ArgumentException(
            builder.ToString(0, builder.Length - 2), nameof(registrations));
    }
}
```

10.2 发布健康报告

健康报告有利于开发人员对服务进行监测，可以定期收集和发布到指定的平台，如开发人员可以自定义发布到控制台中，或者存储到数据库中等。

当开发人员需要定期检查服务状态时，需要通过 IHealthCheckPublisher 接口自定义逻辑，如代码 10-20 所示。

代码10-20

```
var builder = WebApplication.CreateBuilder(args);
builder.Logging.ClearProviders();
builder.Services.AddHealthChecks()
    .AddCheck("Foo", Check)
    .AddCheck("Bar", Check)
    .AddCheck("Baz", Check);

builder.Services.Configure<HealthCheckPublisherOptions>(options =>
{
    options.Delay = TimeSpan.FromSeconds(2);
    options.Period = TimeSpan.FromSeconds(5);
});
builder.Services.AddSingleton<IHealthCheckPublisher, SimplePublisher>();

var app = builder.Build();
app.UseHealthChecks("/health");

app.Run();

HealthCheckResult Check()
{
    var rnd = new Random().Next(1, 4);
    return rnd switch
    {
        1 => HealthCheckResult.Healthy(),
        2 => HealthCheckResult.Unhealthy(),
```

```
            _ => HealthCheckResult.Degraded()
        };
}
```

先调用 WebApplicationBuilder 对象的 Logging 属性的 ClearProviders 扩展方法,再设置发布者的配置选项。其中,Period 属性和 Delay 属性分别表示健康报告的发布时间间隔和健康检查服务启动后收集工作的延后时间。

如代码 10-21 所示,创建一个名为 SimplePublisher 的类,继承 IHealthCheckPublisher 接口,实现 PublishAsync 方法。因为 PublishAsync 方法会被调用并接收健康状态,所以可以在此处对健康状态做出响应。本实例还是比较简单的,直接将健康状态输出到控制台中。

代码10-21

```
public class SimplePublisher : IHealthCheckPublisher
{
    public Task PublishAsync(HealthReport report,
                             CancellationToken cancellationToken)
    {
        if (report.Status == HealthStatus.Healthy)
        {
            Console.ForegroundColor = ConsoleColor.Green;
            Console.WriteLine(
                    $"{DateTime.UtcNow} Prob status: {report.Status}");
        }
        else
        {
            Console.ForegroundColor = ConsoleColor.Red;
            Console.WriteLine(
                    $"{DateTime.UtcNow} Prob status: {report.Status}");
        }
        var sb = new StringBuilder();

        foreach (var name in report.Entries.Keys)
        {
            sb.AppendLine($" {name}: {report.Entries[name].Status}");
        }
        cancellationToken.ThrowIfCancellationRequested();
        Console.WriteLine(sb);
```

```
        Console.ResetColor();
        return Task.CompletedTask;
    }
}
```

发布的健康报告如图 10-3 所示。

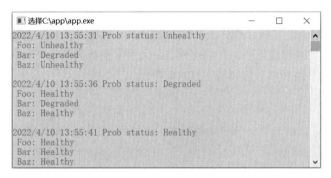

图 10-3　发布的健康报告

如代码 10-22 所示，IHealthCheckPublisher 接口作为发布健康报告的约束，在应用程序中可以注册多样化的健康报告输出服务，如可以发布到控制台、日志文件及数据库中。对于多样化的健康报告，可以通过如上实例自行实现存储服务。

代码10-22

```
public interface IHealthCheckPublisher
{
    Task PublishAsync(HealthReport report, CancellationToken cancellationToken);
}
```

如代码 10-23 所示，HealthCheckPublisherOptions 配置选项作为 HealthCheckPublisherHostedService 服务的配置对象。HealthCheckPublisherOptions 对象的 Delay 属性表示服务启动后收集工作的延后时间，默认为 5 秒；Period 属性表示健康报告的发布时间间隔，默认为 30 秒；Predicate 属性表示为 IHealthCheck 对象进行过滤；Timeout 属性表示 IHealthCheckPublisher 对象的超时时间。

代码10-23

```
public sealed class HealthCheckPublisherOptions
{
    public TimeSpan Delay { get; set; } = TimeSpan.FromSeconds(5);
    public TimeSpan Period { get; set; } = TimeSpan.FromSeconds(30);
```

```
    public Func<HealthCheckRegistration, bool>? Predicate { get; set; }
    public TimeSpan Timeout { get; set; } = TimeSpan.FromSeconds(30);
}
```

如代码 10-24 所示，HealthCheckPublisherHostedService 类实现了 IHostedService 接口。HealthCheckPublisherHostedService 类作为一个后台任务，首先执行 StartAsync 方法，随之创建一个 Timer 对象，并落在 RunAsyncCore 方法上，在该方法内会调用 HealthCheckService 对象的 CheckHealthAsync 方法；然后获取健康报告；最后调用 IHealthCheckPublisher 对象的 PublishAsync 方法发布健康报告，并输出到开发人员实现的健康检查发布服务中。当然，应用程序允许开发人员注册多个健康检查发布服务。

代码10-24

```
internal sealed partial class HealthCheckPublisherHostedService :
                                                    IHostedService
{
    private readonly HealthCheckService _healthCheckService;
    private readonly IOptions<HealthCheckPublisherOptions> _options;
    private readonly ILogger _logger;
    private readonly IHealthCheckPublisher[] _publishers;

    private readonly CancellationTokenSource _stopping;
    private Timer? _timer;
    private CancellationTokenSource? _runTokenSource;

    public HealthCheckPublisherHostedService(
            HealthCheckService healthCheckService,
                IOptions <HealthCheckPublisherOptions> options,
                    ILogger<HealthCheckPublisherHostedService> logger,
                        IEnumerable <IHealthCheckPublisher> publishers)
    {
        _healthCheckService = healthCheckService;
        _options = options;
        _logger = logger;
        _publishers = publishers.ToArray();
        _stopping = new CancellationTokenSource();
    }

    internal bool IsStopping => _stopping.IsCancellationRequested;
```

```csharp
internal bool IsTimerRunning => _timer != null;
public Task StartAsync(CancellationToken cancellationToken = default)
{
    _timer = NonCapturingTimer.Create(Timer_Tick,
      null, dueTime: _options.Value.Delay, period: _options.Value.Period);
    return Task.CompletedTask;
}

public Task StopAsync(CancellationToken cancellationToken = default)
{
    _stopping.Cancel();
    _timer?.Dispose();
    _timer = null;
    return Task.CompletedTask;
}

private async void Timer_Tick(object? state)
{
    await RunAsync();
}
internal void CancelToken()
{
    _runTokenSource!.Cancel();
}
internal async Task RunAsync()
{
    CancellationTokenSource? cancellation = null;
    try
    {
        var timeout = _options.Value.Timeout;
        cancellation =
       CancellationTokenSource.CreateLinkedTokenSource(_stopping.Token);
        _runTokenSource = cancellation;
        cancellation.CancelAfter(timeout);
        await RunAsyncCore(cancellation.Token);
    }
    catch (OperationCanceledException) when (IsStopping) { }
    catch (Exception ex) { }
```

```
        finally
        {
            cancellation?.Dispose();
        }
    }

    private async Task RunAsyncCore(CancellationToken cancellationToken)
    {
        await Task.Yield();
        var report = await _healthCheckService.CheckHealthAsync(
                           _options.Value.Predicate, cancellationToken);
        var publishers = _publishers;
        var tasks = new Task[publishers.Length];
        for (var i = 0; i < publishers.Length; i++)
        {
            tasks[i] = RunPublisherAsync(publishers[i], report,
                                                    cancellationToken);
        }
        await Task.WhenAll(tasks);
    }

    private async Task RunPublisherAsync(IHealthCheckPublisher publisher,
            HealthReport report, CancellationToken cancellationToken)
    {
        try
        {
            await publisher.PublishAsync(report, cancellationToken);
        }
        catch (OperationCanceledException) when (IsStopping) { }
    }
}
```

10.3 可视化健康检查界面

在健康检查中，还提供了一个可视化界面，引用"AspNetCore.HealthChecks.UI" NuGet 包可以启用可视化界面。如代码 10-25 所示，先通过 WebApplicationBuilder 对象的 Services

属性调用 AddHealthChecksUI 扩展方法，注册可视化界面依赖的相关服务，HealthChecksUI 提供了多种存储方式，以内存存储为例，可以调用 AddInMemoryStorage 扩展方法，再通过 WebApplication 对象调用 UseHealthChecksUI 扩展方法注册相关的中间件。

代码10-25

```
var builder = WebApplication.CreateBuilder(args);

builder.Services.AddHealthChecksUI()
   .AddInMemoryStorage();

var app = builder.Build();

app.UseHealthChecksUI();

await app.RunAsync();
```

如代码 10-26 所示，在 appsettings.json 文件中配置 HealthChecksUI 的配置项。HealthChecks 是一个数组，允许定义多个要检查的服务或应用程序；EvaluationTimeinSeconds 表示每隔多少秒轮询一次健康检查；MinimumSecondsBetweenFailureNotifications 表示故障通知每隔指定的秒数触发一次。

代码10-26

```
{
  "HealthChecksUI": {
    "HealthChecks": [
      {
        "Name": "Test Health",
        "Uri": "http://localhost:5179/health"
      },
      {
        "Name": "Test Health1",
        "Uri": "http://localhost:5000/health"
      }
    ],
    "EvaluationTimeinSeconds": 10,
    "MinimumSecondsBetweenFailureNotifications": 60
  }
}
```

通过/healthchecks-ui 路径访问可视化界面，如图 10-4 所示。

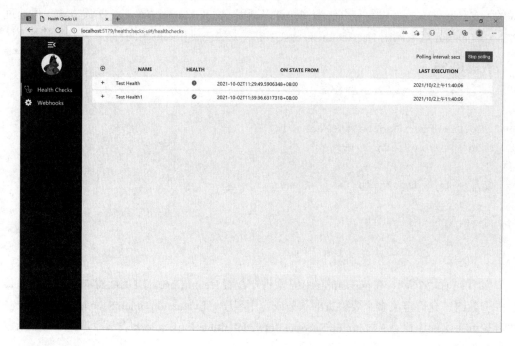

图 10-4　可视化界面

10.4　小结

本章介绍了健康检查的核心内容，主要包括检查当前应用的健康状态和发布健康报告。前者用于对应用程序的存活进行监控；后者支持将健康报告发布到控制台、日志等其他地方，用于开发人员对应用程序的监测。

10.3 节主要介绍如何添加健康检查面板。通过学习本节，读者可以搭建属于自己的健康检查面板。

第 11 章

文件系统

ASP.NET Core 提供的静态文件模块用来管理本地静态文件,其扩展性比较高。利用静态文件模块可以处理静态文件请求,同时提供物理资源、嵌入式文件或其他自定义方式。本章主要介绍静态文件中间件的使用,以及如何扩展文件系统。

11.1 ASP.NET Core 静态文件

静态文件是指 JavaScript、CSS 和图片等格式的文件。利用 ASP.NET Core 中提供的静态文件中间件可以对这些静态文件进行处理,请求可以通过 HTTP 方式获取指定的物理文件,也能获得文件的物理目录结构。

11.1.1 静态文件

笔者采用"结论先行,体验优先"的方式创建一个项目实例。如代码 11-1 所示,首先通过 WebApplication 对象的 CreateBuilder 扩展方法获取 WebApplicationBuilder 对象,然后调用 WebApplicationBuilder 对象的 Build 方法创建 WebApplication 对象,并利用 WebApplication 对象调用 UseStaticFiles 扩展方法,最后调用 Run 方法启动该服务。

代码11-1

```
var builder = WebApplication.CreateBuilder(args);
```

```
var app = builder.Build();

app.UseStaticFiles();

app.Run();
```

图11-1所示为项目目录，静态文件存储在项目的wwwroot目录下，默认目录为{content root}/wwwroot，可以通过wwwroot的相关路径访问静态文件。例如，在wwwroot目录下包含多个文件夹。

- wwwroot。
- css。
- js。
- lib。

如图11-2所示，读取静态文件https://<hostname>/dotnet.png，以css文件夹为例，访问css文件夹的格式为https://<hostname>/css/<css_file_name>。

图11-1 项目目录

图11-2 读取静态文件

11.1.2 新增静态文件目录

在默认情况下，静态文件目录为wwwroot，允许开发人员自定义静态文件所在的目录。如代码11-2所示，只需在调用UseStaticFiles扩展方法时，传递一个StaticFileOptions配置选项，FileProvider属性为文件操作者（该属性为IFileProvider类型），RequestPath属性为该处理程序的请求路径前缀。

代码11-2

```
var builder = WebApplication.CreateBuilder(args);
```

第 11 章 文件系统

```
var app = builder.Build();

app.UseStaticFiles(new StaticFileOptions
{
    FileProvider = new PhysicalFileProvider(
        Path.Combine(builder.Environment.ContentRootPath,
"MyStaticFiles")),
    RequestPath = "/StaticFiles"
});

app.Run();
```

如图 11-3 所示，读取图片的格式为 https://<hostname>/StaticFiles/dotnet.png。

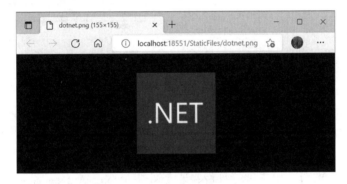

图 11-3 读取图片

11.1.3 自定义 HTTP 响应头

静态文件可以自定义 HTTP 响应头。如代码 11-3 所示，只需通过 StaticFileOptions 对象的 OnPrepareResponse 属性设置 HTTP 响应即可。

代码11-3

```
var builder = WebApplication.CreateBuilder(args);

var app = builder.Build();

const string cacheMaxAge = "604800";
app.UseStaticFiles(new StaticFileOptions
```

```
{
    OnPrepareResponse = ctx =>
    {
        ctx.Context.Response.Headers.Append(
            "Cache-Control", $"public, max-age={cacheMaxAge}");
    }
});

app.Run();
```

以 Cache-Control 头为例,可以通过 StaticFileResponseContext 对象的 Context 属性调用它的 Response 属性设置响应头。如图 11-4 所示,查看自定义响应输出。

11.1.4 静态文件的授权管理

与其他常规中间件一样,UseStaticFiles 扩展方法在 UseAuthorization 调用之前,这样不会对静态文件的执行进行任何授权检查,以便支持对静态文件进行公开访问。

当然,开发人员也可以对静态文件进行授权访问。如代码 11-4 所示,可以在注册

图 11-4 查看自定义响应输出

Authorization 核心服务时,通过 AuthorizationOptions 配置选项设置 FallbackPolicy 属性,回退授权按照策略要求所有用户进行身份验证。通过 RequireAuthenticatedUser 扩展方法将 DenyAnonymousAuthorizationRequirement 添加到当前实例中,这将强制对当前用户进行身份验证。

代码11-4

```
var builder = WebApplication.CreateBuilder(args);
builder.Services.AddAuthentication(authOpt =>
    {
        authOpt.DefaultAuthenticateScheme =
                        CookieAuthenticationDefaults.AuthenticationScheme;
        authOpt.DefaultChallengeScheme =
                        CookieAuthenticationDefaults.AuthenticationScheme;
    })
    .AddCookie(o =>
    {
```

```
      //TODO
   });
builder.Services.AddAuthorization(options =>
{
   options.FallbackPolicy = new AuthorizationPolicyBuilder()
      .RequireAuthenticatedUser()
      .Build();
});

var app = builder.Build();

app.UseStaticFiles();

app.UseAuthentication();
app.UseAuthorization();

app.UseStaticFiles(new StaticFileOptions
{
   FileProvider = new PhysicalFileProvider(
      Path.Combine(builder.Environment.ContentRootPath,
                                              "MyStaticFiles")),
   RequestPath = "/StaticFiles"
});

app.Run();
```

11.1.5 开启目录浏览服务

在 ASP.NET Core 中，出于安全方面的考虑，默认目录浏览处于禁用状态。可以通过如代码 11-5 所示的调用实现目录浏览，首先注册 AddDirectoryBrowser 服务，然后调用 UseDirectoryBrowser 扩展方法，注册 DirectoryBrowserMiddleware 中间件。

代码11-5

```
var builder = WebApplication.CreateBuilder(args);
builder.Services.AddDirectoryBrowser();

var app = builder.Build();
```

```
app.UseStaticFiles(new StaticFileOptions
{
    FileProvider = new PhysicalFileProvider(
        Path.Combine(builder.Environment.ContentRootPath, "MyStaticFiles")),
    RequestPath = "/MyImages"
});

app.UseDirectoryBrowser(new DirectoryBrowserOptions
{
    FileProvider = new PhysicalFileProvider(
      Path.Combine(builder.Environment.ContentRootPath, "MyStaticFiles")),
    RequestPath = "/MyImages"
});

app.Run();
```

在注册中间件时，创建 DirectoryBrowserOptions 配置选项作为该中间件的设置，如设置开启浏览的目录，以及设置该目录引导的路径。图 11-5 所示为文件目录。

图 11-5　文件目录

11.1.6　自定义文件扩展名

FileExtensionContentTypeProvider 对象提供了 Mappings 属性，如在 ASP.NET Core 中默认没有为 .less 扩展文件提供支持，这时可以通过该属性进行映射。如代码 11-6 所示，可以通过 FileExtensionContentTypeProvider 对象进行扩展。

代码11-6

```
var builder = WebApplication.CreateBuilder(args);
```

```
var app = builder.Build();

var provider = new FileExtensionContentTypeProvider();
provider.Mappings[".less"] = "text/css";
provider.Mappings[".htm3"] = "text/html";
provider.Mappings[".image"] = "image/png";

app.UseStaticFiles(new StaticFileOptions
{
    ContentTypeProvider = provider
});

app.Run();
```

11.2 自定义一个简单的文件系统

在 ASP.NET Core 中，允许开发人员自定义文件系统，可以利用 IFileProvider 接口构建一个文件系统。在本节中，笔者将替换默认的物理文件系统，将 Redis 作为文件系统。

11.2.1 文件和目录

要构建一个文件系统，需要先创建 RedisFileInfo 类和 EnumerableDirectoryContents 类，分别用来表示文件的信息和文件的目录。如代码 11-7 所示，RedisFileInfo 类需要继承 IFileInfo 接口，Exists 属性表示判断目录或文件是否真的存在，Length 属性表示文件内容的字节长度，Name 属性表示文件或目录的名称，PhysicalPath 属性表示文件或目录的物理路径，LastModified 属性表示文件或目录最后一次修改的时间，IsDirectory 属性表示是目录还是文件。

代码11-7
```
public class RedisFileInfo : IFileInfo
{
    public RedisFileInfo(string name, string content)
    {
        Name = name;
        LastModified = DateTimeOffset.Now;
        _fileContent = Convert.FromBase64String(content);
```

```
    }

    public RedisFileInfo(string name, bool isDirectory)
    {
        Name = name;
        LastModified = DateTimeOffset.Now;
        IsDirectory = isDirectory;
    }
    public Stream CreateReadStream()
    {
        var stream = new MemoryStream(_fileContent);
        stream.Position = 0;
        return stream;
    }

    public bool Exists => true;

    public long Length => _fileContent.Length;

    public string Name { get; set; }
    public string PhysicalPath { get; }
    public DateTimeOffset LastModified { get; }
    public bool IsDirectory { get; }

    private readonly byte[] _fileContent;
}
```

实现 CreateReadStream 方法，利用该方法返回读取的文件内容，并返回一个 Stream 类型。

如代码 11-8 所示，创建 EnumerableDirectoryContents 类，主要用于表示目录的信息。同样，在目录内部维护一个 IFileInfo 集合，该集合中的每个元素都是一个 RedisFileInfo 对象。

代码11-8

```
public class EnumerableDirectoryContents : IDirectoryContents
{
    private readonly IEnumerable<IFileInfo> _entries;

    public EnumerableDirectoryContents(IEnumerable<IFileInfo> entries)
    {
        _entries = entries;
```

```
    }

    public IEnumerator<IFileInfo> GetEnumerator()
    {
        return _entries.GetEnumerator();
    }

    IEnumerator IEnumerable.GetEnumerator()
    {
        return GetEnumerator();
    }

    public bool Exists => true;
}
```

11.2.2　RedisFileProvider

创建一个应用程序用于处理 Redis 中读取文件的逻辑，新建一个名为 RedisFileProvider 的解析器，主要用于通过指定的名称（也就是指定的路径名称或文件名称）从 Redis 中读取存储的图片内容。

如代码 11-9 所示，创建一个名为 RedisFileOptions 的配置选项，定义 HostAndPort 属性用于配置 Redis 的连接信息。

代码11-9

```
public class RedisFileOptions
{
    public string HostAndPort { get; set; }
}
```

如代码 11-10 所示，RedisFileProvider 类继承了 IFileProvider 接口，并定义了一个 Redis 客户端，通过 RedisFileOptions 配置选项读取 Redis 的连接信息，代码的定义还是比较简单的。使用 GetDirectoryContents 方法可以得到指定的目录，使用 GetFileInfo 方法可以得到指定目录或文件的 IFileInfo 对象，此处暂时先不实现 Watch 方法。

代码11-10

```
public class RedisFileProvider : IFileProvider
{
    private readonly RedisFileOptions _options;
```

```csharp
private readonly ConnectionMultiplexer _redis;

public RedisFileProvider(IOptions<RedisFileOptions> options)
{
    _options = options.Value;
    _redis = ConnectionMultiplexer.Connect(
        new ConfigurationOptions
        {
            EndPoints = { _options.HostAndPort }
        });
}

public IDirectoryContents GetDirectoryContents(string subpath)
{
    var db = _redis.GetDatabase();
    var server = _redis.GetServer(_options.HostAndPort);
    var list = new List<IFileInfo>();
    subpath = NormalizePath(subpath);
    foreach(var key in server.Keys(0, $"{subpath}*"))
    {
        var k = "";
        if(subpath != "")
        {
            k = key.ToString().Replace(subpath, "").Split(":")[0];
        }
        else
        {
            k = key.ToString().Split(":")[0];
        }

        if(list.Find(f => f.Name == k) == null)
        {
            //判断是否存在"."
            if (k.IndexOf('.', StringComparison.OrdinalIgnoreCase) >= 0)
            {
                list.Add(new RedisFileInfo(k, db.StringGet(k)));
            }
            else
```

```csharp
            {
                list.Add(new RedisFileInfo(k, true));
            }
        }
    }

    if(list.Count == 0)
    {
        return NotFoundDirectoryContents.Singleton;
    }

    return new EnumerableDirectoryContents(list);
}

private static string NormalizePath(string path) =>
                    path.TrimStart ('/').Replace('/', ':');

public IFileInfo? GetFileInfo(string subpath)
{
    subpath = NormalizePath(subpath);
    var db = _redis.GetDatabase();
    var redisValue = db.StringGet(subpath);
    return !redisValue.HasValue ? new NotFoundFileInfo(subpath) :
            new RedisFileInfo(subpath, redisValue.ToString());
}

public IChangeToken Watch(string filter) =>
                    throw new NotImplementedException();
}
```

在 GetFileInfo 方法中通过 subpath 参数值在 Redis 客户端读取文件信息，该信息当前保存的是 base64 字符串。如果字符串存在则创建 RedisFileInfo 对象。GetDirectoryContents 方法也是根据相应的 subpath 参数值，在 Redis 中读取对应的文件夹和目录，并读取 lib 文件夹下面的文件夹和目录等。

在应用程序中，扫描 Redis 中的内容，进行模糊匹配，若匹配前缀为 lib 的 key，则先获取指定的服务器，再使用 Keys 命令遍历 key。可以先使用 lib*进行匹配，再循环获取匹配的 key，创建 RedisFileInfo 对象，并将其保存到集合中，最后创建 EnumerableDirectoryContents 对象。

11.2.3 使用自定义文件系统

创建完 RedisFileProvider 类之后就可以将 Redis 作为应用程序的文件系统。如代码 11-11 所示，先调用 UseStaticFiles 扩展方法，再创建 StaticFileOptions 配置选项，利用 FileProvider 属性设置当前构建的 RedisFileProvider 实例对象，通过它可以替换文件提供者，将 Redis 作为应用程序的文件系统。因为 RedisFileProvider 类接收一个 IOptions 配置选项，所以调用 Options 静态类的 Create 方法可以构建一个配置选项对象。

代码11-11

```
var builder = WebApplication.CreateBuilder(args);

var app = builder.Build();

app.UseStaticFiles(new StaticFileOptions()
{
    FileProvider = new RedisFileProvider(Options.Create(new RedisFileOptions
    {
        HostAndPort = "localhost:6379"
    }))
});

app.UseDirectoryBrowser(new DirectoryBrowserOptions
{
    FileProvider = new RedisFileProvider(Options.Create(new RedisFileOptions
    {
        HostAndPort = "localhost:6379"
    }))
});

app.Run();
```

调用 UseDirectoryBrowser 扩展方法启用目录浏览，一般来说，除非在必要情况下，否则不建议开启目录浏览。图 11-6 所示为浏览器中浏览文件的目录。

在 Redis 中，key 的命名用":"隔开不同的层次及命名空间。图 11-7 所示为 Redis 通过命名方式分层，在 Redis 中可以以层次的形式作为文件夹的分层。通过 Redis Desktop Manager（Redis 可视化工具）可以查看存储的内容。

第 11 章　文件系统

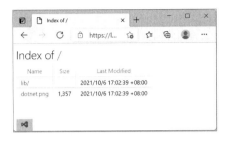

图 11-6　浏览器中浏览文件的目录

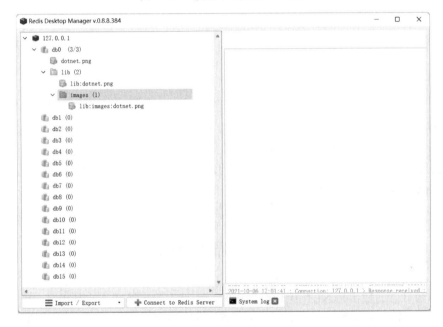

图 11-7　Redis 通过命名方式分层

11.3　小结

通过 ASP.NET Core 提供的静态文件模块和静态文件中间件，开发人员可以轻松地让应用程序拥有访问静态文件的功能，同时可以基于 IFileProvider 对象来扩展新的文件系统，如基于 Redis、SQL Server 等做扩展文件系统。

第 12 章

日志

日志用于快速对应用程序的运行状况进行实时查看。当出现错误时，可以基于日志进行问题定位和调试。在.NET 中提供了多种方式的日志输出，如控制台日志、调试日志、事件日志等。下面介绍日志的输出方式。

12.1　控制台日志

控制台日志的输出是.NET 应用最简单的一种日志输出形式。它利用日志模块抽象的 ILogger 接口创建了一个 ConsoleLogger 对象，同时利用 ILoggerProvider 对象实现一个 ConsoleLoggerProvider 类型，控制台日志输出模块位于"Microsoft.Extensions.Logging.Console"NuGet 包中。

12.1.1　控制台日志的输出

首先创建控制台项目，如代码 12-1 所示，创建 ServiceCollection 对象，先调用 AddLogging 扩展方法注册日志服务，再利用 Action<ILoggingBuilder>对象调用 AddConsole 扩展方法注册 ConsoleLoggerProvider 对象。先调用 BuildServiceProvider 扩展方法返回 ServiceProvider 对象，再调用 GetRequiredService<T>扩展方法获取相应的服务实例，最后调用 CreateLogger<T>扩展方法创建一个 ILogger 对象。

代码12-1

```
var logger = new ServiceCollection()
    .AddLogging(builder =>
        builder.AddConsole())
    .BuildServiceProvider()
    .GetRequiredService<ILoggerFactory>()
    .CreateLogger<Program>();
var levels = Enum.GetValues<LogLevel>()
    .Where(l => l != LogLevel.None);
var eventId = 1;
foreach (var level in levels)
{
    logger.Log(level, eventId++, "LogLevel:{0}", level);
}
```

接下来输出不同级别的类型，以便查看其输出结果，利用 LogLevel 枚举类获取日志级别集合，并在集合中忽略 None 枚举成员。随后循环该集合，通过 ILogger 对象调用 Log 方法，如代码 12-1 所示，日志内容包括日志级别和事件 ID。最终这些日志都会依次被打印到控制台中。图 12-1 所示为控制台日志的输出。

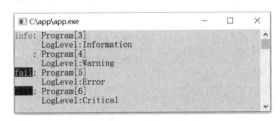

图 12-1　控制台日志的输出

12.1.2　日志级别和事件 ID

日志级别记录了日志的严重性，这是一种简单且强大的区分日志事件的方法。在正确使用日志级别的情况下，可以非常方便地对日志进行排查，并对问题进行定位。一般来说，对日志的写入都会关联一个具体发生的事件（Event），所以每条日志都存在一个事件 ID（EventID），以进行标识。事件 ID 可以标识日志记录，可以对日志记录进行筛选过滤及分组。

如代码 12-2 所示，LogLevel 类是一个枚举类型，定义了 7 种不同的日志级别，此处按照日志的严重性从低到高排序。

代码12-2

```
public enum LogLevel
{
    Trace = 0,
    Debug = 1,
    Information = 2,
    Warning = 3,
    Error = 4,
    Critical = 5,
    None = 6,
}
```

LogLevel 类的属性/值如表 12-1 所示，表中对 LogLevel 类的 7 种日志级别进行了描述，可以参照这些信息来决定如何选择日志消息的等级。

表 12-1 LogLevel 类的属性/值

属性	值	说明
Trace	0	记录一些详细的消息，用于对某个问题或方法进行跟踪调试。由于这些消息可能包含敏感的信息，因此在默认情况下不建议开启此等级
Debug	1	用于记录开发和调试期间的消息，这种日志具有较短的时效性
Information	2	用于跟踪应用程序的一般流程，通常来说可以忽略，这种日志具有较长的时效性
Warning	3	用于记录应用程序发生的意外或异常事件，但不影响应用程序继续工作，可以用于调查错误或其他情况
Error	4	应用程序遇到无法处理的错误和异常，这些消息表示对于当前操作或请求失败
Critical	5	用于记录一些需要立即被关注的故障，通常是难以恢复的灾难性问题，如数据丢失、磁盘空间不足等
None	6	指定日志记录类别不应写入任何消息

如代码 12-3 所示，通过该实例演示不同日志级别的输出形式，日志框架可以通过一系列预定义的扩展方法输出不同等级的日志消息，可以通过 ILogger<T> 对象调用相关的扩展方法，创建一个名为 ApplicationEvents 的类，定义多个常量用于表示事件 ID，表示 Create、Read、Update 和 Delete 操作时对应的日志输出。

代码12-3

```
var logger = new ServiceCollection()
    .AddLogging(builder =>
        builder.AddConsole())
```

```
    .BuildServiceProvider()
    .GetRequiredService<ILoggerFactory>()
    .CreateLogger<Program>();

logger.LogInformation("This information log from Program");
logger.LogError("This error log from Program");
logger.LogCritical("This critical log from Program");
logger.LogWarning("This warning log from Program");

logger.LogInformation(ApplicationEvents.Create, "订单创建");
logger.LogInformation(ApplicationEvents.Delete, "订单删除,订单编号:{0}",
"NO.10000000");
logger.LogInformation(ApplicationEvents.Read, "读取订单信息");
logger.LogInformation(ApplicationEvents.Update, "修改了订单的配送地址:从"{0}"
修改到"{1}"","A小区","B小区");

internal static class ApplicationEvents
{
    internal const int Create = 1000;
    internal const int Read = 1001;
    internal const int Update = 1002;
    internal const int Delete = 1003;
}
```

例如,调用 LogInformation 扩展方法输出日志,分别在这几种操作下以不同的事件 ID 进行输出,方便日志消息的筛选和过滤。图 12-2 所示为控制台日志的输出。

图 12-2 控制台日志的输出

12.1.3 日志过滤

AddFilter 扩展方法通过定义日志级别,并根据指定的日志级别输出消息。当然,可以单独设置过滤类别,如可以为指定的命名空间下输出的日志级别消息设置一种策略,限制在该命名空间的类下允许输出的日志级别消息,这是一种很有价值的方法,这两个限制的深度在一般情况下满足绝大部分的需求。

如代码 12-4 所示,创建控制台项目,调用两次 AddFilter 扩展方法,分别用来对 Error 日志级别进行过滤和对指定名称的对象做 Information 日志级别的过滤。另外,创建两个 ILogger<T>实例,分别根据这两个不同的类创建两个实例对象。

代码12-4

```
var loggerFactory = new ServiceCollection()
    .AddLogging(builder =>
        builder.AddConsole()
            .AddFilter(f => f == LogLevel.Error)
            .AddFilter("Simple", f => f == LogLevel.Information)
    )
    .BuildServiceProvider()
    .GetRequiredService<ILoggerFactory>();

var logger = loggerFactory.CreateLogger<Program>();
var simpleLogger = loggerFactory.CreateLogger<Simple>();

logger.LogInformation("This information log from Program");
logger.LogError("This error log from Program");
simpleLogger.LogInformation("This information log from Simple");
simpleLogger.LogError("This error log from Simple");
```

图 12-3 所示为条件过滤后的日志输出,可以看出,ILogger<Program>实例保留了 Error 日志级别的输出,而 ILogger<Simple>实例也是如期输出 Information 日志级别的消息。

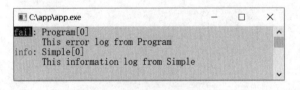

图 12-3　条件过滤后的日志输出

开发人员可以限制日志级别,也可以按条件限制到指定的类。另一种限制方式是设置允许输出的最低日志级别。如代码 12-5 所示,创建控制台项目,添加并调用 SetMinimumLevel 扩展方法,将其值设置为 LogLevel.Warning,也就是说,允许输出的日志级别最低为 Warning。

代码12-5

```
var loggerFactory = new ServiceCollection()
   .AddLogging(builder =>
      builder.AddConsole()
         .SetMinimumLevel(LogLevel.Warning)
   )
   .BuildServiceProvider()
   .GetRequiredService<ILoggerFactory>();

var logger = loggerFactory.CreateLogger<Program>();
var simpleLogger = loggerFactory.CreateLogger<Simple>();

logger.LogInformation("This information log from Program");
logger.LogError("This error log from Program");
logger.LogWarning("This warning log from Program");
simpleLogger.LogInformation("This information log from Simple");
simpleLogger.LogError("This error log from Simple");
simpleLogger.LogWarning("This warning log from Program");
```

如图 12-4 所示，设置最低日志级别的输出，输出的日志始终以 Warning 为最低的级别进行限制并输出。

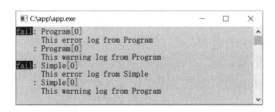

图 12-4　设置最低日志级别的输出

12.1.4　日志范围

使用 BeginScope 方法不仅可以对日志进行追踪，还可以对一个区间范围内的日志定义一个范围内的日志标识。在范围内的日志可以更方便地对日志信息进行标识和定位。

如代码 12-6 所示，创建 ServiceCollection 对象，首先调用 AddLogging 扩展方法注册日志服务，然后通过 Action<ILoggingBuilder> 对象调用 AddSimpleConsole 扩展方法注册 ConsoleLoggerProvider 对象，最后通过 Action<SimpleConsoleFormatterOptions> 配置选项将 IncludeScopes 属性设置为 true 来启用日志范围。首先调用 BuildServiceProvider 扩展方法返回

ServiceProvider 对象，然后调用 GetRequiredService<T>扩展方法获取相应的服务实例，最后调用 CreateLogger<T>扩展方法创建 ILogger 对象。

代码12-6

```
var loggerFactory = new ServiceCollection()
   .AddLogging(builder =>
           builder.AddSimpleConsole(options =>
           {
               options.IncludeScopes = true;
           })
   )
   .BuildServiceProvider()
   .GetRequiredService<ILoggerFactory>();

var logger = loggerFactory.CreateLogger<Program>();
using (logger.BeginScope("Scope Id:{id}", Guid.NewGuid().ToString ("N")))
{
   logger.LogInformation("start get");
   logger.LogInformation("result=1");
   logger.LogInformation("end get");
}
```

需要注意的是，日志范围对象是在 using 代码块中进行的，所以在 using 代码块中执行完成后会调用该对象的 Dispose 方法进行释放。BeginScope 是一个扩展方法，允许传递一个格式化字符串的参数（以字符串占位符的形式传递，如 User{User}）。另外，BeginScope 方法还支持一个 params object[]类型的参数，开发人员可以通过传递多个参数来填充格式化字符串。

注意：Console 日志提供了 3 种预定义格式化设置的支持，包括 Simple、Systemd 和 Json，本节通过调用 AddSimpleConsole 扩展方法来注册 Simple 格式化设置。要注册其他格式化设置，可以使用 Add{Type}Console 扩展方法。

如图 12-5 所示，设置作用域日志的输出（通过设置作用域执行代码并输出到控制台中）。

图 12-5　设置作用域日志的输出

根据控制台中的输出结果来看,如对一个订单做一个操作流程,通过范围设置可以更友好地输出日志。

12.1.5　JSON 配置

开发人员既可以通过代码配置日志,也可以以配置文件的方式配置日志。创建一个控制台项目,如代码 12-7 所示,首先读取配置文件,同时创建并使用 ConfigurationBuilder 对象,调用 AddJsonFile 扩展方法注册配置文件。然后调用 ConfigurationBuilder 对象的 Build 方法创建 IConfiguration 对象。创建一个 ServiceCollection 对象,调用 AddLogging 扩展方法注册日志服务,同时利用 Action<ILoggingBuilder>对象调用 AddConsole 扩展方法注册 ConsoleLoggerProvider 对象,调用 AddConfiguration 扩展方法,并接收 IConfiguration 参数对象,通过前面的 IConfiguration 对象调用 GetSection 方法获取指定键的配置信息。最后调用 BuildServiceProvider 扩展方法返回 ServiceProvider 对象,调用 GetRequiredService<T>扩展方法获取相应的服务实例,调用 CreateLogger<T>扩展方法创建 ILogger 对象。

代码12-7

```
var configuration = new ConfigurationBuilder()
    .AddJsonFile("appsettings.Development.json")
    .Build();

var loggerFactory = new ServiceCollection()
    .AddLogging(builder =>
       builder.AddConsole()
          .AddConfiguration(configuration.GetSection("Logging"))
    )
    .BuildServiceProvider()
    .GetRequiredService<ILoggerFactory>();

var logger = loggerFactory.CreateLogger<Program>();
logger.LogInformation("This information log from Program");
logger.LogError("This error log from Program");
logger.LogWarning("This warning log from Program");

using (logger.BeginScope("Scope Id:{id}", Guid.NewGuid().ToString ("N")))
{
    logger.LogWarning("start get");
```

```
        logger.LogWarning("result=1");
        logger.LogWarning("end get");
}
```

通过 ILogger<Program>实例对象调用相应的日志输出扩展方法，对日志进行输出，之后调用 BeginScope 扩展方法输出范围性的日志消息。

如代码 12-8 所示，在日志配置中通过 LogLevel 指定 Program 类别和 Default 类别，Program 类别在日志级别上定义了 Warning 级别，所以在 Program 类别中匹配的日志会以 Warning 级别或更高的级别输出。另外，在 Console 中通过 FormatterName 设置了 Simple 值，将日志输出格式设置为 Simple 类型,通过 FormatterOptions 的 IncludeScopes 属性启动了范围性的日志。

代码12-8

```
{
  "Logging": {
    "LogLevel": {
      "Program": "Warning",
      "Default": "Information"
    },
    "Console": {
      "FormatterName": "Simple",
      "FormatterOptions": {
        "IncludeScopes": true
      }
    }
  }
}
```

如图 12-6 所示，在设置 JSON 配置文件后，运行应用程序会输出预期的内容。

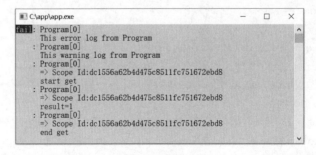

图 12-6　输出结果

12.1.6 控制台日志格式化

Console 日志记录提供了 3 个预定义格式化设置选项（见表 12-2）可供选择，分别为 Simple、Systemd 和 Json。如果注册格式化设置选项，则可以调用 Add{Type}Console 扩展方法。

表 12-2 格式化设置选项

可用类型	注册类型的方法
ConsoleFormatterNames.Json	ConsoleLoggerExtensions.AddJsonConsole
ConsoleFormatterNames.Simple	ConsoleLoggerExtensions.AddSimpleConsole
ConsoleFormatterNames.Systemd	ConsoleLoggerExtensions.AddSystemdConsole

如代码 12-9 所示，创建一个控制台项目，使用 Simple 方式格式化日志内容，通过 AddSimpleConsole 扩展方法注册 ConsoleFormatterNames.Simple 格式化程序。

代码12-9

```
using ILoggerFactory loggerFactory =
   LoggerFactory.Create(builder =>
      builder.AddSimpleConsole(options =>
      {
          options.IncludeScopes = true;
          options.SingleLine = true;
          options.TimestampFormat = "hh:mm:ss ";
      }));

ILogger<Program> logger = loggerFactory.CreateLogger<Program>();
using (logger.BeginScope("[scope is enabled]"))
{
   logger.LogInformation("Hello World!");
   logger.LogInformation("Logs contain timestamp and log level.");
   logger.LogInformation("Each log message is fit in a single line.");
}
```

AddSimpleConsole 扩展方法支持 SimpleConsoleFormatterOptions 配置选项，使用 IncludeScopes 属性可以开启范围日志，使用 SingleLine 属性可以将日志进行缩进，使用 TimestampFormat 属性可以设置日期格式化，在设置 Simple 格式化以后，运行应用程序输出的结果如图 12-7 所示，以 Simple 方式输出。

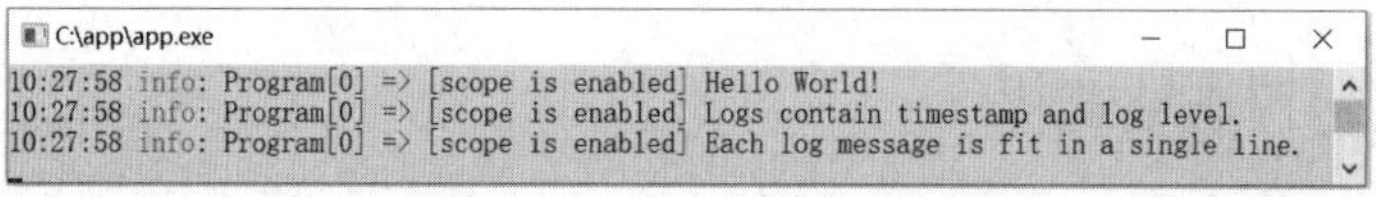

图 12-7　Simple 方式的输出结果

如代码 12-10 所示，通过调用 AddSystemdConsole 扩展方法注册 ConsoleFormatterNames.Systemd 格式化程序。通常来说对于容器环境非常有用，因为在容器中经常使用 Systemd 控制台日志记录。另外，Systemd 控制台记录还可以实现以单行方式进行日志记录（见图 12-8），并且允许禁用颜色。

代码12-10

```
using ILoggerFactory loggerFactory =
    LoggerFactory.Create(builder =>
        builder.AddSystemdConsole(options =>
        {
            options.IncludeScopes = true;
            options.TimestampFormat = "hh:mm:ss ";
        }));

ILogger<Program> logger = loggerFactory.CreateLogger<Program>();
using(logger.BeginScope("[scope is enabled]"))
{
    logger.LogInformation("Hello World!");
    logger.LogInformation("Logs contain timestamp and log level.");
    logger.LogInformation(
                "Systemd console logs never provide color options.");
    logger.LogInformation(
                "Systemd console logs always appear in a single line.");
}
```

图 12-8　Systemd 方式的输出结果

对于默认的几种输出方式，开发人员还可以输出 JSON 格式的日志，如代码 12-11 所示，可以使用 AddJsonConsole 扩展方法进行注册。

代码12-11

```
using ILoggerFactory loggerFactory =
```

```
LoggerFactory.Create(builder =>
    builder.AddJsonConsole(options =>
    {
        options.TimestampFormat = "hh:mm:ss ";
        options.JsonWriterOptions = new JsonWriterOptions
        {
            Indented = true
        };
    }));

ILogger<Program> logger = loggerFactory.CreateLogger<Program>();

logger.LogInformation("Hello World!");
```

如图 12-9 所示，在默认情况下，JSON 控制台格式化程序将每条消息以单行的形式进行记录，但是为了保证日志的可读性，可以将 JsonWriterOptions.Indented 属性设置为 true，最终以 JSON 格式化的形式输出。

图 12-9　JSON 方式的输出结果

12.1.7　实现自定义格式化程序

下面介绍一个简单的自定义格式化程序。如代码 12-12 所示，定义一个名为 CustomOptions 的配置选项，继承 ConsoleFormatterOptions 对象。理解了 CustomOptions 类的定义，读者应该能明白期望这个程序可以达到的功能。先不要继续往后看，如果让你来实现，你该怎么做？

代码12-12

```
public class CustomOptions : ConsoleFormatterOptions
{
    public string CustomPrefix { get; set; }
```

```
    public string CustomSuffix { get; set; }
}
```

创建一个名为 CustomFormatter 的类，该类继承 ConsoleFormatter 类和 IDisposable 接口。

如代码 12-13 所示，实现并重写 ConsoleFormatter 类中的 Write<TState>方法，使用该方法可以对日志执行写入操作。另外，在 CustomOptions 配置选项中定义了两个自定义属性，表示日志记录的前缀和后缀，此处用 WritePrefix 方法和 WriteSuffix 方法来包装前缀字符串和后缀字符串。

代码12-13

```
public sealed class CustomFormatter : ConsoleFormatter, IDisposable
{
    private readonly IDisposable _optionsReloadToken;
    private CustomOptions _formatterOptions;
    public CustomFormatter(IOptionsMonitor<CustomOptions> options)
        //Case insensitive
        : base("customName") =>
        (_optionsReloadToken, _formatterOptions) =
        (options.OnChange(ReloadLoggerOptions), options.CurrentValue);

    private void ReloadLoggerOptions(CustomOptions options) =>
        _formatterOptions = options;
    public override void Write<TState>(
                in LogEntry<TState> logEntry,
                IExternalScopeProvider scopeProvider,
                TextWriter textWriter)
    {
        string message =
            logEntry.Formatter(
                logEntry.State, logEntry.Exception);

        if(message == null)
        {
            return;
        }
        WritePrefix(textWriter);
        textWriter.Write(message);
        WriteSuffix(textWriter);
```

```
    }
    private void WritePrefix(TextWriter textWriter)
    {
        DateTime now = _formatterOptions.UseUtcTimestamp
            ? DateTime.UtcNow
            : DateTime.Now;
        textWriter.Write(
                $"{_formatterOptions.CustomPrefix}
                    {now.ToString(_formatterOptions.TimestampFormat)} ");
    }

    private void WriteSuffix(TextWriter textWriter)
    {
        textWriter.WriteLine(_formatterOptions.CustomSuffix);
    }

    public void Dispose() => _optionsReloadToken?.Dispose();
}
```

在 CustomFormatter 对象中,通过 base 关键字可以调用基类的构造函数,利用 IOptionsMonitor<CustomOptions>对象可以获取实时变更的信息,调用该对象的 OnChange 方法可以返回一个 Tuple(元祖),利用返回值可以获取 CustomOptions 配置选项和 IDisposable 接口。

接下来为自定义格式化程序创建一个扩展方法,如代码 12-14 所示。

代码12-14

```
public static class ConsoleLoggerExtensions
{
    public static ILoggingBuilder AddCustomFormatter(
            this ILoggingBuilder builder, Action<CustomOptions> configure)
    {
        return builder.AddConsole(options =>
                options.FormatterName = "customName")
            .AddConsoleFormatter<CustomFormatter, CustomOptions>(configure);
    }
}
```

如代码 12-15 所示，通过调用 AddCustomFormatter 扩展方法注册自定义格式化程序，通过 CustomOptions 配置选项设置日志的前缀和后缀。

代码12-15

```
using ILoggerFactory loggerFactory =
    LoggerFactory.Create(builder =>
        builder.AddCustomFormatter(options =>
        {
            options.CustomPrefix = "<|";
            options.CustomSuffix = "|>";
        }));

ILogger<Program> logger = loggerFactory.CreateLogger<Program>();

logger.LogInformation("Hello World!");
```

图 12-10 所示为自定义输出，在自定义的日志格式设置好之后，将结果输出到控制台中。

12.1.8 日志的设计与实现

日志模型中提供了 3 个核心对象，分别为 ILogger 对象、ILoggerProvider 对象和 ILoggerFactory 对象。

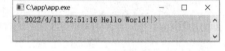

图 12-10 自定义输出

图 12-11 所示为 3 个核心对象的关系图，开发人员可以将 ILoggerFactory 对象和 ILoggerProvider 对象作为 ILogger 对象的辅助方法。

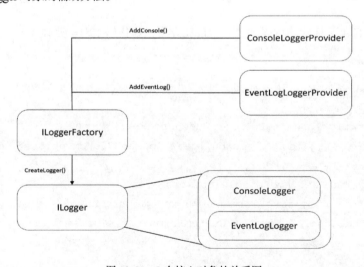

图 12-11 3 个核心对象的关系图

如代码 12-16 所示，ILogger 对象包含的方法有 BeginScope、IsEnabled 和 Log<TState>。使用 IsEnabled 方法可以判断当前 Logger 是否支持指定的日志级别，如果不支持则忽略这条日志，导致该日志不会被输出。使用 BeginScope 方法不仅可以跟踪日志，还可以对一个范围内的日志进行标识。使用 Log<TState>方法可以对日志进行记录。

代码12-16

```
public interface ILogger
{
    void Log<TState>(LogLevel logLevel,
                     EventId eventId, TState state, Exception?
                     exception, Func<TState, Exception?, string> formatter);
    bool IsEnabled(LogLevel logLevel);
    IDisposable BeginScope<TState>(TState state);
}
public interface ILogger<out TCategoryName> : ILogger
{
}
```

可以利用 ILoggerProvider 对象创建一个 ILogger 实例对象，如代码 12-17 所示，ILoggerProvider 对象可以利用日志的类型（categoryName）调用 CreateLogger 方法获取 ILogger 对象。

代码12-17

```
public interface ILoggerProvider : IDisposable
{
    ILogger CreateLogger(string categoryName);
}
```

如代码 12-18 所示，ILoggerFactory 对象包含 CreateLogger 方法和 AddProvider 方法，这两个方法分别用于创建 ILogger 对象和注册 ILoggerProvider 对象。

代码12-18

```
public interface ILoggerFactory : IDisposable
{
    ILogger CreateLogger(string categoryName);
    void AddProvider(ILoggerProvider provider);
}
```

如代码 12-19 所示，通过简化 LoggerFactory 类中的代码，可以帮助读者理解日志模型中的 3 个核心对象。LoggerFactory 类会利用一个 Dictionary<string,Logger>字典对象存储 Logger 对象并将其缓存起来，此处说是缓存是因为 ILoggerFactory 对象的生命周期为 Singleton 模式，LoggerFactory 类会利用一个 List<ProviderRegistration>集合来保存 ProviderRegistration 对象（该对象主要为 ILoggerProvider 类提供一个定义）。

代码12-19

```csharp
public class LoggerFactory : ILoggerFactory
{
    private readonly Dictionary<string, Logger> _loggers =
                new Dictionary<string, Logger>(StringComparer.Ordinal);
    private readonly List<ProviderRegistration> _providerRegistrations =
                                new List<ProviderRegistration>();
    private readonly object _sync = new object();
    public LoggerFactory(IEnumerable<ILoggerProvider> providers,
        IOptionsMonitor<LoggerFilterOptions> filterOption, IOptions
                            <LoggerFactoryOptions> options = null)
    {
        foreach(ILoggerProvider provider in providers)
        {
            AddProviderRegistration(provider, dispose: false);
        }
        ...
    }
    public ILogger CreateLogger(string categoryName)
    {
        lock(_sync)
        {
            if(!_loggers.TryGetValue(categoryName, out Logger logger))
            {
                logger = new Logger
                {
                    Loggers = CreateLoggers(categoryName),
                };
                _loggers[categoryName] = logger;
            }
            return logger;
        }
```

```csharp
}
public void AddProvider(ILoggerProvider provider)
{
    AddProviderRegistration(provider, dispose: true);
    ...
}
private void AddProviderRegistration(
                        ILoggerProvider provider, bool dispose)
{
    _providerRegistrations.Add(new ProviderRegistration
    {
        Provider = provider,
        ShouldDispose = dispose
    });
}
private LoggerInformation[] CreateLoggers(string categoryName)
{
    var loggers = new LoggerInformation[_providerRegistrations.Count];
    for (int i = 0; i < _providerRegistrations.Count; i++)
    {
        loggers[i] =
            new LoggerInformation(
                _providerRegistrations[i].Provider, categoryName);
    }
    return loggers;
}
private struct ProviderRegistration
{
    public ILoggerProvider Provider;
    public bool ShouldDispose;
}
...
}
```

AddProvider 方法接收一个 ILoggerProvider 对象，并且在方法内部会调用 AddProviderRegistration 方法将该对象添加到 List<ProviderRegistration>集合中。在 CreateLogger 方法中，利用 Dictionary<string,Logger>字典，将 categoryName 参数作为字典的 Key 进行获取。

如果在字典中获取到对象则直接返回该对象；如果在字典中没有获取到对象则创建一个 Logger 对象实例，在创建实例时会调用 CreateLoggers 方法创建其存储的对象。在 CreateLoggers 方法中，首先循环 List<ProviderRegistration> 集合为每个日志提供者创建一个 LoggerInformation 对象并保存到 LoggerInformation 数组中，然后返回该对象，将该对象存储在 Logger 对象实例中，最后将实例对象存储到 Dictionary<string,Logger> 字典中，并返回 Logger 对象。

如代码 12-20 所示，日志框架核心服务的注册是通过 AddLogging 扩展方法进行的，在该方法中利用 IServiceCollection 对象注册 ILoggerFactory 对象和 ILogger<>对象。另外，还可以看出，在 AddLogging 方法中可以调用 AddOptions 扩展方法注册 Options 模式的核心服务，以及 IConfigurations<LoggerFilterOptions>服务，该方法被注册为 Singleton 模式。IConfigurations<LoggerFilterOptions>服务具体的实例是一个 DefaultLoggerLevelConfigureOptions 对象。

代码12-20

```
public static class LoggingServiceCollectionExtensions
{
    public static IServiceCollection AddLogging(
                                    this IServiceCollection services)
    {
        return AddLogging(services, builder => { });
    }

    public static IServiceCollection AddLogging(
        this IServiceCollection services, Action<ILoggingBuilder> configure)
    {
        if (services == null)
        {
            throw new ArgumentNullException(nameof(services));
        }
        services.AddOptions();
        services.TryAdd(
            ServiceDescriptor.Singleton<ILoggerFactory, LoggerFactory>());
        services.TryAdd(
          ServiceDescriptor.Singleton(typeof(ILogger<>), typeof(Logger<>)));
        services.TryAddEnumerable(
        ServiceDescriptor.Singleton<IConfigureOptions<LoggerFilterOptions>>(
            new DefaultLoggerLevelConfigureOptions(LogLevel.Information)));
        configure(new LoggingBuilder(services));
        return services;
```

    }
}
```

### 12.1.9 控制台日志的设计与实现

对于控制台日志来说，ILogger 对象的实现是 ConsoleLogger 类型，对应的 ILoggerProvider 对象的实现是 ConsoleLoggerProvider 对象。代码 12-21 展示了 ConsoleLogger 类型的定义，该类型将 ConsoleLoggerProcessor 对象作为控制台日志输出的处理者。

代码12-21

```csharp
internal sealed class ConsoleLogger : ILogger
{
    internal ConsoleLogger(
             string name, ConsoleLoggerProcessor loggerProcessor);

    public void Log<TState>(LogLevel logLevel, EventId eventId,
                  TState state, Exception exception,
                     Func<TState, Exception, string> formatter);

    public bool IsEnabled(LogLevel logLevel)
       => logLevel != LogLevel.None;
    public IDisposable BeginScope<TState>(TState state) =>
                  ScopeProvider?. Push(state) ?? NullScope.Instance;
}
```

如代码 12-22 所示，在 ConsoleLoggerProcessor 对象中利用 BlockingCollection<LogMessageEntry>集合对象存储日志记录信息，可以发现它的容量被初始化为 1024 条。另外，在 ConsoleLoggerProcessor 对象中创建并通过一个后台线程处理日志记录，先循环 BlockingCollection<LogMessageEntry>集合，再利用 IConsole 接口将日志记录输出到控制台中。

代码12-22

```csharp
internal class ConsoleLoggerProcessor : IDisposable
{
    private const int _maxQueuedMessages = 1024;
    private readonly BlockingCollection<LogMessageEntry> _messageQueue =
          new BlockingCollection<LogMessageEntry>(_maxQueuedMessages);
```
```

```csharp
private readonly Thread _outputThread;
public IConsole Console;
public IConsole ErrorConsole;
public ConsoleLoggerProcessor()
{
    _outputThread = new Thread(ProcessLogQueue)
    {
        IsBackground = true,
        Name = "Console logger queue processing thread"
    };
    _outputThread.Start();
}
public virtual void EnqueueMessage(LogMessageEntry message)
{
    if(!_messageQueue.IsAddingCompleted)
    {
        _messageQueue.Add(message);
        return;
    }
    WriteMessage(message);
}
internal virtual void WriteMessage(LogMessageEntry entry)
{
    IConsole console = entry.LogAsError ? ErrorConsole : Console;
    console.Write(entry.Message);
}
private void ProcessLogQueue()
{
    try
    {
        foreach(LogMessageEntry message in 
                        _messageQueue.GetConsumingEnumerable())
        {
            WriteMessage(message);
        }
    }
    catch
```

```
        {
            _messageQueue.CompleteAdding();
        }
    }
    public void Dispose()
    {
        _messageQueue.CompleteAdding();
        try
        {
            _outputThread.Join(1500);
        }
        catch (ThreadStateException) { }
    }
}
```

如代码 12-23 所示,在 IConsole 接口中定义一个 Write 方法,当调用 Write 方法时会输出指定的字符串。

代码12-23

```
internal interface IConsole
{
    void Write(string message);
}
```

如代码 12-24 所示,在 ConsoleLoggerOptions 类中,定义了 FormatterName 属性和 LogToStandardErrorThreshold 属性,分别表示格式化处理程序的名称和日志输出的最低级别。

代码12-24

```
public class ConsoleLoggerOptions
{
    public string FormatterName { get; set; }
    public LogLevel LogToStandardErrorThreshold { get; set; } = LogLevel.None;
}
```

代码 12-25 展示了 ConsoleLoggerProvider 类的定义,该类标注了 ProviderAlias 特性,利用该特性可以将别名设置为 Console。

代码12-25

```
[ProviderAlias("Console")]
```

```csharp
public class ConsoleLoggerProvider : ILoggerProvider, ISupportExternalScope
{
    public ConsoleLoggerProvider(
        IOptionsMonitor<ConsoleLoggerOptions> options)
            : this(options, Array.Empty<ConsoleFormatter>());

    public ConsoleLoggerProvider(
        IOptionsMonitor<ConsoleLoggerOptions> options,
            IEnumerable<ConsoleFormatter> formatters);

    public ILogger CreateLogger(string name);
    public void Dispose();
    public void SetScopeProvider(IExternalScopeProvider scopeProvider);
}
```

代码 12-26 展示了注册控制台日志的扩展方法。控制台日志的核心服务会通过 AddConsole 扩展方法进行注册，配置选项会通过 Action<ConsoleLoggerOptions>委托对象定义。

代码12-26

```csharp
public static class ConsoleLoggerExtensions
{
    public static ILoggingBuilder AddConsole(this ILoggingBuilder builder)
    {
        builder.AddConfiguration();
        builder.AddConsoleFormatter<JsonConsoleFormatter,
                                    JsonConsoleFormatterOptions>();
        builder.AddConsoleFormatter<SystemdConsoleFormatter,
                                    ConsoleFormatterOptions>();
        builder.AddConsoleFormatter<SimpleConsoleFormatter,
                                    SimpleConsoleFormatterOptions>();
        builder.Services.TryAddEnumerable(ServiceDescriptor.Singleton
                    <ILoggerProvider, ConsoleLoggerProvider>());
        LoggerProviderOptions.RegisterProviderOptions
          <ConsoleLoggerOptions, ConsoleLoggerProvider>(builder.Services);
        return builder;
    }
}
```

```
public static ILoggingBuilder AddConsole(this ILoggingBuilder builder,
                Action<ConsoleLoggerOptions> configure)
{
    builder.AddConsole();
    builder.Services.Configure(configure);
    return builder;
}
public static ILoggingBuilder AddConsoleFormatter<
        TFormatter, TOptions> (this ILoggingBuilder builder)
        where TOptions : ConsoleFormatterOptions where
            TFormatter : ConsoleFormatter
{
    builder.AddConfiguration();
    builder.Services.TryAddEnumerable(
            ServiceDescriptor.Singleton<ConsoleFormatter, TFormatter>());
    builder.Services.TryAddEnumerable(
             ServiceDescriptor.Singleton<IConfigureOptions <TOptions>,
        ConsoleLoggerFormatterConfigureOptions<TFormatter, TOptions>>());
    builder.Services.TryAddEnumerable(
        ServiceDescriptor.Singleton<IOptionsChange TokenSource<TOptions>,
ConsoleLoggerFormatterOptionsChangeTokenSource<TFormatter, TOptions>>());
    return builder;
}
}
```

12.1.10 格式化的设计与实现

控制台日志是可以格式化的，默认提供了 3 种预定义格式，分别为 Simple、Json 和 Systemd，如代码 12-27 所示。

代码12-27

```
public static class ConsoleFormatterNames
{
public const string Simple = "simple";
public const string Json = "json";
```

```
public const string Systemd = "systemd";
}
```

代码 12-28 展示了 ConsoleFormatter 类的定义，应用程序最终会通过 Write 方法格式化并输出日志。

代码12-28

```
public abstract class ConsoleFormatter
{
    protected ConsoleFormatter(string name)
    {
        Name = name ?? throw new ArgumentNullException(nameof(name));
    }
    public string Name { get; }

    public abstract void Write<TState>(in LogEntry<TState> logEntry ,
            IExternalScopeProvider scopeProvider, TextWriter textWriter);
}
```

如代码 12-29 所示，ConsoleFormatterOptions 类作为控制台日志格式化的定义，如果需要自定义一个格式化程序则可以继承该类。

代码12-29

```
public class ConsoleFormatterOptions
{
    public ConsoleFormatterOptions() { }
    public bool IncludeScopes { get; set; }
    public string TimestampFormat { get; set; }
    public bool UseUtcTimestamp { get; set; }
}
```

代码 12-30 所示，在 ConsoleLoggerExtensions 类中，还包括 AddSimpleConsole 扩展方法和 AddJsonConsole 扩展方法，最终核心对象都会通过该类进行注册。

代码12-30

```
public static class ConsoleLoggerExtensions
{
    public static ILoggingBuilder AddConsole(this ILoggingBuilder builder)
    {
        ...
    }
```

```csharp
public static ILoggingBuilder AddConsole(this ILoggingBuilder builder,
                            Action<ConsoleLoggerOptions> configure)
{
    ...
}
public static ILoggingBuilder AddSimpleConsole(
    this ILoggingBuilder builder,
      Action<SimpleConsoleFormatterOptions> configure) =>
      builder.AddConsoleWithFormatter<SimpleConsoleFormatterOptions>
                        (ConsoleFormatterNames.Simple, configure);
public static ILoggingBuilder AddJsonConsole(
    this ILoggingBuilder builder,
      Action<JsonConsoleFormatterOptions> configure) =>
      builder.AddConsoleWithFormatter<JsonConsoleFormatterOptions>
                        (ConsoleFormatterNames.Json,configure);
public static ILoggingBuilder AddSystemdConsole(
    this ILoggingBuilder builder,
      Action<ConsoleFormatterOptions> configure) =>
      builder.AddConsoleWithFormatter<ConsoleFormatterOptions>
                        (ConsoleFormatterNames.Systemd,configure);
public static ILoggingBuilder AddSystemdConsole(
    this ILoggingBuilder builder) =>
        builder.AddFormatterWithName(ConsoleFormatterNames.Systemd);
internal static ILoggingBuilder AddConsoleWithFormatter<TOptions>(
  this ILoggingBuilder builder, string name, Action<TOptions> configure)
                    where TOptions : ConsoleFormatterOptions
{
    builder.AddFormatterWithName(name);
    builder.Services.Configure(configure);
    return builder;
}
private static ILoggingBuilder AddFormatterWithName(
        this ILoggingBuilder builder, string name) =>
            builder.AddConsole((ConsoleLoggerOptions options) =>
                                    options. FormatterName = name);
public static ILoggingBuilder AddConsoleFormatter
  <TFormatter, TOptions> (this ILoggingBuilder builder)
```

```
                where TOptions : ConsoleFormatterOptions
                    where TFormatter : ConsoleFormatter
{
    ...
}
```

12.1.11　Source Generator

在"Microsoft.Extensions.Logging" NuGet 包下定义一个 LoggerMessageAttribute 特性，该特性可以用于在编译时生成"高性能日志 API"方法。使用 LoggerMessageAttribute 特性可以对方法进行标记，通过 Source Generator 在编译时生成相应的日志调用逻辑代码，同时由 Source Generator 生成代码可以减少代码量。

如代码 12-31 所示，在使用 LoggerMessageAttribute 特性时，方法和类需要添加 partial 关键字，这样才会触发编译时代码生成。

代码12-31

```
public static partial class Log
{
    [LoggerMessageAttribute(EventId = 0,Level = LogLevel.Information,
                                        Message = "Hello {Name}")]
    public static partial void SayHello(this ILogger logger, string name);
}
```

例如，笔者通过控制台项目，先利用上面的 SayHello 日志输出方法进行调用并输出，通过 LoggerFactory 类调用 Create 方法，再利用 Action<ILoggingBuilder> 对象调用 AddConsole 扩展方法注册 ConsoleLoggerProvider 对象，如代码 12-32 所示。

代码12-32

```
var loggerFactory = LoggerFactory.Create(
    builder => builder.AddConsole());
var logger = loggerFactory.CreateLogger<Program>();

logger.SayHello("HueiFeng");
```

通过 LoggerFactory 实例对象，先调用 CreateLogger<T> 扩展方法创建一个 ILogger 对象，再调用 SayHello 扩展方法。控制台日志的输出结果如图 12-12 所示。

图 12-12 控制台日志的输出结果

12.2 调试日志

在.NET 日志系统中，还提供了调试器 Debug 输出的方式。采用这种方式有利于开发人员将日志输出到调试窗口中，但这种方式仅在 Debug 模式下生效。

12.2.1 调试日志的输出

下面以控制台项目为例展开介绍。如代码 12-33 所示，创建 ServiceCollection 对象，首先调用 AddLogging 扩展方法注册日志服务，然后利用 Action<ILoggingBuilder>对象调用 AddConsole 扩展方法注册 ConsoleLoggerProvider 对象（非必需调用，该调用的主要目的是开启控制台日志输出），并调用 AddDebug 扩展方法注册 DebugLoggerProvider 对象，最后调用 SetMinimumLevel 扩展方法将最低日志级别设置为 LogLevel.Trace。首先调用 BuildServiceProvider 扩展方法返回 ServiceProvider 对象，然后调用 GetRequiredService<T>扩展方法获取相应的服务实例，最后调用 CreateLogger<T>扩展方法创建 ILogger 对象。

代码12-33

```
var logger = new ServiceCollection()
    .AddLogging(builder =>
        builder.AddConsole()
            .AddDebug()
            .SetMinimumLevel(LogLevel.Trace))
    .BuildServiceProvider()
    .GetRequiredService<ILoggerFactory>()
    .CreateLogger<Program>();

logger.LogTrace("Trace...");
logger.LogDebug("Debug...");
```

```
logger.LogInformation("Information...");
logger.LogWarning("Warning...");
logger.LogError("Error...");
logger.LogCritical("Critical...");
```

运行控制台项目，调试日志的输出结果如图 12-13 所示，将内容输出到调试窗口中。

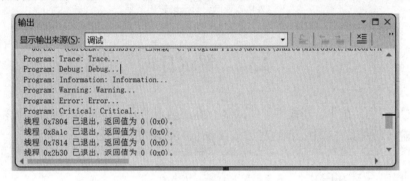

图 12-13　调试日志的输出结果

12.2.2　调试日志的设计与实现

如代码 12-34 所示，将日志记录输出到调试窗口中（调试日志的输出结果如图 12-14 所示）。

代码12-34

```
System.Diagnostics.Debug.WriteLine("Hello,World!");
```

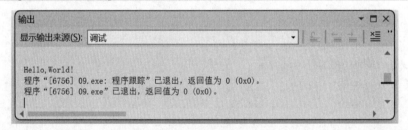

图 12-14　调试日志的输出结果

代码 12-35 展示了 DebugLoggerProvider 类的定义，该类标注了 ProviderAlias 特性，并且利用该特性将别名设置为 Debug，同时在 CreateLogger 方法中创建一个 DebugLogger 实例对象。

代码12-35

```
[ProviderAlias("Debug")]
public class DebugLoggerProvider : ILoggerProvider
{
    public ILogger CreateLogger(string name)
    {
        return new DebugLogger(name);
    }
    public void Dispose()
    {
    }
}
```

DebugLogger 实例对象继承 ILogger 接口，对应调试窗口日志的实现。代码 12-36 展示了 DebugLogger 类的定义，在日志字符串被包装后，可以通过调用 DebugWriteLine 方法记录输出日志。

代码12-36

```
internal sealed partial class DebugLogger : ILogger
{
    private readonly string _name;
    public DebugLogger(string name)
    {
        _name = name;
    }
    public IDisposable BeginScope<TState>(TState state)
        => NullScope.Instance;

    public bool IsEnabled(LogLevel logLevel)
        => Debugger.IsAttached && logLevel != LogLevel.None;

    public void Log<TState>(LogLevel logLevel, EventId eventId, TState state,
            Exception exception, Func<TState, Exception, string> formatter)
    {
        if (!IsEnabled(logLevel))
        {
            return;
        }
```

```
        if(formatter == null)
        {
            throw new ArgumentNullException(nameof(formatter));
        }
        string message = formatter(state, exception);
        if(string.IsNullOrEmpty(message))
        {
            return;
        }
        message = $"{ logLevel }: {message}";
        if(exception != null)
        {
            message += Environment.NewLine + Environment.NewLine + exception;
        }
        DebugWriteLine(message, _name);
    }
}
```

如代码 12-37 所示,在 DebugWriteLine 方法中调用了 System.Diagnostics.Debug.WriteLine 方法,用于将日志记录输出到调试窗口中。对 DebugWriteLine 方法而言,可以选择复制到项目中,查看其输出情况,日志框架的"调试日志模块"背后是 System.Diagnostics.Debug.WriteLine 方法输出的。

代码12-37

```
internal sealed partial class DebugLogger
{
    private void DebugWriteLine(string message, string name)
    {
        System.Diagnostics.Debug.WriteLine(message, category: name);
    }
}
```

AddDebug 是调试日志核心服务注册的方法,如代码 12-38 所示,在该方法中注册了 ILoggerProvider 服务,对应的实现类是 DebugLoggerProvider。

代码12-38

```
public static class DebugLoggerFactoryExtensions
{
    public static ILoggingBuilder AddDebug(this ILoggingBuilder builder)
```

```
    {
        builder.Services.TryAddEnumerable(ServiceDescriptor.Singleton
                        <ILoggerProvider, DebugLoggerProvider>());
        return builder;
    }
}
```

12.3　事件日志

在日志模块中，还提供了 Windows 事件日志的记录，并且被定义在"Microsoft.Extensions.Logging.EventLog" NuGet 包中。

12.3.1　事件日志的输出

以控制台项目为例，先引入"Microsoft.Extensions.Logging.EventLog" NuGet 包，如代码 12-39 所示，创建 ServiceCollection 对象，先调用 AddLogging 扩展方法注册日志服务，再利用 Action<ILoggingBuilder>对象调用 AddEventLog 扩展方法注册 EventLogLoggerProvider 对象。先调用 BuildServiceProvider 扩展方法返回 ServiceProvider 对象，再调用 GetRequiredService <T>扩展方法获取相应的服务实例，最后调用 CreateLogger<T>扩展方法创建一个 ILogger 对象。

代码12-39

```
var logger = new ServiceCollection()
    .AddLogging(builder =>
        builder
            .AddEventLog())
    .BuildServiceProvider()
    .GetRequiredService<ILoggerFactory>()
    .CreateLogger<Program>();

logger.LogInformation("Information...");
logger.LogWarning("Warning...");
logger.LogError("Error...");
```

如图 12-15 所示，运行控制台项目，打开"事件查看器"窗口，查看事件日志的输出结果。

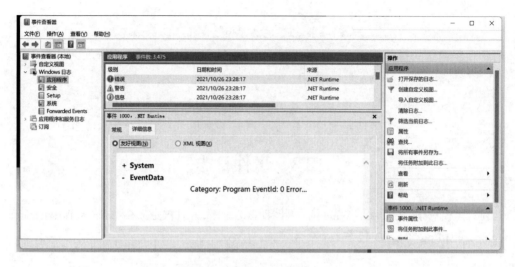

图 12-15　事件日志的输出结果

12.3.2　事件日志的设计与实现

代码 12-40 展示了事件日志内部的实现，最终通过 EventLog 对象将日志输出到"事件查看器"窗口中。

代码12-40

```
var eventLog = new EventLog();
eventLog.Source = "MySource";
eventLog.WriteEntry("Test EventLog");
```

代码 12-41 展示了 EventLogLoggerProvider 类的定义，该类标注了 ProviderAlias 特性，并且利用该特性将别名设置为 EventLog。另外，在 CreateLogger 方法中创建一个 EventLogLogger 实例对象，用于事件日志的记录。

代码12-41

```
[ProviderAlias("EventLog")]
public class EventLogLoggerProvider : ILoggerProvider, ISupportExternalScope
{
    internal readonly EventLogSettings _settings;

    private IExternalScopeProvider _scopeProvider;
    public EventLogLoggerProvider()
```

```csharp
            : this(settings: null)
    {
    }
    public EventLogLoggerProvider(EventLogSettings settings)
    {
        _settings = settings ?? new EventLogSettings();
    }

    public EventLogLoggerProvider(IOptions<EventLogSettings> options)
        : this(options.Value)
    {
    }

    public ILogger CreateLogger(string name)
        => new EventLogLogger(name, _settings, _scopeProvider);

    public void Dispose()
    {
        if(_settings.EventLog is WindowsEventLog windowsEventLog)
        {
#if NETSTANDARD
            Debug.Assert(RuntimeInformation.IsOSPlatform(OSPlatform.Windows));
#endif
            windowsEventLog.DiagnosticsEventLog.Dispose();
        }
    }

    public void SetScopeProvider(IExternalScopeProvider scopeProvider)
    {
        _scopeProvider = scopeProvider;
    }
}
```

如代码 12-42 所示，EventLogLogger 类继承 ILogger 接口，在该实现类中，笔者简化了类的实现，保留了一些关键性信息。首先调用 Log<TState>方法，在方法内部对日志字符串进行包装，然后调用 WriteMessage 方法进行日志的输出。

代码12-42

```csharp
internal sealed class EventLogLogger : ILogger
{
    private readonly string _name;
    private readonly EventLogSettings _settings;
    private readonly IExternalScopeProvider _externalScopeProvider;
    private const string ContinuationString = "...";
    private readonly int _beginOrEndMessageSegmentSize;
    private readonly int _intermediateMessageSegmentSize;

    public EventLogLogger(string name,
                          EventLogSettings settings,
                          IExternalScopeProvider externalScopeProvider);

    public IEventLog EventLog { get; }
    public IDisposable BeginScope<TState>(TState state)
                                => _externalScopeProvider?.Push(state);

    public bool IsEnabled(LogLevel logLevel)
        => logLevel != LogLevel.None &&
           (_settings.Filter == null || _settings.Filter(_name, logLevel));

    public void Log<TState>(LogLevel logLevel, EventId eventId, TState state,
        Exception exception, Func<TState, Exception, string> formatter)
    {
        if (!IsEnabled(logLevel)) return;
        string message = formatter(state, exception);
        if (string.IsNullOrEmpty(message)) return;
        StringBuilder builder =
            new StringBuilder().Append("Category: ").AppendLine(_name)
              .Append("EventId: ").Append(eventId.Id).AppendLine();
        ...
        builder.AppendLine().AppendLine(message);
        WriteMessage(builder.ToString(),
                GetEventLogEntryType(logLevel),
                    EventLog.DefaultEventId ?? eventId.Id);
    }
}
```

```csharp
private void WriteMessage(string message,
                EventLogEntryType eventLogEntryType, int eventId)
{
    int startIndex = 0;
    string messageSegment = null;
    while(true)
    {
        ...
        EventLog.WriteEntry(messageSegment,
                        eventLogEntryType, eventId, category: 0);
    }
}
private EventLogEntryType GetEventLogEntryType(LogLevel level)
{
    //...
}
}
```

如代码 12-43 所示，EventLogLogger 抽象了 IEventLog 接口，在 IEventLog 接口中定义的 WriteEntry 方法用于日志的写入，定义的 DefaultEventId 属性和 MaxMessageSize 属性分别表示默认的事件 ID（eventID）和日志消息文本长度的最大限制。

代码12-43

```csharp
internal interface IEventLog
{
    int? DefaultEventId { get; }
    int MaxMessageSize { get; }
    void WriteEntry(string message,
            EventLogEntryType type, int eventID, short category);
}
```

如代码 12-44 所示，在 WindowsEventLog 类中实现了 IEventLog 接口，在该类的构造函数中通过指定的参数 logName、machineName 和 sourceName 来创建 EventLog 实例对象，在 WriteEntry 方法中调用了 EventLog 实例对象的 WriteEvent 方法，通过调用该方法可以将日志记录到 Windows 事件中。需要注意的是，WindowsEventLog 对象实现了 IEventLog 接口的 MaxMessageSize 属性，并且定义了一个 int 类型的常量，默认值为 31839，也就是说，日志消息文本不能超过该长度。

代码12-44

```csharp
[SupportedOSPlatform("windows")]
internal sealed class WindowsEventLog : IEventLog
{
    private const int MaximumMessageSize = 31839;
    private bool _enabled = true;
    public WindowsEventLog(string logName,
                                    string machineName, string sourceName)
    {
        DiagnosticsEventLog = new System.Diagnostics.EventLog(
                                    logName, machineName, sourceName);
    }
    public System.Diagnostics.EventLog DiagnosticsEventLog { get; }
    public int MaxMessageSize => MaximumMessageSize;
    public int? DefaultEventId { get; set; }
    public void WriteEntry(string message, EventLogEntryType type,
                                    int eventID, short category)
    {
        try
        {
            if(_enabled)
            {
                DiagnosticsEventLog.WriteEvent(
                    new EventInstance(eventID, category, type), message);
            }
        }
        catch(SecurityException sx)
        {
            _enabled = false;
            try
            {
                using(var backupLog = new System.Diagnostics.EventLog(
                            "Application", ".", "Application"))
                {
                    backupLog.WriteEvent(new EventInstance(
```

```
                    instanceId: 0, categoryId: 0, EventLogEntryType.Error),
                $"Unable to log .NET application events.{sx.Message}");
            }
        }
        catch(Exception)
        {
        }
    }
}
```

如代码 12-45 所示，在 EventLoggerFactoryExtensions 类中定义的 3 个扩展方法用于注册事件日志核心对象。

代码12-45

```
public static class EventLoggerFactoryExtensions
{
    public static ILoggingBuilder AddEventLog(this ILoggingBuilder builder)
    {
        builder.Services.TryAddEnumerable(ServiceDescriptor.Singleton
                    <ILoggerProvider, EventLogLoggerProvider>());
        return builder;
    }

    public static ILoggingBuilder AddEventLog(this ILoggingBuilder builder,
                                    EventLogSettings settings)
    {
        builder.Services.TryAddEnumerable(ServiceDescriptor.Singleton
                <ILoggerProvider>(new EventLogLoggerProvider(settings)));
        return builder;
    }

    public static ILoggingBuilder AddEventLog(this ILoggingBuilder builder,
                                    Action<EventLogSettings> configure)
    {
        builder.AddEventLog();
```

```
        builder.Services.Configure(configure);
        return builder;
    }
}
```

12.4　EventSource 日志

EventSource 采用订阅发布的设计模式，事件日志开发人员可以通过 EventListener 对象来订阅感兴趣的日志事件，EventSource 日志模块被定义在 "Microsoft.Extensions. Logging. EventSource" NuGet 包中。

12.4.1　EventSource 日志的输出

笔者以控制台项目为例展开介绍。先引入 "Microsoft.Extensions.Logging.EventSource" NuGet 包，如代码 12-46 所示，创建 ServiceCollection 对象，首先调用 AddLogging 扩展方法注册日志服务，然后利用 Action<ILoggingBuilder>对象调用 AddEventSourceLogger 扩展方法注册 EventSourceLoggerProvider 对象。首先调用 BuildServiceProvider 扩展方法返回 ServiceProvider 对象，然后调用 GetRequiredService<T>扩展方法获取相应的服务实例，最后调用 CreateLogger<T>扩展方法创建一个 ILogger 对象。

代码12-46

```
var logger = new ServiceCollection()
    .AddLogging(builder =>
        builder
            .AddEventSourceLogger())
    .BuildServiceProvider()
    .GetRequiredService<ILoggerFactory>()
    .CreateLogger<Program>();

logger.LogInformation("Information...");
logger.LogWarning("Warning...");
logger.LogError("Error...");
```

接下来利用 PerfView 查看写入的事件（开启 PerfView，选择 Collect→Run 命令，打开如

图 12-16 所示的窗口）。通过 PerfView 启动应用程序，并收集运行时的性能数据。

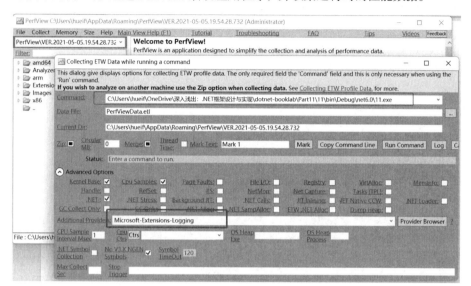

图 12-16　查看写入的事件

在 Command 文本框中输入的是 .exe 应用程序的地址，在 Additional Providers 文本框中输入 Microsoft-Extensions-Logging。单击 Run Command 按钮，启动应用程序，产生的 ETW 相关的性能数据将被 PerfView 收集起来，收集的数据最终会被生成一个 .etl.zip 文件，文件路径显示在 Data File 文本框中。EventSource 日志的输出结果如图 12-17 所示。

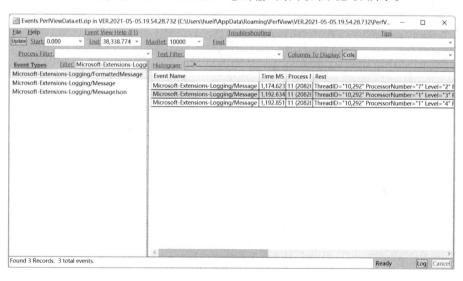

图 12-17　EventSource 日志的输出结果

12.4.2 EventSource 日志的设计与实现

如代码 12-47 所示，LoggingEventSource 类实现了 EventSource 类，通过 EventSource 特性标注 EventSource 日志的名称为 Microsoft-Extensions-Logging。

代码12-47

```
[EventSource(Name = "Microsoft-Extensions-Logging")]
public sealed class LoggingEventSource : EventSource
{
    public static class Keywords
    {
        public const EventKeywords Meta = (EventKeywords)1;
        public const EventKeywords Message = (EventKeywords)2;
        public const EventKeywords FormattedMessage = (EventKeywords)4;
        public const EventKeywords JsonMessage = (EventKeywords)8;
    }
    Private LoggingEventSource() : base(
                    EventSourceSettings.EtwSelfDescribingEventFormat);

    [Event(1, Keywords = Keywords.FormattedMessage, Level = EventLevel.LogAlways)]
    internal unsafe void FormattedMessage(LogLevel Level, int FactoryID,
     string LoggerName, int EventId, string EventName, string FormattedMessage);

    [Event(2, Keywords = Keywords.Message, Level = EventLevel.LogAlways)]
    [DynamicDependency(DynamicallyAccessedMemberTypes.PublicProperties,
                    typeof(KeyValuePair<string, string>))]
    internal void Message(LogLevel Level, int FactoryID, string LoggerName,
                int EventId, string EventName, ExceptionInfo Exception,
                IEnumerable <KeyValuePair<string, string>> Arguments);
    [Event(5, Keywords = Keywords.JsonMessage, Level = EventLevel.LogAlways)]
    internal unsafe void MessageJson(LogLevel Level, int FactoryID,
        string LoggerName, int EventId, string EventName, string ExceptionJson,
                    string ArgumentsJson, string FormattedMessage);
}
```

如代码 12-48 所示，EventSourceLogger 类实现了 ILogger 接口，并且利用 Log<TState>

方法输出日志。另外，在 Log<TState>方法中会调用 LoggingEventSource 对象的方法（LoggingEventSource 对象主要提供 3 个日志事件方法）。

代码12-48

```
internal sealed class EventSourceLogger : ILogger
{
    public void Log<TState>(LogLevel logLevel, EventId eventId, TState state,
            Exception exception, Func<TState, Exception, string> formatter)
    {
        if(!IsEnabled(logLevel))
        {
            return;
        }
        string message = null;
        if(_eventSource.IsEnabled(EventLevel.Critical,
                        LoggingEventSource.Keywords.FormattedMessage))
        {
            message = formatter(state, exception);
            _eventSource.FormattedMessage(logLevel, _factoryID,
                    CategoryName, eventId.Id, eventId.Name, message);
        }
        if(_eventSource.IsEnabled(EventLevel.Critical,
                        LoggingEventSource.Keywords.Message))
        {
            ExceptionInfo exceptionInfo = GetExceptionInfo(exception);
            IReadOnlyList<KeyValuePair<string, string>> arguments =
                                            GetProperties(state);
            _eventSource.Message(logLevel, _factoryID, CategoryName,
                    eventId.Id, eventId.Name, exceptionInfo, arguments);
        }
        if(_eventSource.IsEnabled(EventLevel.Critical,
                        LoggingEventSource.Keywords.JsonMessage))
        {
            string exceptionJson = "{}";
            if(exception != null)
            {
                ExceptionInfo exceptionInfo = GetExceptionInfo(exception);
```

```csharp
            KeyValuePair<string, string>[] exceptionInfoData = new[]
            {
                new KeyValuePair<string, string>(
                    "TypeName", exceptionInfo.TypeName),
                new KeyValuePair<string, string>(
                    "Message", exceptionInfo.Message),
                new KeyValuePair<string, string>(
                    "HResult", exceptionInfo.HResult.ToString()),
                new KeyValuePair<string, string>(
                    "VerboseMessage", exceptionInfo.VerboseMessage),
            };
            exceptionJson = ToJson(exceptionInfoData);
        }
        IReadOnlyList<KeyValuePair<string, string>> arguments =
            GetProperties(state);
        message ??= formatter(state, exception);
        _eventSource.MessageJson(logLevel, _factoryID, CategoryName,
    eventId.Id, eventId.Name, exceptionJson, ToJson(arguments), message);
    }
}
```

EventSourceLoggerProvider 类实现了 ILoggerProvider 接口，如代码 12-49 所示，在实现的 CreateLogger 方法中会创建一个 EventSourceLogger 对象，之后利用 CreateLogger 扩展方法创建一个 ILogger 对象。

代码12-49

```csharp
[ProviderAlias("EventSource")]
public class EventSourceLoggerProvider : ILoggerProvider
{
    private static int _globalFactoryID;
    private readonly int _factoryID;

    private EventSourceLogger _loggers;
    private readonly LoggingEventSource _eventSource;

    public EventSourceLoggerProvider(LoggingEventSource eventSource)
    {
```

```csharp
        _eventSource = eventSource;
        _factoryID = Interlocked.Increment(ref _globalFactoryID);
    }

    public ILogger CreateLogger(string categoryName)
    {
        return _loggers = new EventSourceLogger(
                    categoryName, _factoryID, _eventSource, _loggers);
    }

    public void Dispose()
    {
        for (EventSourceLogger logger = _loggers; logger != null;
                                                    logger = logger.Next)
        {
            logger.Level = LogLevel.None;
        }
    }
}
```

如代码12-50所示，EventSourceLoggerProvider对象的注册被定义在AddEventSourceLogger扩展方法中，EventSource日志的核心对象都是通过该扩展方法注册的。

代码12-50

```csharp
public static class EventSourceLoggerFactoryExtensions
{
    public static ILoggingBuilder AddEventSourceLogger(
                                    this ILoggingBuilder builder)
    {
        builder.Services.TryAddSingleton(LoggingEventSource.Instance);
        builder.Services.TryAddEnumerable(ServiceDescriptor.Singleton
                <ILoggerProvider, EventSourceLoggerProvider>());
        builder.Services.TryAddEnumerable(ServiceDescriptor
                .Singleton<IConfigureOptions<LoggerFilterOptions>,
                        EventLogFiltersConfigureOptions>());
        builder.Services.TryAddEnumerable(ServiceDescriptor.Singleton
                <IOptionsChangeTokenSource<LoggerFilterOptions>,
                    EventLogFiltersConfigureOptionsChangeSource>());
        return builder;
```

```
    }
}
```

12.5 TraceSource 日志

TraceSource 顾名思义是跟踪日志，在默认情况下跟踪输出会利用 DefaultTraceListener 对象，本节的代码直接采用 ConsoleTraceListener 对象输出跟踪结果。在跟踪日志的模块中定义了 TraceSourceLoggerProvider 对象，该对象实现了 ILoggerProvider 接口，跟踪日志模块的代码被定义在 "Microsoft.Extensions.Logging.TraceSource" NuGet 包中。

12.5.1 TraceSource 日志的输出

创建一个控制台项目，并引入 "Microsoft.Extensions.Logging.TraceSource" NuGet 包。如代码 12-51 所示，创建一个 ServiceCollection 对象，先调用 AddLogging 扩展方法注册日志服务，再利用 Action<ILoggingBuilder> 对象调用 AddTraceSource 扩展方法注册 TraceSourceLoggerProvider 对象，并且传入 SourceSwitch 对象和 ConsoleTraceListener 对象，SourceSwitch 对象用于控制日志输出的等级，ConsoleTraceListener 对象用于将追踪日志输出到控制台中。接下来调用 BuildServiceProvider 扩展方法返回 ServiceProvider 对象，调用 GetRequiredService<T> 扩展方法获取相应的服务实例，调用 CreateLogger<T> 扩展方法创建一个 ILogger<T> 对象。

代码12-51

```
var firstSwitch = new SourceSwitch("FirstSwitch")
{
    Level = SourceLevels.All
};
var logger = new ServiceCollection()
    .AddLogging(builder =>
        builder
            .AddTraceSource(firstSwitch, new ConsoleTraceListener()))
    .BuildServiceProvider()
    .GetRequiredService<ILoggerFactory>()
    .CreateLogger<Program>();
logger.LogInformation("Information...");
```

```
logger.LogWarning("Warning...");
logger.LogError("Error...");
```

利用 ILogger<T>对象输出日志,如图 12-18 所示,将结果输出到控制台中。

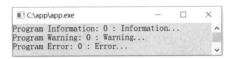

图 12-18　将结果输出到控制台中

12.5.2　TraceSource 日志的设计与实现

TraceSourceLogger 类实现了 ILogger 接口,并利用 Log<TState>方法输出日志。在 Log<TState>方法中会调用 TraceSource 对象的 TraceEvent 方法写入日志消息,如代码 12-52 所示。

代码12-52

```
internal sealed class TraceSourceLogger : ILogger
{
    private readonly DiagnosticsTraceSource _traceSource;
    public TraceSourceLogger(DiagnosticsTraceSource traceSource);

    public void Log<TState>(LogLevel logLevel, EventId eventId, TState state,
            Exception exception, Func<TState, Exception, string> formatter)
    {
        if(!IsEnabled(logLevel))
            return;
        string message = string.Empty;
        if(formatter != null)
        {
            message = formatter(state, exception);
        }
        else if(state != null)
        {
            message += state;
        }
        if(exception != null)
        {
```

```
            string exceptionDelimiter = string.IsNullOrEmpty(message)
                                            ? string.Empty : " ";
            message += exceptionDelimiter + exception;
        }
        if(!string.IsNullOrEmpty(message))
        {
            _traceSource.TraceEvent(GetEventType(logLevel),
                                            eventId.Id, message);
        }
    }

    public bool IsEnabled(LogLevel logLevel)
    {
        if(logLevel == LogLevel.None)
            return false;
        TraceEventType traceEventType = GetEventType(logLevel);
        return _traceSource.Switch.ShouldTrace(traceEventType);
    }

    private static TraceEventType GetEventType(LogLevel logLevel);
    public IDisposable BeginScope<TState>(TState state)
    {
        return new TraceSourceScope(state);
    }
}
```

TraceSourceLoggerProvider 类实现了 ILoggerProvider 接口，如代码 12-53 所示，在实现的 CreateLogger 方法中创建一个 TraceSourceLogger 对象，不难看出它会根据指定的名称将创建的 TraceSource 对象缓存起来，所以，当调用时会根据一个指定的名称到缓存中查询是否存在，如果存在则直接返回缓存中的 TraceSource 对象，如果不存在则根据指定的名称创建一个 TraceSourceLogger 对象。

代码12-53

```
[ProviderAlias("TraceSource")]
public class TraceSourceLoggerProvider : ILoggerProvider
{
    private readonly SourceSwitch _rootSourceSwitch;
    private readonly TraceListener _rootTraceListener;
```

```csharp
private readonly ConcurrentDictionary<string, DiagnosticsTraceSource>
    _sources = new ConcurrentDictionary<string, DiagnosticsTraceSource> (
                                        StringComparer.OrdinalIgnoreCase);

public TraceSourceLoggerProvider(SourceSwitch rootSourceSwitch)
                            : this(rootSourceSwitch, null);

public TraceSourceLoggerProvider(SourceSwitch rootSourceSwitch,
                            TraceListener rootTraceListener);

public ILogger CreateLogger(string name)
  => new TraceSourceLogger(GetOrAddTraceSource(name));

private DiagnosticsTraceSource GetOrAddTraceSource(string name)
   => _sources.GetOrAdd(name, InitializeTraceSource);

private DiagnosticsTraceSource InitializeTraceSource(
                                        string traceSourceName)
{
    var traceSource = new DiagnosticsTraceSource(traceSourceName);
    string parentSourceName = ParentSourceName(traceSourceName);
    if(string.IsNullOrEmpty(parentSourceName))
    {
        if(HasDefaultSwitch(traceSource))
        {
            traceSource.Switch = _rootSourceSwitch;
        }

        if(_rootTraceListener != null)
        {
            traceSource.Listeners.Add(_rootTraceListener);
        }
    }
    else
    {
        if(HasDefaultListeners(traceSource))
```

```
        {
            DiagnosticsTraceSource parentTraceSource =
                        GetOrAddTraceSource (parentSourceName);
            traceSource.Listeners.Clear();
            traceSource.Listeners.AddRange(parentTraceSource.Listeners);
        }
        if(HasDefaultSwitch(traceSource))
        {
            DiagnosticsTraceSource parentTraceSource =
                        GetOrAddTraceSource (parentSourceName);
            traceSource.Switch = parentTraceSource.Switch;
        }
    }
    return traceSource;
  }
}
```

如代码 12-54 所示，TraceSourceFactoryExtensions 类中提供了 AddTraceSource 扩展方法，开发人员可以调用 AddTraceSource 方法注册 TraceSource 的核心对象。

代码12-54

```
public static class TraceSourceFactoryExtensions
{
    public static ILoggingBuilder AddTraceSource(
        this ILoggingBuilder builder, string switchName)
          => builder.AddTraceSource(new SourceSwitch(switchName));

    public static ILoggingBuilder AddTraceSource(
        this ILoggingBuilder builder, string switchName,
        TraceListener listener)
          => builder.AddTraceSource(new SourceSwitch(switchName), listener);

    public static ILoggingBuilder AddTraceSource(
        this ILoggingBuilder builder, SourceSwitch sourceSwitch)
    {
        builder.Services.AddSingleton<ILoggerProvider>(_
                => new TraceSourceLoggerProvider(sourceSwitch));
        return builder;
```

```
    }
    public static ILoggingBuilder AddTraceSource(
        this ILoggingBuilder builder,
        SourceSwitch sourceSwitch, TraceListener listener)
    {
        builder.Services.AddSingleton<ILoggerProvider>(_
                => new TraceSourceLoggerProvider(sourceSwitch, listener));
        return builder;
    }
}
```

12.6 DiagnosticSource 日志

DiagnosticSource 采用观察者模式设计日志模块，并且将日志写入 DiagnosticSource 中，会被订阅者消费。DiagnosticSource 只是一个抽象类，定义了事件所需的方法，主要核心对象是 DiagnosticListener，而每个 DiagnosticListener 对象都具备一个独立的名称标识。DiagnosticListener 对象充当发布者角色，通过 Write 方法向 DiagnosticSource 发布日志，同时使用 Subscribe 方法设置订阅者来消费 DiagnosticSource 中的日志。

12.6.1 DiagnosticSource 日志的输出

笔者以控制台项目为例，利用 DiagnosticListener 对象的 AllListeners 属性存储当前进程创建的所有 IObservable<DiagnosticListener>订阅者对象，如代码 12-55 所示，通过调用 Subscribe 方法将 IObservable<DiagnosticListener>对象以订阅者的形式注册到 AllListeners 属性上。先创建一个 DiagnosticListener 对象，并将其命名为 App，再通过 DiagnosticSource 对象的 IsEnabled 方法判断指定的订阅者是否存在，如果存在，那么调用该对象的 Write 方法发送一个名为 App.Log 的日志事件。

代码12-55
```
DiagnosticListener.AllListeners.Subscribe(new MyObserver());
var diagnosticSource = new DiagnosticListener("App.Log");
var name = "App.Log";
if (diagnosticSource.IsEnabled(name))
{
    diagnosticSource.Write(name, new { Name = "Mr.A", Age = "18" });
```

}

如代码 12-56 所示，订阅消息需要实现 IObserver 接口，定义一个名为 MyObserver 的对象，作为订阅者的实现类。在 OnNext 方法中通过 DiagnosticListener 对象的 Subscribe 方法订阅指定的订阅者进行消费，在该对象中会订阅 OnNext (KeyValuePair <string,object?> value)方法进行消费。

代码12-56

```csharp
public class MyObserver : IObserver<DiagnosticListener>,
IObserver<KeyValuePair<string, object?>>
{
    public void OnCompleted()
    {
    }

    public void OnError(Exception error)
    {
    }

    public void OnNext(KeyValuePair<string, object?> value)
    {
        Console.WriteLine($"{value.Key}:{value.Value}");
    }

    public void OnNext(DiagnosticListener value)
    {
        if (value.Name == "App.Log")
        {
            value.Subscribe(this);
        }
    }
}
```

如图 12-19 所示，运行该项目，将结果输出到控制台中。

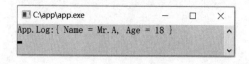

图 12-19　输出结果

第 12 章 日志

如图 12-19 所示，获取到该数据后也需要对数据进行转换，所以，是否有方法可以将数据直接定义为强类型呢？答案是有的，可以使用 DiagnosticName 特性设置订阅者的名称。如代码 12-57 所示，创建一个方法定义参数，用于接收参数的内容。

代码12-57

```
public class MyDiagnosticListener
{
    [DiagnosticName("App.Log")]
    public void Log(string name, string age)
    {
        System.Console.WriteLine($"DiagnosticName:App.Log");
        System.Console.WriteLine($"Name:{name},Age:{age}");
    }
}
```

如代码 12-58 所示，修改 MyObserver 类只需要定义一个简单的格式即可。

代码12-58

```
public class MyObserver : IObserver<DiagnosticListener>
{
    public void OnCompleted()
    {
    }

    public void OnError(Exception error)
    {
    }

    public void OnNext(DiagnosticListener value)
    {
    }
}
```

下面修改 Program 类，如代码 12-59 所示，利用 SubscribeWithAdapter 扩展方法注册订阅者对象即可。

代码12-59

```
var diagnosticListener = new DiagnosticListener("App");
DiagnosticSource diagnosticSource = diagnosticListener;
diagnosticListener.SubscribeWithAdapter(new MyDiagnosticListener());
```

```
var name = "App.Log";
if (diagnosticSource.IsEnabled(name))
{
    diagnosticSource.Write(name, new { Name = "Mr.A", Age = "18" });
}
```

12.6.2　DiagnosticSource 日志的设计与实现

DiagnosticSource 日志通过标准化的信息向应用程序传达内部的"消息"，并且采用事件的订阅发布模式。

如代码 12-60 所示，定义了两个接口，分别表示该模式中的订阅者和发布者。在 IObservable<T>接口中定义了一个名为 Subscribe 的方法，用于注册订阅者；在 IObserver<T> 接口中定义了 3 个方法，OnNext 表示事件回调方法，在观察者接收到通知后进行相关操作，OnCompleted 表示事件结束后调用的方法，OnError 表示事件异常调用的方法。

代码12-60

```
public interface IObserver<in T>
{
    void OnNext(T value);
    void OnError(Exception error);
    void OnCompleted();
}

public interface IObservable<out T>
{
    IDisposable Subscribe(IObserver<T> observer);
}
```

DiagnosticSource 是 DiagnosticListener 的基类，如代码 12-61 所示，DiagnosticSource 为日志事件的抽象类，用于定义事件日志的基本操作，此处定义了 IsEnabled 方法和 Write 方法。

代码12-61

```
public abstract partial class DiagnosticSource
{
    public abstract void Write(string name, object? value);
```

```
    public abstract bool IsEnabled(string name);

    public virtual bool IsEnabled(string name, object? arg1, object? arg2 = null)
    {
        return IsEnabled(name);
    }
}
```

如代码 12-62 所示,DiagnosticListener 类定义的 Write 方法用来发送日志事件,同时定义 3 个 Subscribe 方法用于注册订阅者,定义 IsEnabled 方法用于判断指定名称的订阅者是否存在。

代码12-62

```
public partial class DiagnosticListener : DiagnosticSource,
    IObservable<KeyValuePair<string, object?>>, IDisposable
{
    public static IObservable<DiagnosticListener> AllListeners;

    public virtual IDisposable Subscribe(
        IObserver<KeyValuePair<string, object?>> observer,
                                        Predicate<string>? isEnabled);
    public virtual IDisposable Subscribe(
        IObserver<KeyValuePair<string, object?>> observer,
                    Func<string, object?, object?, bool>? isEnabled);
    public virtual IDisposable Subscribe(
                IObserver<KeyValuePair<string, object?>> observer)
        => SubscribeInternal(observer, null, null, null, null);

    public bool IsEnabled();
    public override bool IsEnabled(string name);
    public override bool IsEnabled(
                    string name, object? arg1, object? arg2 = null);

    public override void Write(string name, object? value);
    public virtual void Dispose();

    public string Name { get; private set; }
    public override string ToString()
```

```
{
    return Name ?? string.Empty;
}
}
```

使用 DiagnosticNameAttribute 特性有助于开发人员完成一个强类型的订阅者，该特性的定义如代码 12-63 所示。

代码12-63

```
public class DiagnosticNameAttribute : Attribute
{
    public DiagnosticNameAttribute(string name);
    public string Name { get; }
}
```

如代码 12-64 所示，DiagnosticListenerExtensions 类定义了 SubscribeWithAdapter 方法，该方法用于注册带有 DiagnosticNameAttribute 特性的类，可以注册一个或多个订阅事件。

代码12-64

```
public static class DiagnosticListenerExtensions
{
    public static IDisposable (
                    this DiagnosticListener diagnostic, object target)
    {
        var adapter = new DiagnosticSourceAdapter(target);
        return diagnostic.Subscribe(
                    adapter, (Predicate<string>)adapter.IsEnabled);
    }

    public static IDisposable SubscribeWithAdapter(
        this DiagnosticListener diagnostic, object target,
        Func<string, bool> isEnabled)
    {
        var adapter = new DiagnosticSourceAdapter(target, isEnabled);
        return diagnostic.Subscribe(adapter,
                        (Predicate<string>)adapter.IsEnabled);
    }

    public static IDisposable SubscribeWithAdapter(
```

```
    this DiagnosticListener diagnostic,object target,
    Func<string, object, object, bool> isEnabled)
{
    var adapter = new DiagnosticSourceAdapter(target, isEnabled);
    return diagnostic.Subscribe(adapter,
            (Func<string, object, object,bool>)adapter.IsEnabled);
}
}
```

12.7 小结

日志不仅方便开发人员排查和定位问题，还是记录一些关键信息的一种手段。本章详细介绍了日志相关的知识点，可以通过.NET 提供的多种日志实现更丰富的应用功能。

第 13 章

多线程与任务并行

对当今的.NET 技术而言，异步已经成为.NET 技术体系中多线程的代名词，但由于多线程的概念非常重要，因此本章将对该话题进行深入探讨。目前，线程处理是最基础的技术点，而多线程机制是.NET 框架最核心的功能之一。充分利用多线程机制，可以极大地提高程序吞吐量，带来更高的应用性能。本章先讲述多线程的多个细节，包括线程的概念、线程的使用方式，以及线程管理等，再介绍基于任务的异步编程，使读者深刻理解异步任务。

13.1 线程简介

13.1.1 线程的概念

在操作系统中，进程（Process）是资源分配的基本单位，而线程（Thread）是运算调度的基本单位。当操作系统创建进程后，便会注入线程用于执行进程中的代码。创建进程后由操作系统启动第一个线程（即进程的主线程），而应用程序亦可自行创建多个线程，主线程的生命周期与应用程序的生命周期一致。如果主线程结束，则进程会退出，被操作系统销毁。操作系统在创建进程时，会为进程分配虚拟内存等资源，进程中的多个线程共享进程所拥有的资源、环境。

如果使用 C 语言开发应用程序，在 UNIX 操作系统下则可以使用 Posix 中的 pthread.h 来创建线程，而在 Windows 操作系统下则可以使用 processthreadsapi.h。通过这种系统的接口创建的线程一般称为系统线程或原生线程。如图 13-1 所示，在 Windows 任务管理器中可以查看系统的线程数量。

在 .NET 中，可以使用 Thread 类创建一个线程对象，创建线程对象后，CLR 会做一些初始工作，如设置默认状态、优先级等。当调用 Start 方法时，CLR 才会真正创建一个系统线程，因此，.NET 中的线程与系统线程是一一对应的。通过.NET 创建的线程和使用其他方式创建但会进入.NET 托管环境的线程称为托管线程。当使用托管线程时，开发人员不必关注如何管理系统线程，.NET 通过一系列接口，屏蔽了 UNIX 操作系统和 Windows 操作系统的接口的差异，开发人员也不需要关注线程栈内存的分配和回收。

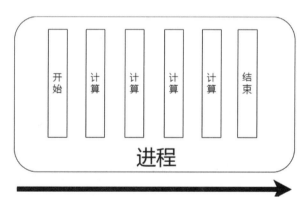

图 13-1　Windows 任务管理器

通常来说，开发人员学习一门新语言，会从 Hello World 应用程序开始。Hello World 应用程序是顺序型的，也就是说，在这个过程中它具备一个开始和一个结束。

图 13-2 所示为单线程执行图，线程类似于描述的顺序程序，单线程有一个开始、一个结束和一个执行点。线程本身不是程序，不能独立运行，需要在程序内运行。进程相当于线程的容器。

图 13-2　单线程执行图

上面介绍的是单线程的相关内容。但单线程不是最主要的，真正吸引我们的是使用多个线程同时执行，并在单个程序中执行不同的任务。图 13-3 所示为多线程执行图。

多线程的功能很多，如可以对 UI 进行渲染，利用工作线程可以对一些任务进行计算，从而分开执行这些任务。

进程是操作系统分配资源的基本单位，而线程具有和进程类似的各种状态，因此，在一些地方线程被称为轻量级进程。线程与进程相似，两者都有一个连续的控制流。在一些地方线程被认为是轻量级的，因为它运行在应用程序的上下文中，并利用为该程序和程序环境分配的资源。

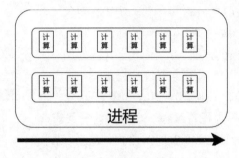

图 13-3　多线程执行图

13.1.2　线程的使用方式

操作系统中提供的线程的各种非托管 API 用来创建和管理线程，而在.NET 中封装了这些非托管线程。在 System.Threading 命名空间中定义了一个名为 Thread 的类，表示托管线程，可以通过对这个类的引用轻松地进行线程的创建与管理工作。

如代码 13-1 所示，Thread 类的构造函数有 4 个，可以通过这 4 个函数创建一个新的线程。

代码13-1

```
public Thread(ParameterizedThreadStart start)
public Thread(ParameterizedThreadStart start, int maxStackSize)
public Thread(ThreadStart start)
public Thread(ThreadStart start, int maxStackSize)
```

如代码 13-2 所示，在控制台实例中，利用多线程可以并发地从网络中下载网页文件。

代码13-2

```
Console.WriteLine("输入一行地址，按 Enter 键进行下载");
while (true)
{
    var url = Console.ReadLine();
    Uri uri = new Uri(url);
    new Thread(() =>
    {
        Request(uri);
    }).Start();
```

```
}
static void Request(Uri uri)
{
    using HttpClient client = new HttpClient();
    HttpResponseMessage response;
    try
    {
        response = client.GetAsync(uri).Result;
        var content = response.Content.ReadAsByteArrayAsync().Result;
        if(content != null)
        {
            using var fileStream = File.Create(
                $"./{DateTime.Now.ToString("yyyyMMdd-HHmmssff")}.html");
            fileStream.Write(content);
            fileStream.Flush();
        }
        Console.WriteLine($"{uri} 下载任务成功");
    }
    catch(Exception ex)
    {
        Console.WriteLine($"{uri} 下载任务失败，错误：{ex}");
    }
}
```

程序启动后，输入多行地址，输出结果如下。

```
输入一行地址，按 Enter 键进行下载
https://www.******.com
https://www.******.com
https://www.******.com
https://www.******.com/ 下载任务成功
https://www.******.com/ 下载任务成功
https://www.******.com/ 下载任务成功
```

由此可知，使用多线程可以并发地执行多个任务，而多个线程之间不会相互干扰。通过多线程同时执行工作，程序能够更快地完成任务。

为了使代码可以在不同的上下文中运行，需要实例化一个 Thread 对象。Thread 类接收一个参数需要设置为 ThreadStart 类型或 ParameterizedThreadStart 类型的委托标识要执行的代码，Thread 类被定义在 System.Threading 命名空间中。实例化后可以调用 thread.Start 方法来

启动该线程。Thread 类提供了 Join 方法，可以通过 Join 方法让子线程开始执行，主线程暂停执行，等待子线程执行完后，主线程再往下执行。调用 Join 方法使一个线程等待另一个线程，当子线程执行完毕，主线程才获取到执行权，开始执行。

代码 13-3 所示，通过 lambda 表达式来简写。

代码13-3

```
static void Main()
{
    Thread t = new Thread(() => Console.WriteLine("Hello!"));
    t.Start();
}
```

13.1.3 线程管理

- **IsAlive**：判断当前线程是否处于活动状态。活动状态是指线程已经启动且尚未终止。如果线程处于正在运行或准备开始运行的状态，则认为线程是"存活"的。
- **IsBackground**：应用程序的主线程及通过 Thread 对象创建的线程默认为前台线程。在进程中只要前台线程不退出，进程就不会终止，因为主线程是前台线程，所以无论后台线程是否退出，都不会影响进程。但是如果前台线程退出，进程就会自动终止。开发人员可以通过 Thread 对象将 IsBackground 属性设置为 true，从而将线程标记为后台线程。后台线程不会阻止进程的终止操作，但是进程的前台线程终止后，该进程也会终止。
- **Priority**：获取或设置线程调度的优先级。Priority 属性为 ThreadPriority 枚举类型（见表 13-1），优先级的默认值为 Normal，优先级较高的线程优先执行，当执行完毕，才会执行优先级较低的线程。如果线程的优先级相同，则采取顺序执行的方式。
- **ThreadState**：获取当前线程的状态，初始值为 Unstarted。ThreadState 是一个枚举类型。ThreadState 属性比 IsAlive 属性更具体一些。
- **Sleep**：暂停当前线程的执行，使当前线程进入休眠状态，直至被重新唤醒。暂停时间通过 Sleep 方法的参数进行设置，单位为毫秒。其实，Sleep 方法还有一些其他用处，就是将线程剩下的时间片送给其他线程，直到休眠结束，休眠的线程将继续执行。

表 13-1　ThreadPriority 类的属性

属性	值	说明
AboveNormal	3	在 Highest 优先级之后，在 Normal 优先级之前

续表

属性	值	说明
BelowNormal	1	在 Normal 优先级之后，在 Lowest 优先级之前
Highest	4	最高优先级，在任何优先级之前
Lowest	0	最低优先级，在任何优先级之后
Normal	2	默认优先级，在 AboveNormal 优先级之后，在 BelowNormal 优先级之前

13.1.4 线程池

线程池（Thread Pool）是一种线程使用的模式。虽然线程是一种比较稀缺的资源，但线程过多也会带来一定的开销，包括创建销毁线程的开销、调度线程的开销等。为了解决资源分配问题，线程池采用了池化（Pooling）思想，基础类库中提供了线程池，通过统一地进行管理和调度，可以合理地分配内部资源，解决资源不足的问题，根据操作系统当前的情况调整线程的数量。如代码 13-4 所示，以控制台项目为例，创建一个 for 循环，同时将一个名为 DoWork 的方法放在线程池中排队，通过该方法打印并输出线程信息。

代码13-4

```
for(int i = 0; i < 20; i++)
{
    ThreadPool.QueueUserWorkItem(state => DoWork());
}

static void DoWork()
{
    Console.WriteLine(
      $"ThreadId: {Thread.CurrentThread.ManagedThreadId}, 
      ThreadPoolThread: {Thread.CurrentThread.IsThreadPoolThread}, 
      Background: {Thread.CurrentThread.IsBackground}");
}
```

注意：托管线程池中的线程是后台线程，也就是说，它的 IsBackground 属性为 true，这意味着，在所有前台线程退出后，ThreadPool 线程也会因此退出。

图 13-4 所示为线程信息的输出结果。

调用 ThreadPool 对象的 QueueUserWorkItem 方法，该方法将一个 WaitCallback 类型的委托作为回调方法，在 WaitCallback 中有一个 object 类型的委托。通过 QueueUserWorkItem 方

法将任务加到队列中，并在 DoWork 方法中打印线程 ID、是否是后台线程，以及该线程是否来自线程池中。通过 ThreadPool.QueueUserWorkItem 方法调用永远都是线程池中的线程，而线程池中的工作线程与 I/O 线程的最大线程数也是可以设置的，如代码 13-5 所示。

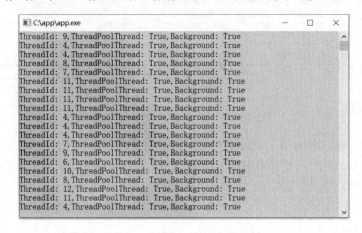

图 13-4　线程信息的输出结果

代码 13-5

```
ThreadPool.GetMaxThreads(out int workerThreads, out int completionPortThreads);
ThreadPool.SetMaxThreads(int workerThreads, int completionPortThreads);
```

表 13-2 所示为 ThreadPool 对象的方法。

表 13-2　ThreadPool 对象的方法

方法	说明
GetAvailableThread	获取由 ThreadPool.GetMaxThreads(out int,out int)方法返回线程数和当前活动线程数之间的差值
GetMaxThreads	获取线程池中最大的线程数和异步 I/O 线程数
GetMinThreads	获取线程池中最小的线程数和异步 I/O 线程数
QueueUserWorkItem	启动线程池中的一个线程（以队列的方式，如果线程池中暂时没有空闲线程，则进入队列排队）
SetMaxThreads	设置线程池中最大的线程数
SetMinThreads	设置线程池最少需要保留的线程数

13.2　基于任务的异步编程

通过学习 13.1 节，读者可以基本了解线程和线程池。创建线程有额外的开销，会占用大量的虚拟内存。使用线程池则可以在需要时进行线程分配，运行结束后，再为后续的异步工作做线程的重用，而不是即用即销毁。

13.2.1 异步任务

从.NET Framework 4.0 开始，TPL（Task Parallel Library，任务并行库）不是每次异步工作时都会创建一个线程，而是创建一个 Task。Task 是 TPL 的抽象级，可以简化异步编程。在 Task 中可以让开发人员忽略一些实现的细节，将注意力集中在业务逻辑编写上，使异步代码的编写变得更加容易。

Thread 是系统级别的线程，而 ThreadPool 和 Task 是通过任务调度器（TaskScheduler）执行的。任务调度器可以采取多种策略，但默认从线程池中请求一个工作线程。线程池会对其进行伸缩判断，也就是说，线程池会选择最优的方式，判断是创建新线程还是采用之前已经结束工作的现有线程。

如代码 13-6 所示，对主线程而言，依然面向过程执行，通过 Task.Run 方法创建 Task 对象后选择继续执行代码。

代码13-6

```
Task task = Task.Run(() =>
{
   Console.WriteLine("任务");
});
Console.WriteLine("主线程执行");
task.Wait();
```

调用 Task.Run 方法在新线程上运行，该方法接收一个委托类型，该委托以 Lambda 表达式的形式将结果输出到控制台中（见图 13-5）。

在调用 Task.Run 方法后，委托方法便开始运行，在调用 Wait 方法后就意味着强制主线程等待任务的完成。

如果 Task 有返回值，那么开发人员可以使用 Task<T>类型执行一个有返回值的任务。如代码 13-7 所示，通过一个控制台实例进行演示。

代码13-7

```
Task<int> task = Task.Run(() =>
{
   return new Random().Next();
});
```

```
Console.WriteLine(task.Result);
```

Task<int>是该任务返回的对象，如图 13-6 所示，获取返回值，通过 Task<int>对象的 Result 属性获取类型为 int 的返回值。在调用 Result 属性后主线程会进行等待，此时线程等待会造成当前线程的堵塞，一直到结果变为可用。

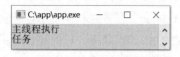

图 13-5　异步任务的输出结果

图 13-6　获取返回值

13.2.2　Task 生命周期

无论是什么技术，掌握其生命周期有助于对该技术的理解，同时能更好地应对错误的排查。

在 Task 对象中有一个 TaskStatus 枚举类型的属性，TaskStatus 表示当前 Task 的状态。TaskStatus 枚举值列表如表 13-3 所示。

表 13-3　TaskStatus 枚举值列表

枚举值	说明
Canceled	表示任务已取消，Task 对象的 IsFaulted 属性被设置为 true
Created	已初始化，但尚未被计划
Faulted	表示因为未处理的异常而完成的任务
RunToCompletion	已成功完成执行的任务
Running	正在执行，但尚未完成
WaitingForActivation	正在等待依赖的任务完成对其进行调度
WaitingForChildrenToComplete	已完成执行，正在隐式等待附加的子任务完成
WaitingToRun	已被计划执行，但尚未开始执行

任务可以通过 Task.Run 方法或 Task.Factory.StartNew 方法创建。如果状态为 WaitingToRun，则意味着任务已经与任务调度程序进行了关联，后续等待轮流运行。

Task 对象的 Status 属性主要用于跟踪其生命周期。图 13-7 所示为任务的生命周期。

任务初始化之后的状态为 Created，在任务启动后，TaskStatus 的状态变为 WaitingToRun，随后，任务调度器 TaskScheduler 实例开始在它指定的线程上执行任务。此时任务的状态为 Running，在任务开始后会产生 3 种可能的结果。如果在执行时正常退出，则任务的状态为 RunToCompletion；如果引发了未处理的异常，则任务的状态为 Faulted；任务也可以以 Canceled 结束，为此，必须将 CancellationToken（取消令牌）对象作为方法参数，在 Task 对象创建任务时传递，取消任务的相关内容请参考 13.2.3 节。如果该令牌在任务开始之前发出了取消请求的信号，则阻止该任务运行。如果该令牌在任务执行后发出了取消请求的信号，则任务的 Status 属性将转换为 Canceled，如果执行时引发了 OperationCanceledException 异常，则 OperationCanceledException 对象的 CancellationToken 属性会返回创建任务时传递的令牌，随后任务进入取消状态。

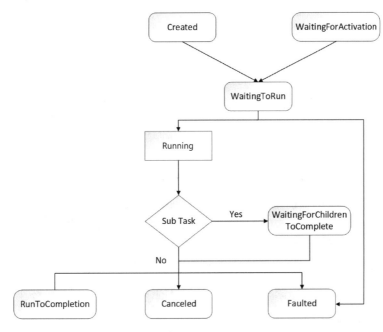

图 13-7　任务的生命周期

13.2.3　取消任务

CancellationToken 比较重要，因为合理应用 CancellationToken 可以提升应用程序的性能。当应用程序启动 Task 后，如果想关闭一项耗时的任务或一组连续的任务，则可以使用 CancellationToken 让应用程序主动在这些场景中完成线程的取消操作。因此，充分利用 CancellationToken 可以完美地避免此处不必要的资源浪费。

下面通过一个控制台实例来演示令牌（Token）的使用方式。如代码 13-8 所示，首先创建一个 CancellationTokenSource 对象，Task 对象可以接收取消令牌参数，所以要将创建的 CancellationTokenSource 对象的 CancellationToken 属性作为参数传递给 Task 对象。

代码13-8

```
var cts = new CancellationTokenSource();
```

```csharp
var task = new Task<int>(() => TaskMethod(10, cts.Token), cts.Token);
Console.WriteLine(task.Status);
cts.Cancel();
Console.WriteLine(task.Status);

static int TaskMethod(int count, CancellationToken token)
{
    for (int i = 0; i < count; i++)
    {
        Thread.Sleep(1000);
        if (token.IsCancellationRequested) return -1;
    }
    return count;
}
```

创建一个名为 TaskMethod 的方法,也可以以委托的形式将方法传递给 Task 构造函数,因为在 Task 类中存在 Action 委托参数的构造函数。将创建的 CancellationTokenSource 对象的 Token 属性以参数的形式传递给 Task 构造函数,并将 Task 对象的 Status 属性先后打印输出到控制台中。图 13-8 所示为输出的 Task 任务的状态,在输出的语句中调用 CancellationTokenSource 对象的 Cancel 方法,将任务关闭,以最后一条输出语句为例输出 Task 对象的状态。

如图 13-8 所示,先后输出 Created 和 Canceled,第一个输出结果为 Created,第二个输出结果为 Canceled。因为调用了 Cancel 方法,所以该任务被关闭,输出结果也是符合预期的。

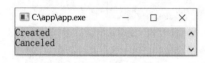

图 13-8 输出的 Task 任务的状态

Cancel 方法不会强制关闭运行时的 Task 任务,所以通过 CancellationToken 参数对象的 IsCancellationRequested 属性进行判断,IsCancellationRequested 属性会返回一个 Boolean 值,该值表示是否调用了 CancellationTokenSource 对象的 Cancel 方法,如代码 13-9 所示。

代码13-9

```csharp
var cts = new CancellationTokenSource();
var task = new Task<int>(() => TaskMethod(10, cts.Token), cts.Token);
Console.WriteLine(task.Status);
task.Start();
for (int i = 0; i < 10; i++)
{
    Thread.Sleep(100);
```

```
    Console.WriteLine(task.Status);
}
cts.Cancel();
for (int i = 0; i < 10; i++)
{
    Thread.Sleep(100);
    Console.WriteLine(task.Status);
}
Console.WriteLine(task.Result);

static int TaskMethod(int count, CancellationToken token)
{
    for (int i = 0; i < count; i++)
    {
        Thread.Sleep(1000);
        if (token.IsCancellationRequested) return -1;
    }
    return count;
}
```

调用 Task 对象的 Start 方法，启动 Task 任务，并以循环的形式打印输出到控制台中，同时在每次打印时等待 100 毫秒，循环结束后调用 CancellationTokenSource 对象的 Cancel 方法，在 TaskMethod 方法中通过 IsCancellationRequested 属性判断是否调用了 Cancel 方法，如果调用了 Cancel 方法，则结束该方法并返回-1。调用 Task 对象的 Result 属性，并打印输出返回值。图 13-9 所示为启动 Task 任务的输出结果。

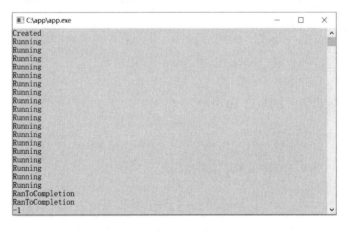

图 13-9　启动 Task 任务的输出结果

13.2.4 使用关键字 await 和 async 获取异步任务的结果

上面介绍了 TPL 异步操作实例的 Task 类和 Task<T>类，前者无返回值，后者返回一个 T 类型的返回值。对于异步模式来说，开发人员还需要掌握 await 关键字和 async 关键字。这两个可以用于异步任务处理的关键字，大大简化了 TPL 的使用，使异步编程更简单、便捷。本节主要介绍 await 关键字和 async 关键字的工作方式。

下面通过控制台实例来演示。如代码 13-10 所示，调用 HttpClient.GetByteArrayAsync 方法返回 Task<byte[]>对象，该实例先调用 DownloadDocsMainPageAsync 方法，再等待下载完成，获取返回内容的大小。

代码13-10

```
Task<int> downloading = DownloadDocsMainPageAsync();
Console.WriteLine($"{nameof(Program)}: Launched downloading.");
int bytesLoaded = await downloading;
Console.WriteLine($"{nameof(Program)}: Downloaded {bytesLoaded} bytes.");

static async Task<int> DownloadDocsMainPageAsync()
{
    Console.WriteLine(
        $"{nameof(DownloadDocsMainPageAsync)}: About to start downloading.");

    var client = new HttpClient();
    byte[] content = await client.GetByteArrayAsync(
                                "https://docs.******.com/en-us/");
    Console.WriteLine(
            $"{nameof(DownloadDocsMainPageAsync)}: Finished downloading.");
    return content.Length;
}
```

异步任务的输出结果如图 13-10 所示。

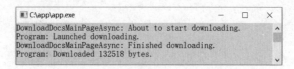

图 13-10　异步任务的输出结果

由此可以看出，主线程调用异步 DownloadDocsMainPageAsync 方法后执行了内部的输出，输出 About to start downloading 信息，输出完成后主线程并没有等待该异步方法完成，而是执行并输出 Main 函数内的 Launched downloading 信息，之后通过 await 关键字等待 downloading 实例，应用程序不会导致线程堵塞，当 Task 任务完成之后继续执行 await 关键字后面的代码，最终打印输出该文件的长度。

13.2.5　处理异步操作中的异常

相信读者对异常处理并不陌生。但是异步和同步中的异常有一定的差异，异常通常会传播到处理的 try/catch 语句中，如果任务是附加的子任务的父级，则会引发多个异常。

下面通过控制台实例来演示。如代码 13-11 所示，通过 TaskMethod 方法抛出异常（This exception is expected!），将该方法以委托参数的形式创建一个 Task 对象。

代码13-11

```
class Program
{
    static void Main(string[] args)
    {
        try
        {
            var task = Task.Run(() => TaskMethod());
            task.Wait();
        }
        catch (AggregateException ae)
        {
            foreach (var e in ae.InnerExceptions)
            {
                Console.WriteLine(e.Message);
            }
        }
    }
    static Task TaskMethod()
    {
        throw new Exception("This exception is expected!");
    }
}
```

上述代码使用 try/catch 语句处理异常。为了使所有的异常都正常地传回调用的线程，笔者通过 AggregateException 对象来捕获。AggregateException 对象通过 InnerExceptions 属性来确定异常源，InnerExceptions 为集合类型，所以以循环的形式输出捕获的异常信息。另外，对于不同的异常类，可以通过判断异常类型检查该异常的原始异常，并选择不同的处理方式，这是可行的。图 13-11 所示为异步操作的异常捕获，并输出到控制台中。

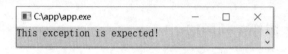

图 13-11 异步操作的异常捕获

13.2.6 上下文延续

在异步编程中，使用 ContinueWith 方法可以使当前 Task 对象在执行完成后，再执行 ContinueWith 方法内部的代码，并且 ContinueWith 方法内部的代码还可以调用当前 Task 对象的结果。

如代码 13-12 所示，创建一个 Task 对象，该对象返回当前时间，获取到当前时间之后，执行 ContinueWith 方法中的委托代码。

代码13-12

```
class Program
{
    public static async Task Main()
    {
        Task<DateTime> taskA = Task.Run(() => DateTime.Now);
        await taskA.ContinueWith(antecedent
                => Console.WriteLine(
                $"Today is {antecedent.Result.DayOfWeek}."));
    }
}
```

在一个任务结束执行后，通过调用 Task.ContinueWith 方法执行委托代码，这样可以做到任务的延续，通过调用 Task 对象的 Result 属性返回一个 DateTime 类型的结果，调用 DateTime 对象的 DayOfWeek 属性指示当天为星期几，输出打印字符串内容，输出结果如图 13-12 所示。

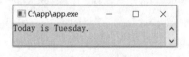

图 13-12 输出结果

13.2.7 TaskScheduler

TaskScheduler 是 TPL 的一部分，负责 Task 任务的调度和管理。不仅如此，并行、数据流的代码等都使用 TaskScheduler.Default。

理解 TaskScheduler 最好的方法是实现一个自定义的 TaskScheduler，并且通过自定义的 TaskScheduler 运行，下面通过实例来演示。

首先需要重写 TaskScheduler，创建一个名为 CustomTaskScheduler 的类，如代码 13-13 所示，要重写 TaskScheduler 就需要对下面这些方法进行重写。

代码13-13

```
public class CustomTaskScheduler : TaskScheduler, IDisposable
{
    protected override IEnumerable<Task>? GetScheduledTasks()
    {
        throw new NotImplementedException();
    }
    protected override void QueueTask(Task task)
    {
        throw new NotImplementedException();
    }
    protected override bool TryExecuteTaskInline(
                         Task task, bool taskWasPreviouslyQueued)
    {
        throw new NotImplementedException();
    }
    public void Dispose()
    {
        throw new NotImplementedException();
    }
}
```

QueueTask 方法返回 void，并将 Task 类型对象作为参数，在调用 TaskScheduler 时调用此方法，GetScheduledTasks 方法返回已调度的所有任务的列表，TryExecuteTaskInline 方法用于以内联方式（即在当前线程上）执行的任务，在这种情况下，无须排队就可以执行任务。如代码 13-14 所示，使用自定义的 TaskScheduler 对象。

代码13-14

```
var scheduler = new CustomTaskScheduler();
```

```
List<Task> tasks = new List<Task>();
Task task1 = new Task(() =>
{
    Write("Running 1 seconds");
    Thread.Sleep(1000);
});
tasks.Add(task1);
Task task2 = new Task(() =>
{
    Write("Running 2 seconds");
    Thread.Sleep(2000);
});
tasks.Add(task2);
foreach (var t in tasks)
{
    t.Start(scheduler);
}
Write("Press any key to quit..");
Console.ReadKey();

static void Write(string msg)
{
Console.WriteLine($"{DateTime.Now:HH:mm:ss} on Thread
 {Thread.CurrentThread.ManagedThreadId} -- {msg}");
}
```

首先创建 CustomTaskScheduler 类，然后创建一个 List<Task> 集合用于保存任务集，最后将任务添加并保存到集合中，以循环的形式调用每个 Task 对象的 Start 方法进行启动。另外，Start 方法接收一个 TaskScheduler 参数，此时传递 CustomTaskScheduler 类。自定义 TaskScheduler 的输出结果如图 13-13 所示。

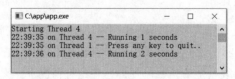

图 13-13　自定义 TaskScheduler 的输出结果

通过调用 Write 方法输出信息，分别打印当前时间、线程 ID 及信息内容，Thread 1 为主线程，Thread 4 由自定义的 CustomTaskScheduler 类创建的线程处理。代码 13-15 展示了

CustomTaskScheduler 类的内部实现。

代码13-15

```csharp
public class CustomTaskScheduler : TaskScheduler, IDisposable
{
    private BlockingCollection<Task> _tasks = new BlockingCollection<Task>();
    private readonly Thread _mainThread;
    public CustomTaskScheduler()
    {
        _mainThread = new Thread(this.Execute);
    }
    private void Execute()
    {
        Console.WriteLine(
            $"Starting Thread {Thread.CurrentThread.ManagedThreadId}");
        foreach (var t in _tasks.GetConsumingEnumerable())
        {
            TryExecuteTask(t);
        }
    }
    protected override IEnumerable<Task>? GetScheduledTasks()
    {
        return _tasks.ToArray<Task>();
    }
    protected override void QueueTask(Task task)
    {
        _tasks.Add(task);
        if (!_mainThread.IsAlive)
        {
            _mainThread.Start();
        }
    }
    protected override bool TryExecuteTaskInline(
                        Task task, bool taskWasPreviouslyQueued)
    {
        return false;
    }
    public void Dispose()
```

```
    {
        _tasks.CompleteAdding();
    }
}
```

如上所示，QueueTask 方法将 Task 对象作为参数，并将其添加到 BlockingCollection 集合中。BlockingCollection 为线程安全集合，拥有阻塞功能。通过 CustomTaskScheduler 构造函数，创建一个线程并执行 Execute 方法，在 Execute 方法中首先对当前线程的 ID 进行打印输出，然后循环 BlockingCollection 集合，并调用 TryExecuteTask 方法，该方法将始终处于堵塞状态，会通过调度程序运行任务，并在任务完成后返回。

13.3 线程并行

现代计算机的 CPU 通常都是多核的，而多线程可以充分利用 CPU 资源。其实，单核 CPU 并不能真正并行执行程序，只是同一个时间内处理器切换得比较快，为每个线程提供的 CPU 时间片非常短。为了发挥多核 CPU 的优势，可以利用并行执行。本节主要介绍并行开发编程。

当任务并行时会将应用程序分割成一组任务，使用不同的线程来运行这些任务。通过 Parallel 库，开发人员无须创建线程，无须主动加锁，也无须主动分割工作的细节。

下面通过控制台实例来演示并行的使用方式。创建 Parallel.ForEach 方法，执行循环操作，如代码 13-16 所示。

代码13-16

```
var cts = new CancellationTokenSource();
Parallel.ForEach(Enumerable.Range(1, 10), new ParallelOptions
{
    CancellationToken = cts.Token,
    MaxDegreeOfParallelism = Environment.ProcessorCount,
    TaskScheduler = TaskScheduler.Default
}, (i, state) =>
{
    Process(i);
});
Console.WriteLine("Done");

static void Process(int i)
{
```

```
Console.WriteLine(
    $"Id: {i}, ThreadId: {Thread.CurrentThread. ManagedThreadId}");
}
```

ForEach 方法的第一个参数接收一个集合,通过 Enumerable.Range 方法进行模拟,ParallelOptions 对象允许开发人员对并行循环进行一些配置,使用 CancellationToken 属性可以取消循环,使用 MaxDegreeOfParallelism 属性可以限制最大并行数,使用 TaskScheduler 属性允许通过自定义 TaskScheduler 类来调度任务,Action 可以接收一个附加的 ParallelLoopState 参数,可以用于循环中跳出或检查当前循环的状态。

ParallelLoopState 参数有两种方法可以停止并行循环,分别为 Break 方法和 Stop 方法。Stop 方法会通知循环停止处理任何工作,并设置 IsStopped 属性为 true;Break 方法用于阻止之后的迭代,但是并不影响已经开始的迭代。并行循环的输出结果如图 13-14 所示。

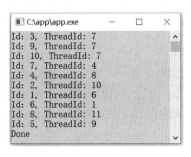

图 13-14　并行循环的输出结果

13.4　小结

本章主要介绍线程、异步、并行的相关内容。掌握这些知识点,有助于读者合理地利用资源。为了使读者更好地了解 TaskScheduler,笔者创建了一个自定义的 TaskScheduler,但是建议不要使用自定义的调度器,因为默认的调度器通常是最合适的。在一般情况下,线程并不是越多越好,要有一个度,合理利用线程能让应用程序的性能最大化。并行编程也会带来相应的开销,所以需要对其进行合理应用。

第 14 章

线程同步机制和锁

在使用多线程进行任务处理的过程中,当多个线程同时操作同一份共享资源时,可能会发生冲突,出现资源争抢的现象,这就会引发线程同步问题,在.NET 中可以引入多种方式来解决这个问题,包括原子操作和各种锁(如自旋锁、混合锁、互斥锁、信号量、读写锁)等。这些机制相当于为线程同步制定了一系列用于维护"秩序"的规则,线程要遵守这个规则来对共享资源进行操作。

14.1 原子操作

原子操作(Atomic Operation)是指不能被分割的操作,不会因为任务的调度等原因被打断(要么执行完,要么不执行,不会被其他线程打断)而影响执行结果。在.NET 中提供了 Interlocked 类,它的原子操作基于 CPU 本身,并非堵塞行为。

14.1.1 无锁编程

如代码 14-1 所示,以控制台实例为例,先创建一个名为 value 的 int 类型的静态变量,再通过 Parallel 类执行并行操作,并创建 10000 次循环任务,每次循环操作通过 Parallel 类的委托方法执行 value + 1 将静态变量 value 进行累加。因为循环是并行操作,所以避免不了多个线程同时并行操作。而静态变量在内存中只有一份,只占用一个内存区域,也就是采用的共享内存,所以,多个线程同时共享一个内存区域,在不对内存进行保护的情况下操作,就需要注意内存安全。

代码14-1

```
class Program
{
    static int value = 0;
    static void Main(string[] args)
    {
        Parallel.For(0, 10000,
            _ =>
            {
                value += 1;
            });
        Console.WriteLine(value);
    }
}
```

运行上面的代码，可以发现由于使用多线程并行操作，因此总会在共享内存操作时出现一些问题，导致定义的 value 值有误，以至于在运行后会出现 value 值小于 10000 的情况。

如代码 14-2 所示，可以通过 Interlocked 类对数据进行操作。

代码14-2

```
class Program
{
    static int value = 0;
    static void Main(string[] args)
    {
        Parallel.For(0, 10000, _ =>
        {
            Interlocked.Add(ref value, 1);
        });
        Console.WriteLine(value);
    }
}
```

运行后可以发现，结果始终为预期的 10000，Interlocked 类的每个方法都执行一个原子级别的读取/写入操作。

表 14-1 所示为 Interlocked 类的方法。使用.NET 提供的 Interlocked 类有助于对数据进行原子操作，看起来似乎跟 lock 一样，但它并不是 lock，它的原子操作是基于 CPU 本身的，非堵塞的，所以比 lock 的效率高。

表 14-1　Interlocked 类的方法

方法	说明
Add	使计数器增加指定的值
CompareExchange	先把计数器与某个值进行比较，如果相等，则把计数器设定为指定的值
Decrement	使计数器减少 1
Exchange	将计数器设定为指定的值
Increment	使计数器增加 1
Read	读取计数器的值

14.1.2　无锁算法

要实现一个原子的功能，可以通过无锁算法（Lock Free Algorithm）来实现。CAS（Compare And Swap，比较并替换）就是一种无锁算法。在 .NET 中可以通过 Interlocked 类实现 CAS 算法。

代码 14-3 展示了使用 Interlocked.CompareExchange 的无锁算法。

代码 14-3

```
class Program
{
    static void Main()
    {
        var location = 1;
        var value = 3;
        var compared = 1;
        Interlocked.CompareExchange(ref location, value, compared);
        Console.WriteLine(location);
    }
}
```

运行后可以发现，当前控制台中输出的是 3。需要特别说明的是，当原始值 location 与比较值 compared 相等时，当前的 value 值会被替换为原始值。

如代码 14-4 所示，可以看到使用了 Interlocked.Exchange 方法，从本质上来说 Exchange 方法是一个赋值操作，传递给它的参数为要更改的值。在 CompareExchange 方法中，如果该值等于第二个参数，则将其更改为第三个参数，long 类型的返回值为第二个参数的值。

代码 14-4

```
class Program
{
    static long _value;
```

```
static void Main()
{
    Thread thread1 = new Thread(new ThreadStart(DoWork));
    thread1.Start();
    thread1.Join();
    Console.WriteLine(Interlocked.Read(ref _value));
}

static void DoWork()
{
    Interlocked.Exchange(ref _value, 10);
    long result = Interlocked.CompareExchange(ref _value, 30, 10);
    Console.WriteLine(result);
}
```

14.2 自旋锁

在多线程编程中,当自旋锁(SpinLock)在一个线程中获取到锁(也就是锁被占用)时,在当前线程中就无法获取锁,该线程会处于等待状态。自旋锁具备一定的间隔时间,继续获取锁。

14.2.1 SpinWait 实现自旋锁

自旋等待(SpinWait)是一个轻量级的线程等待类型。如代码 14-5 所示,以控制台实例为例,定义一个名为 DoWork 的方法,在 Main 方法内创建一个 Thread 对象,用该线程执行 DoWork 方法,直到 50 毫秒之后主线程将静态变量_isCompleted 设置为 true,DoWork 循环被停止。通过 SpinWait 实例对象调用 SpinOnce 方法执行单一的自旋,同时输出一条控制台消息,判断 spinWait.SpinOnce 方法的下一次调用是否触发上下文切换和内核转换主要使用 NextSpinWillYield 属性。

代码14-5

```
class Program
{
    static void Main(string[] args)
    {
```

```csharp
    var thread = new Thread(DoWork);
    thread.Start();
    Thread.Sleep(50);
    isCompleted = true;
    Console.ReadKey();
}

private static volatile bool _isCompleted = false;

static void DoWork()
{
    SpinWait spinWait = new SpinWait();
    while (!_isCompleted)
    {
        spinWait.SpinOnce();
        Console.WriteLine($"自旋次数：{spinWait.Count}，下一次调用触发上
                    下文切换和内核转换：{spinWait.NextSpinWillYield}");
    }
    Console.WriteLine("Waiting is complete");
}
```

当使用 volatile 关键字来声明静态变量 _isCompleted 时，读取这个变量的值的时候每次都是从内存中读取而不是从缓存中读取，这样可以保证变量的信息总是最新的，变量可以同时执行多个线程的修改。

使用 volatile 关键字来声明，变量不会被编译器和处理器优化，如图 14-1 所示，将结果输出到控制台中。

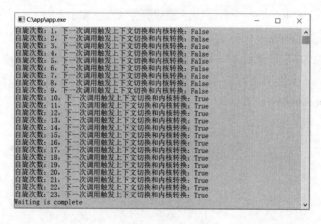

图 14-1　输出结果

代码 14-6 展示了 SpinWait 内部的实现,读者需要知道的是 SpinWait 是一个结构体（struct）类型。

代码14-6

```
public struct SpinWait
{
    public int Count;
    private int _count;
    public bool NextSpinWillYield =>
                this._count >= 10 || Environment.IsSingleProcessor;

    public void SpinOnce();
    public void SpinOnce(int sleep1Threshold);

    private void SpinOnceCore(int sleep1Threshold)
    {
        if (this._count >= 10 &&
            (this._count >= sleep1Threshold &&
              sleep1Threshold >= 0 || (this._count - 10) % 2 == 0) ||
                Environment.IsSingleProcessor)
        {
            if (this._count >= sleep1Threshold && sleep1Threshold >= 0)
                Thread.Sleep(1);
            else if ((this._count >= 10
                        ? (this._count - 10) / 2 : this._count) % 5 ==4)
                Thread.Sleep(0);
            else
                Thread.Yield();
        }
        else
        {
            int iterations = Thread.OptimalMaxSpinWaitsPerSpinIteration;
            if (this._count <= 30 && 1 << this._count < iterations)
                iterations = 1 << this._count;
            Thread.SpinWait(iterations);
        }
        this._count = this._count == int.MaxValue ? 10 : this._count + 1;
    }
```

```
public void Reset() => this._count = 0;

public static void SpinUntil(Func<bool> condition) =>
                            SpinWait.SpinUntil(condition, -1);
public static bool SpinUntil(Func<bool> condition, TimeSpan timeout);
public static bool SpinUntil(Func<bool> condition,
                                          int millisecondsTimeout);
}
```

需要注意 SpinOnceCore 方法，SpinWait 方法的内部调用了 Thread.SpinWait 方法，可以看到，当 NextSpinWillYield 属性在 _count 私有属性大于或等于 10，或者运行在单核计算机上时，NextSpinWillYield 属性会返回 true，也就是说，它总是进行上下文切换。

SpinUntil 方法提供了 Func<bool> 委托参数，也就是说，可以通过该参数确定一个条件，满足该条件后才会进入自旋锁。

14.2.2 SpinLock 实现自旋锁

自旋锁（SpinLock）是一个轻量级的线程锁，基于原子操作实现。如代码 14-7 所示，以控制台实例来实现，先创建一个 SpinLock 实例对象，再通过 Parallel 类调用 For 方法进行并行循环，因为并行循环会出现非预期值 10000000 的情况，所以通过 SpinLock 来实现。

代码14-7

```
class Program
{
    static void Main()
    {
        var spinLock = new SpinLock();
        var list = new List<int>();
        Parallel.For(0, 10000000, r =>
        {
            bool lockTaken = false;          //释放成功
            try
            {
                spinLock.Enter(ref lockTaken);   //进入锁
                list.Add(r);
            }
```

```
        finally
        {
            if (lockTaken) spinLock.Exit(false);  //释放
        }
    });
    Console.WriteLine(list.Count);
    //输出 10000000
}
```

代码 14-8 展示了 SpinLock 的实现。SpinLock 不可重入，在线程进入锁之后，必须先正确退出锁才能重入锁。通过任何的重入锁都会导致死锁，如果在调用 Exit 前没有调用 Enter，那么 SpinLock 的状态可能会被破坏。

代码14-8

```
public struct SpinLock
{
    public void Enter(ref bool lockTaken);

    public void TryEnter(ref bool lockTaken);
    public void TryEnter(TimeSpan timeout, ref bool lockTaken);
    public void TryEnter(int millisecondsTimeout, ref bool lockTaken);

    public void Exit();
    public void Exit(bool useMemoryBarrier);

    public bool IsHeld;

    public bool IsHeldByCurrentThread;
    public bool IsThreadOwnerTrackingEnabled;
}
```

14.3　混合锁

在 .NET 中提供了一个混合锁 Monitor，混合锁的性能还是比较高的。使用 Monitor 锁定对象（引用类型），可以保证共享资源的线程互斥执行。在 Monitor 类中提供了 Enter 方法和 Exit 方法，分别用于获取锁和释放锁。Monitor 类会利用 try{}finally{}语法糖执行锁的释放操作。

代码 14-9 展示了 Monitor 类的使用。Monitor 作为一个对象的同步锁，可以获取一个锁，并允许一个线程可以访问该代码段的内容。Monitor 类提供了一些有用的方法用于开发人员对锁的处理，还可以对引用对象执行锁操作，但不要对值类型进行锁操作，因为这会不可避免地使值类型装箱，导致装箱成新对象，无法做到线程同步。

代码14-9

```csharp
class Program
{
    private static readonly object _object = new object();

    public static void Print()
    {
        bool _lock = false;
        Monitor.Enter(_object, ref _lock);
        try
        {
            for (int i = 0; i < 5; i++)
            {
                Thread.Sleep(100);
                Console.Write(i + ",");
            }
            Console.WriteLine();
        }
        finally
        {
            if (_lock)
            {
                Monitor.Exit(_object);
            }
        }
    }

    static void Main()
    {
        Thread[] threads = new Thread[3];
```

```
        for (int i = 0; i < 3; i++)
        {
            threads[i] = new Thread(Print);
        }
        foreach (var t in threads)
        {
            t.Start();
        }
        Console.ReadLine();
    }
}
```

通常来说，先通过 Monitor.Enter 获取锁，再通过 Monitor.Exit 释放锁。为了避免这个过程中会出现异常，导致锁无法释放，可以使用 try{}finally{}语句块来释放锁，如图 14-2 所示，运行程序并将结果输出到控制台中。

图 14-2　输出结果

另外，Monitor 类还提供了其他方法：Wait 方法用于释放对象锁并阻止当前线程，直到它重新获取锁；Pulse 方法用于通知等待队列中的线程锁定对象状态的更改；PulseAll 方法用于通知所有的等待线程对象状态的更改；IsEntered 方法用于确定当前线程是否保留指定的对象锁。

除此之外，还可以将 Print 方法中的 Monitor 锁对象替换为 lock 关键字，如代码 14-10 所示。

代码14-10

```
public static void Print()
{
    lock (_object)
    {
        for (int i = 0; i < 5; i++)
        {
            Thread.Sleep(100);
            Console.Write(i + ",");
        }

        Console.WriteLine();
```

lock 关键字实际上是一个语法糖，对 Monitor 锁对象做了一个封装。lock 语法糖背后的语法大概如代码 14-11 所示。

代码14-11

```
try
{
    Monitor.Enter(obj);
}
finally
{
    Monitor.Exit(obj);
}
```

14.4 互斥锁

在获取共享资源时，获取到一个线程后就会对资源加锁，使用完成后会对其解锁，在使用过程中，其他线程如果想访问该资源就会进入等待状态。

Mutex 也像锁一样工作。通过 Mutex 对象，当多个线程同时访问一个资源时保证一次只能有一个线程访问。另外，在并发访问时，使用 Mutex 对象可以获取共享资源的排它锁。Mutex 对象是系统级别的，所以可以跨进程工作。

如代码 14-12 所示，笔者创建了一个控制台实例，通过创建多个线程来观察 Mutex 对象，并在 Print 方法中通过 mutex.WaitOne 方法堵塞当前线程，等待资源被释放，在 finally 中通过调用 ReleaseMutex 方法解决堵塞来释放锁。

代码14-12

```
class Program
{
    private static Mutex mutex = new Mutex(false);
    static void Main()
    {
        for(int i = 0; i < 5; i++)
        {
            Thread thread = new Thread(Print);
            thread.Name = "Thread:" + i;
            thread.Start();
```

```csharp
    }
    Console.ReadKey();
}

static void Print()
{
    Console.WriteLine($"{Thread.CurrentThread.Name} 即将开始进行等待处理");
    try
    {
        //阻塞当前线程，直到 WaitOne 方法接收到信号
        mutex.WaitOne();
        Console.WriteLine($"开始: {Thread.CurrentThread.Name} 处理中...");
        Thread.Sleep(2000);
        Console.WriteLine($"完成: {Thread.CurrentThread.Name}");
    }
    finally
    {
        mutex.ReleaseMutex();
    }
}
```

运行程序，将结果输出到控制台中，如图 14-3 所示。

图 14-3 输出结果

14.5 信号量

信号量类似于互斥锁，但是它允许多个线程同时访问一个资源。信号量具有一个计数器，利用计数器来控制是等待还是访问，如计数器的值为 5，如果一个线程调用了信号量则计数器减 1，直到这个计数器变为 0，这时不允许其他的线程再访问，使它们进入等待状态。如

果线程调用 Release 方法则释放信号量资源，计数器加 1，此时可以进来一个线程进行访问。

14.5.1 Semaphore

信号量主要用于限制共享资源并发访问的线程数，可以通过 Semaphore 允许一个或多个线程进入临界区，在线程安全的情况下并发执行任务。因此，可以在资源数量有限的情况下限制线程数量，如代码 14-13 所示。

代码14-13

```
class Program
{
    private static Semaphore _semaphore = new Semaphore(1, 2);
    static void Main()
    {
        for (int i = 0; i < 3; i++)
        {
            Thread thread = new Thread(Print);
            thread.Name = "Thread:" + i;
            thread.Start();
        }
    }

    private static void Print()
    {
        Console.WriteLine($"{Thread.CurrentThread.Name} 即将开始进行等待处理");
        try
        {
            _semaphore.WaitOne();
            Console.WriteLine($"开始: {Thread.CurrentThread.Name} 处理中...");
            Thread.Sleep(5000);
            Console.WriteLine($"完成: {Thread.CurrentThread.Name}");
        }
        finally
        {
            _semaphore.Release();
            Console.WriteLine("Semaphore Release");
        }
```

    }
}
```

如上所示，实例化一个 Semaphore 对象，并通过构造函数传递两个参数值，这两个值分别代表 InitialCount（初始请求数）和 MaximumCount（最大请求数）。

WaitOne 方法表示阻止当前线程，直到当前线程收到信号；当线程退出临界区时需要调用 Release 方法。

上述代码的执行结果如下。

```
Thread:1 即将开始进行等待处理
Thread:0 即将开始进行等待处理
Thread:2 即将开始进行等待处理
开始: Thread:1 处理中...
完成: Thread:1
Semaphore Release
开始: Thread:0 处理中...
完成: Thread:0
Semaphore Release
开始: Thread:2 处理中...
完成: Thread:2
Semaphore Release
```

如代码 14-14 所示，通过 Semaphore 对象指定系统信号量名称，因此，可以通过信号量在多个进程间使用同步和互斥。

代码14-14

```
class Program
{
    private static Semaphore _semaphore = new Semaphore(5, 5, "Semaphore");
    static void Main()
    {
        for (int i = 0; i < 3; i++)
        {
            Thread thread = new Thread(Print);
            thread.Name = "Thread:" + i;
            thread.Start();
        }
    }
```
```

```csharp
private static void Print()
{
    Console.WriteLine($"{Thread.CurrentThread.Name} 即将开始进行等待处理");
    try
    {
        _semaphore.WaitOne();
        Console.WriteLine($"开始: {Thread.CurrentThread.Name} 处理中...");
    }
    finally
    {
        Console.ReadLine();
        _semaphore.Release();
    }
}
```

使用 Semaphore 对象可以做到进程间的共享。接下来启动两个应用程序进程，如图 14-4 所示，为了验证跨进程信号量，将结果输出到控制台中，可以清晰地发现线程被执行了 5 次，同样对应 MaximumCount（最大请求数）限制为 5。

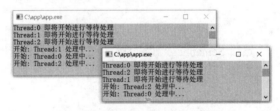

图 14-4 跨进程信号量

14.5.2 SemaphoreSlim

.NET 中还提供了轻量级的信号量（SemaphoreSlim），该类是受托管的。如代码 14-15 所示，SemaphoreSlim 的使用方式和 Semaphore 的使用方式基本类似，并且也可以限制同时访问一个资源的线程数。

代码14-15

```csharp
class Program
{
    private static SemaphoreSlim _semaphore = new SemaphoreSlim(1, 2);
    static void Main()
```

```csharp
{
    for (int i = 0; i < 3; i++)
    {
        Thread thread = new Thread(Print);
        thread.Name = "Thread:" + i;
        thread.Start();
    }
}

private static void Print()
{
    Console.WriteLine($"{Thread.CurrentThread.Name} 即将开始进行等待处理");
    try
    {
        _semaphore.Wait();
        Console.WriteLine($"开始: {Thread.CurrentThread.Name} 处理中...");
        Thread.Sleep(5000);
        Console.WriteLine(Thread.CurrentThread.Name + " 退出.");
    }
    finally
    {
        _semaphore.Release();
        Console.WriteLine("Semaphore Release");
    }
}
```

实例化一个 SemaphoreSlim 对象，并通过构造函数传递两个参数值，限制最大的线程数为 3。

上述代码的执行结果如下。

```
Thread:0 即将开始进行等待处理
Thread:1 即将开始进行等待处理
Thread:2 即将开始进行等待处理
开始: Thread:0 处理中...
Thread:0 退出.
Semaphore Release
开始: Thread:1 处理中...
Thread:1 退出.
```

```
Semaphore Release
开始：Thread:2 处理中...
Thread:2 退出.
Semaphore Release
```

14.6　读写锁

　　读写锁（ReaderWriterLock）允许多个线程同时获取锁，但同一时间只允许一个线程获得写锁，因此也被称为共享独占锁。

　　ReaderWriterLockSlim 作为一个读写锁，代表一个管理资源访问的锁。简单来说，它允许多个线程同时读取锁，但只允许一个线程独占写锁。

　　如代码 14-16 所示，通过一个控制台实例来演示。首先创建 Reader 方法和 Writer 方法，一个用于通过锁读取，一个用于通过锁写入，执行等待模拟写入或读取操作，然后利用 try/finally 代码块来确保锁的释放工作。

代码14-16

```csharp
private static ReaderWriterLockSlim readerWriterLock =
                                new ReaderWriterLockSlim();
private static List<int> items = new List<int>();
private static void Reader()
{
    Console.WriteLine("读取内容");
    while(true)
    {
        try
        {
            readerWriterLock.EnterReadLock();
        }
        finally
        {
            readerWriterLock.ExitReadLock();
        }
    }
}
```

```csharp
private static void Writer()
{
    while(true)
    {
        try
        {
            int number = new Random().Next(20);
            readerWriterLock.EnterWriteLock();
            items.Add(number);
            Console.WriteLine(
                $"Thread: {Thread.CurrentThread.ManagedThreadId} added {number}");
        }
        finally
        {
            readerWriterLock.ExitWriteLock();
        }
    }
}
```

代码 14-17 展示了 Reader 方法和 Writer 方法的调用。

代码14-17

```csharp
static void Main(string[] args)
{
    Thread[] threads = new Thread[3];
    for(int i = 0; i < 3; i++)
    {
        threads[i] = new Thread(Reader);
    }
    foreach(var t in threads)
    {
        t.Start();
    }

    var wThread1 = new Thread(Writer);
    wThread1.Start();
    var wThread2 = new Thread(Writer);
    wThread2.Start();
```

```
Console.ReadKey();
}
```

在 Main 方法中，创建了 3 个线程用于执行读取和写入操作，利用 ReaderWriterLockSlim 类来实现线程的安全操作，在运行过程中，一旦得到写锁，就会阻止数据的读取。也就是说，获取到写锁之后线程就会进入堵塞状态，为了更小化堵塞浪费的时间，可以利用 EnterReadLock 方法和 ExitReadLock 方法进入读模式获取读取锁和释放读取锁，利用 EnterWriteLock 方法和 ExitWriteLock 方法进入写模式锁定和释放锁状态。

14 7　小结

本章介绍了.NET 中多线程间的同步机制，如锁，在多线程编程中对共享资源的"保护"，可以根据具体的应用场景采取不同的锁策略。

第 15 章

内存管理

本章将从内存管理的两个重要部分展开介绍，即内存分配和垃圾回收器，读者可以由此了解与内存分配和垃圾回收相关的基本知识，以及.NET 应用程序对内存的申请、分配和回收。15.4 节会对 GC（Garbage Collection，垃圾回收器）参数调优展开介绍，以方便读者对内存管理有更深入的理解。

15.1 内存分配

通常来说，在应用程序中会将内存分成堆区和栈区，应用程序运行期间可以主动从堆区申请内存空间。这些内存由内存分配器负责分配，由垃圾回收器负责回收。

在 C#中，new 关键字用于分配对象，对象可以是值类型或引用类型。值类型和引用类型的内部分配机制是不同的。

15.1.1 栈空间和堆空间

当声明一个对象后，这个对象的地址会存放在栈上。栈内存是非托管的，不需要垃圾回收器回收，也不需要开发人员管理，当实例化后，会在堆内存中为它分配空间。

图 15-1 所示为 Stack 栈空间和 Heap 堆空间。例如，通过 ClassA a = new ClassA()创建一个对象后，这个对象在堆中，而变量 a 则指向实例，也就是说代表的是指向堆上的引用，所以变量 a 保存在栈空间中。将变量 a 的引用赋值给其他的变量时，其他的变量会保存引用地

址，对变量 a 指向的实例进行任何修改，都会影响其他的变量指向的实例，因为它们都指向同一个实例，这就是共享堆空间。

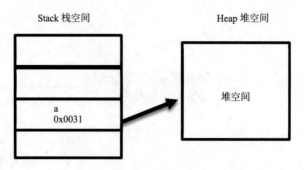

图 15-1　Stack 栈空间和 Heap 堆空间

正如上面提到的，执行 Main 方法，在栈内存中开辟一个新的空间用于存放变量 a，而 a 是局部变量，同时在堆内存中开辟一个空间用于存放 new 的对象。例如，地址 0x0031，对象在堆内存中的地址值会赋给变量 a，这样变量 a 也有地址值，所以变量 a 就指向了这个对象。

15.1.2　值类型和引用类型

如图 15-2 所示，在.NET 中，类型可以分为两大类，分别是值类型（Value Type）和引用类型（Reference Type），值类型和引用类型都有各自的特点。值类型和引用类型的相同点是最终都继承自 System.Object 类，简单来说，继承 System.ValueType 的是值类型，但是 System.ValueType 继承自 System.Object 类，所以它们最终继承自 System.Object 类，如果没有间接类型，则直接继承自 System.Object 类的为引用类型。

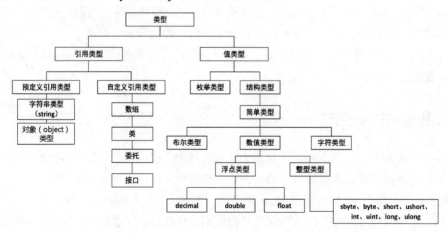

图 15-2　值类型和引用类型

第 15 章 内存管理

值类型和引用类型的内存分配方式有一个比较大的差异,值类型存储在栈上,引用类型则被分配在堆上。

值类型的变量直接存储数据。另外,值类型主要包括基本类型,如整型类型、浮点类型、布尔类型和字符类型这四种。值类型在栈中分配,因此,值类型的效率很高;引用类型的变量保存的是数据的引用,数据存储在堆中,C#中预定义了 object 和 string 等引用类型。如代码 15-1 所示,通过一个控制台实例来演示对象分配。

代码15-1

```
class Program
{
    static void Main()
    {
        ClassA a = new ClassA();
        a.Name = "Test";
        a.Age = 18;
        ClassA a1 = a;
    }

    public class ClassA
    {
        public string Name { get; set; }   //定义引用类型
        public int Age { get; set; }       //定义值类型
    }
}
```

在 ClassA 类中,Age 字段为值类型,但是 class 是一个引用类型,所以 ClassA 作为引用类型实例的一部分,也被分配到托管堆中。图 15-3 所示为引用类型栈和堆。

如代码 15-2 所示,每个变量都有其栈地址,并且不同变量的栈地址也不同,值存储在栈中(见图 15-4)。

代码15-2

```
class Program
{
    static void Main()
    {
        int a = 5;
        int a1 = a;
    }
}
```

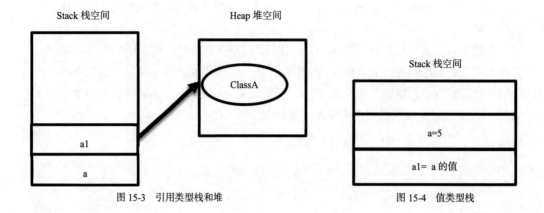

图 15-3 引用类型栈和堆　　　　图 15-4 值类型栈

15.1.3 创建一个新对象

在 C#中，new 关键字用于分配对象，而对象的类型可以是值类型、引用类型，类型的分配的方式也是不同的。如代码 15-3 所示，创建一个引用类型对象和值类型对象。

代码15-3

```
var obj = new SomeClass();
var obj1=new SomeStruct();
```

最终编译器生成 IL 代码，如代码 15-4 所示，而 new 关键字转换为 newobj 指令。需要注意的是，newobj 指令用于分配和初始化类型，而 initobj 指令用于初始化值类型。

代码15-4

```
newobj instance void SomeClass::.ctor()
initobj SomeStruct
```

newobj 指令告知 CLR 执行如下操作。
- 计算该类型要分配的内存大小。
- 检查托管堆上是否有足够的空间，如果有，则调用这个类型的构造方法（.ctor），构造方法会返回一个指向内存的新对象的引用地址，这个地址也是下一个对象指针上一次所指向的位置。

将引用地址返回给调用者之前，让下一个对象指针指向托管堆中下一个可用的位置。

图 15-5 所示为对象压栈。

如果要继续分配新的对象，则放在 NextObjPtr 指针的位置，并且 NextObjPtr 指针向后移动，准备继续接受新对象的分配。

为了使读者对内存分配机制有一个更好的认识,笔者通过以下几种分配器模式展开介绍。如图 15-6 所示,线性分配器（Sequential Allocator）也被称为撞针分配（Bump-the-Pointer）,是一种简单且高效的分配方法。在使用线性分配器时,只需要使用一个指针来记录可用空间的起始地址即可,在向分配器申请内存时,分配器先检查可用空间是否足够分配本次的请求。如果足够,则分配空间,并且将指针的值修改为新对象的地址,分配器会将指针按照相应的字节进行移动;如果不够,则进行垃圾回收。

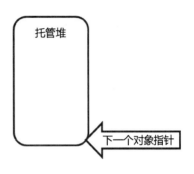

图 15-5　对象压栈

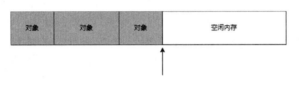

图 15-6　线性分配器

图 15-7 所示为线性分配器回收内存。线性分配器虽然高效,但是也存在问题,如无法在内存被释放时复用内存。如果分配的内存已经被清除回收,则会导致内存碎片化。显然,这种方式会消耗很多内存,但是如果配合垃圾回收器算法来使用,那么这又是一种比较智能的方式,垃圾回收器会定期整理这些空闲空间,这样就能再次进行利用。

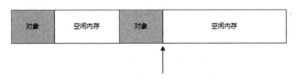

图 15-7　线性分配器回收内存

如图 15-8 所示,空闲链表分配器（Free-List Allocator）可以复用已经释放的内存,可以维护一个类似于链表的数据结构,用于记录那些被释放的内存地址。在分配对象时优先从空闲列表中检查是否有足够的内存空间,如果有则分配内存空间,随后修改链表。

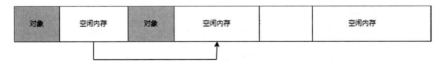

图 15-8　空闲链表分配器

15.1.4　小对象堆分配

小对象（浅表大小小于 85 000 字节）分配在小对象堆（Small Object Heap，SOH）上。为了高效地管理小对象堆，CLR 基于生存期策略将小对象堆分为 3 代，分别为第 0 代、第 1 代和第 2 代，对象根据它们的年龄向上移动这些代。新对象放在第 0 代中，当触发.NET 垃圾回收器后，如果对象幸存下来，就会被移至第 1 代中，如果在下一次触发垃圾回收器后还能幸存下来，则它们将从第 1 代移至第 2 代中。

在第 2 代中，会发生完整 GC（Full GC）运行，清除不需要的第 2 代对象，将第 1 代对象移至第 2 代中，并将第 0 代对象移至第 1 代中，同时清除任何未引用的对象，每次垃圾回收器运行后，受影响的堆都会被压缩，这也就意味着当收集未使用的对象时，垃圾回收器将活动的对象移至间隙中以消除碎片并确保可用的内存是连续的。当然，压缩会涉及一定的开销，但压缩的好处大于开销的成本，所以压缩是在小对象堆上自动执行的。

15.1.5　大对象堆分配

大对象（浅表大小大于或等于 85 000 字节）分配在大对象堆（Large Object Heap，LOH）上。大对象始终分配在第 2 代中，大对象属于第 2 代，因为只有在第 2 代回收期间才能回收它们，并且回收它前面的所有代。例如，执行第 1 代垃圾回收器时，将同时回收第 0 代。当执行第 2 代垃圾回收器时，将回收整个堆。因此，第 2 代垃圾回收器还被称为完整 GC，用于清理整个堆，包括年轻代空间和老年代空间。由于复制大块内存涉及一定的开销，并且压缩、移动的成本也是比较高的，因此它们没有被压缩。

在分配对象时，会检查空闲列表中是否有足够大的空间来容纳该对象。如果存在一块足够大的空闲空间，那么对象就在那里分配；如果不存在，那么对象就在下一个空闲空间中分配。

因为对象不太可能与空闲空间进行精准匹配，所以几乎总是在对象之间留下小块内存，导致内存碎片化。

此外，在分配大对象时倾向于将对象添加到末尾，而不是运行的第 2 代垃圾回收器，这样对性能有好处，但也是产生内存碎片的重要原因。

因此，尽量避免采用 LOH 分配，LOH 分配的开销远远大于 SOH 分配的开销。

如代码 15-5 所示，.NET Core 和.NET Framework（自.NET Framework 4.5.1 开始）提供了 GCSettings.LargeObjectHeapCompactionMode 属性，可以按需压缩 LOH。使用 GCSettings.LargeObjectHeapCompactionMode 属性可以让用户指定并且在下一次完整 GC 期间压缩 LOH。

代码15-5

```
GCSettings.LargeObjectHeapCompactionMode=GCLargeObjectHeapCompactionMode
.CompactOnce;
//FULL GC
GC.Collect();
```

通过设置 GCSettings.LargeObjectHeapCompactionMode 属性的值，在下一次完整 GC 期间时压缩 LOH，设置该属性并执行只会起到一次的效果，如果将该设置变成持久化的开关则会造成不必要的压缩，并且会对性能造成一些影响。GCSettings.LargeObjectHeapCompactionMode 属性的值会重置为 GCLargeObjectHeapCompactionMode.Default。

如代码 15-6 所示，在 .NET Core 3.0 及更高版本中，LOH 会自动压缩，可以在 runtimeconfig.json 文件内指定 LOH 中的阈值大小（以字节为单位），阈值必须大于默认值 85 000 字节。

代码15-6

```
{
    "runtimeOptions":{
        "configProperties":{
            "System.GC.LOHThreshold":120000
        }
    }
}
```

15.1.6　固定对象堆分配

从 .NET 5.0 开始，.NET 垃圾回收器增加了一个新的特性，即固定对象堆（Pinned Object Heap，POH），而在此之前会利用 fixed 关键字来做到短时间的固定。这是一种将特定的局部变量标记为固定的方法，只要不进行垃圾回收，就不会带来额外的开销。如果长时间固定对象，在这种情况下，垃圾回收器会将固定对象移至第 2 代中，由于第 2 代发生垃圾回收的频率较低，因此固定对象的影响可以降到最低，在此之前会利用 GCHandle.Alloc(obj, GCHandleType. Pinned)来做，但是需要更多的开销，因为开发人员还需要分配和解除分配 GCHandle。虽说这样可以，但仍然无法避免堆碎片的产生，这取决于固定对象的数量、时间，以及固定对象在内存中的位置等其他因素。

因此，POH 中包含 SOH/LOH 内的固定对象，这样垃圾回收器在压缩时就会忽略这个位置。从 .NET 5.0 开始，.NET 中提供了 POH 和一个配套的对象分配 API。如代码 15-7 所示，可以通过两种方式来分配数组。

代码15-7

```
class GC
{
    static T[] AllocateUninitializedArray<T>(int length, bool pinned = false);
    static T[] AllocateArray<T>(int length, bool pinned = false);
}
```

如代码 15-7 所示，两个内存分配的 API 利用它们指定想要创建的固定对象，在这里则是直接在 POH 中分配对象，而不是在 SOH/LOH 中分配对象。

POH 中的内存分配比正常的 SOH 中的内存分配慢一些，因为它并不是基于每个线程创建分配上下文，而是向 LOH 中的单个空闲链表分配，所以，当在 POH 中分配内存时，需要注意应具有足够的空闲空间。

POH 中还有一个非常重要的限制，即只允许固定 Blittable 类型的对象，也就是说，还可以固定非托管类型的缓冲区，如 int 或 byte。另外，POH 具备一定的性能优势，也就是说，垃圾回收器在标记可达对象时可以跳过 POH，但是无法固定住一个持有其他引用的对象。

如代码 15-8 所示，相应的检查是在运行时完成的，因此，POH 取决于判断 pinned。

代码15-8

```
public static T[] AllocateArray<T>(int length, bool pinned = false)
{
    GC_ALLOC_FLAGS flags = GC_ALLOC_FLAGS.GC_ALLOC_NO_FLAGS;

    if (pinned)
    {
        if (RuntimeHelpers.IsReferenceOrContainsReferences<T>())
            ThrowHelper.ThrowInvalidTypeWithPointersNotSupported(typeof(T));

        flags = GC_ALLOC_FLAGS.GC_ALLOC_PINNED_OBJECT_HEAP;
    }

    return Unsafe.As<T[]>(AllocateNewArray(
                    typeof(T[]).TypeHandle.Value, length, flags));
}
```

如代码 15-9 所示，可以尝试分配到 POH 中。

代码15-9

```
var array = GC.AllocateArray<string>(10, pinned: true);
```

但是在执行过程中抛出了异常,如代码 15-10 所示。

代码15-10

```
System.ArgumentException:"Cannot use type 'System.String'.
Only value types without pointers or references are supported."
```

15.2　垃圾回收器

在计算机科学中,垃圾回收是一种自动内存管理机制。通常来说,编程语言利用手动和自动两种方式管理内存,现代的高级语言几乎都具有垃圾回收机制,如 Java、Go 和 .NET 等语言采用的是自动内存管理系统,所谓的要回收的垃圾,其实就是应用程序中不再使用的对象,垃圾回收器会把应用程序中不再使用的内存视为"垃圾",在"合适"的时间回收或重用不再被对象占用的内存。在应用程序中如果不回收这些已分配但不使用的内存,就会一直占用并消耗,因此有了垃圾回收机制。

.NET 垃圾回收器主要包含标记阶段(Mark Phase)、计划阶段(Plan Phase)、重定位阶段(Relocale Phase)、清扫阶段(Sweep Phase)和压缩阶段(Compact Phase)。

15.2.1　分代

"代"是 .NET 垃圾回收器采用的一种机制,按照对象的存活时间进行划分。代也被分为第 0 代、第 1 代和第 2 代,这 3 代中的每一代都保存了一定"范围内"的对象,利用分代机制可以提升程序的性能。

CLR 的垃圾回收器算法是基于代的,需要注意以下几点。

- 较新的对象生存期较短,相反,先创建的对象生存期较长。
- 压缩托管堆的部分内存比压缩整个托管堆速度快。

对象的生存期是比较重要的一环,在对象生存期中包括存活较短的对象和存活较长的对象。具有不同的生存期的对象存储在一个区域中,称为代。堆上有 3 代对象。第 0 代是最年轻的,包含生命周期较短的对象,如临时变量。垃圾回收在这一代发生得比较频繁。通常来说,大多数的对象会在第 0 代中被回收,不能回收的被移至下一代中,即第 1 代,这一代包含短寿命的对象,并作为短寿命对象与长期存在的对象之间的一个缓冲区。第 2 代包含长期存在的对象，如存活在进程中的静态数据。

.NET 中的 SOH 分为如下 3 代。

- 第 0 代：包含新创建的对象，到目前为止还没有对其执行任何回收操作。
- 第 1 代：存储在单次垃圾回收中幸存下来的对象（由于仍在使用，因此在第 0 代中没有被回收）。
- 第 2 代：保留在两次或多次垃圾回收中幸存下来的对象。

幸存规则：如果在垃圾回收期间没有被回收的，则说明它仍然被某些东西引用，所以这些就存活下来。

- 如果对象在第 0 代中存活下来，那么对象被提升到第 1 代中。
- 如果对象在第 1 代中存活下来，那么对象被提升到第 2 代中。
- 如果对象在第 2 代中存活下来，那么对象依旧在第 2 代中。

代阈值：在代中所有对象的大小超过阈值就会被触发回收。代阈值在不同垃圾回收模式及不同 LLC 缓存等条件下也会有所不同（见表 15-1）。

表 15-1 以"内存占用"模式为例，即二维数组中的"第一维"（假设为 12MB LLC 缓存）

代	阈值下限	阈值上限	系数下限	系数上限
第 0 代	6MB	6MB	工作站模式：9 服务器模式：20	工作站模式：20 服务器模式：40
第 1 代	160KB	6MB	2	7
第 2 代（小对象）	256KB	SSIZE_T_MAX	1.2	1.8
第 2 代（大对象）	3MB	SSIZE_T_MAX	1.25	4.5

尽管如此，这些只是定义的初始值，在运行时由垃圾回收器动态调整阈值，增加特定阈值的条件之一是一代中的存活率很高（来自特定代的更多对象要么被提升到下一代中，要么留在第 2 代中），使垃圾回收器运行的频率降低（不会经常超过阈值条件）。

15.2.2 标记阶段

标记阶段通常是垃圾回收器的第一阶段，主要目的是找到所有存活的对象。从垃圾回收器根对象（线程栈局部根，垃圾回收器句柄表根，FinalizeQueue 根）开始，垃圾回收器沿着所有的对象引用进行遍历，并将所见的对象都做上标记。分代的垃圾回收器的好处是，可以只遍历回收内存堆的部分对象，而不需要一次性遍历所有对象，只要访问那些需要回收的部分即可。当回收年轻代时，垃圾回收器需要知道这些代中还存活着哪些对象。

下面通过一个简单的实例来介绍标记阶段，如代码 15-11 所示。

代码15-11

```cpp
class Obj
{
    public:
    //表示标记状态
    bool mark;
    //表示对象之间的引用关系
    std::list<Obj> children;
};

std::list<Obj>* roots = new std::list<Obj>();

void mark(Obj obj) {
    //检查对象是否已被标记
    if (obj.mark == false)
    {
        obj.mark = true;
        for (auto child : obj.children)
        {
            mark(child);
        }
    }
}

void mark_phase() {
    //给活动的对象打标记
    for (auto r : *roots)
    {
        mark(r);
    }
}

int main()
{
    mark_phase();
}
```

图 15-9 所示为垃圾回收器标记，在 main 方法中调用 mark_phase 方法，并利用 roots 集

合定义对象，接着通过 mark_phase 方法循环 roots 集合，利用 mark 属性进行标记。在 mark 方法中，children 属性有值也需进行标记，children 属性存储了根对象具有引用关系的对象，这只是一个大概的标记流程，当然，在垃圾回收器中的标记逻辑实际上远远比这些多，但笔者相信，了解基本逻辑可以对项目起到重要作用。

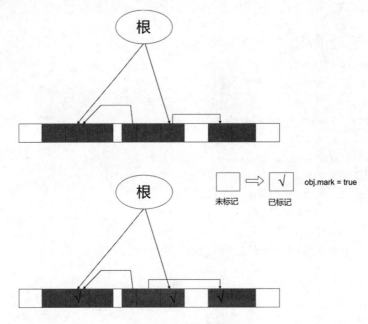

图 15-9　垃圾回收器标记

15.2.3　计划阶段

在标记阶段之后，所有的对象都已经被标记为可到达和不可到达。此时垃圾回收器会得到运行时所需的信息，接下来就是压缩或清扫。代码 15-12 展示了 dotnet/runtime 仓库中垃圾回收器计划阶段的一段代码。

代码15-12

```
plan_phase()
{
    //实际的计划阶段，判断压缩还是清扫
    if (compact)
    {
        relocate_phase();
        compact_phase();
```

```
    }
    else
        make_free_lists();
}
```

15.2.4 重定位阶段

在计划阶段已经收集了相应的数据，垃圾回收器会在移动对象引用之前，对这些对象完成引用的重定位操作，也就是说，重定位阶段会找到被回收对象的所有引用，而标记阶段是找到那些影响对象生命周期的引用，不需要考虑弱引用。

15.2.5 清扫阶段

如果计划阶段不执行压缩，则直接进入清扫阶段，并在该阶段找出那些占用的空间。清扫阶段会在这些存活的对象所占的空间中创建自由对象（Free Object），并将其加入自由列表（Free List）中，这些相邻的对象也会合并成一个自由对象。图 15-10 所示为标记清除算法。

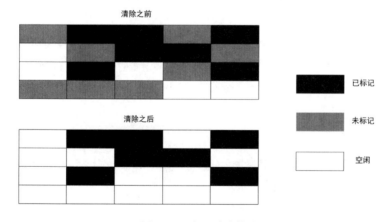

图 15-10　标记清除算法

在标记完可以存活的对象后，会清扫未标记的对象。在清扫后，内存也会变得不连续，这就会造成内存碎片化，在内存碎片越来越多以后，如果分配较大的对象，就无法找到足够大的内存碎片，从而触发垃圾回收器。

15.2.6 压缩阶段

在对象回收之后，内存空间会变得不连续，在计划阶段已经计算出对象需要移动的新位置，而压缩阶段只需要将它们复制到目标地址即可，这样可以使内存变得连续起来。图 15-11 所示为标记整理算法。

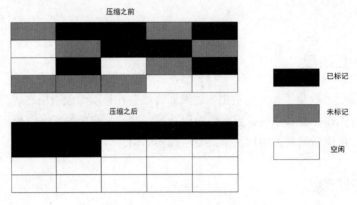

图 15-11　标记整理算法

15.3　资源释放

.NET 中的非托管资源（Unmanaged Resource）不受垃圾回收器管理（如数据库连接和网络连接等），需要开发人员手动释放；但对于托管资源，垃圾回收器会主动管理及完成自动释放工作。通常来说，非托管资源都会实现一个 IDisposable 接口。

15.3.1　Dispose

在.NET 中，非托管资源可以通过 Dispose 方法编写释放逻辑来清理资源，因此，需要实现 IDisposable 接口，如代码 15-13 所示。

代码15-13

```
public class FileManager : IDisposable
{
    FileStream fileStream = new FileStream(@"Test.txt",
                                            FileMode.Append);
    public async Task Write(string text)
    {
        byte[] buffer = Encoding.Unicode.GetBytes(text);
        int offset = 0;
```

```
        try
        {
            await fileStream.WriteAsync(buffer, offset,
                buffer.Length);
        }
        catch
        {
        }
    }
    public void Dispose()
    {
        if(fileStream != null)
        {
            Console.WriteLine("FileManager is dispose");
            fileStream.Dispose();
        }
    }
}
```

也可以显示调用 Dispose 方法,或者通过 using 代码块来释放资源。当然,这也是为了确保 Dispose 方法一直会被释放,如代码 15-14 所示。

代码15-14

```
class Program
{
    static async Task Main(string[] args)
    {
        using(FileManager fileManager = new FileManager())
        {
            await fileManager.Write("This is a text");
        }
    }
}
```

开发人员不需要记住调用的是哪个 Dispose 方法,因为 Dispose 方法在 .NET 中会以安全的方式通过 using 代码块自动执行。

在正常情况下,应该调用 Dispose 方法以确保释放逻辑被执行到,如果开发人员不调用它会怎么样呢?读者可以带着疑问来了解 using 代码块背后的底层机制,最简单的方式是查看 IL 代码，可以利用反编译器打开 .dll 文件（笔者使用的是 JetBrains 的 dotPeek 工具）。IL 代码如代码 15-15 所示。

代码 15-15

```
.class private auto ansi beforefieldinit
  Program
    extends [System.Runtime]System.Object
{
  .method private hidebysig static void
    Main(
      string[] args
    ) cil managed
  {
    .entrypoint
    .custom instance void System.Runtime.CompilerServices.NullableContextAttribute::.ctor([in] unsigned int8)
      = ( 01 00 01 00 00 ) // ...
      //unsigned int8(1) //0x01
    .maxstack 1
    .locals init (
      [0] class FileManager fileManager
    )
    //[7 5 - 7 6]
    IL_0000: nop
    //[8 16 - 8 59]
    IL_0001: newobj      instance void FileManager::.ctor()
    IL_0006: stloc.0     //fileManager
    .try
    {
      //[9 9 - 9 10]
      IL_0007: nop
      //[10 9 - 10 10]
      IL_0008: nop
      IL_0009: leave.s    IL_0016
    } //end of .try
    finally
    {
      IL_000b: ldloc.0    //fileManager
```

```
      IL_000c: brfalse.s    IL_0015
      IL_000e: ldloc.0      //fileManager
      IL_000f: callvirt
            instance void [System.Runtime]System.IDisposable::Dispose()
      IL_0014: nop
      IL_0015: endfinally
  } //end of finally

  //[11 5 - 11 6]
    IL_0016: ret

} //end of method Program::Main
.method public hidebysig specialname rtspecialname instance void
  .ctor() cil managed
{
  .maxstack 8

  IL_0000: ldarg.0       //this
  IL_0001: call    instance void [System.Runtime]System.Object::.ctor()
  IL_0006: nop
  IL_0007: ret

}
}
```

也就是说，using 代码块是一个语法糖，在编译器中编译后会生成 try/finally 块，如代码 15-16 所示。可以通过 try/finally 块调用 Dispose 方法。

代码15-16

```
FileManager fileManager = new FileManager();
try
{
    //TODO
}
finally
{
    if(fileManager != null)
    {
        fileManager.Dispose();
```

IDisposable 接口的定义如代码 15-17 所示，此处只定义了一个 Dispose 方法。

代码15-17

```
public interface IDisposable
{
    void Dispose();
}
```

15.3.2　DisposeAsync

从 C# 8.0 开始，在 .NET 中增加了一个新的接口，即 IAsyncDisposable，允许开发人员以异步方式释放资源。IAsyncDisposable 接口在异步编程中具有重要作用。在此之前，对于资源的释放只能通过同步的 IDisposable 接口来执行，随着 .NET Core 的诞生，它从底层开始几乎无处不在地异步化，这中间大量使用异步的 API。通常来说，在当下使用 .NET 框架更偏重于使用异步接口，因为它们不会堵塞线程的执行，所以在"异步化"的编程世界中，.NET 团队决定引入 IAsyncDisposable 接口。

简化 FileManager 的实现，并实现 IAsyncDisposable 接口，如代码 15-18 所示。

代码15-18

```
public class FileManager : IAsyncDisposable
{
    ...
    public async ValueTask DisposeAsync()
    {
        if (fileStream != null)
        {
            await fileStream.DisposeAsync();
        }
    }
}
```

接下来介绍如何调用 IAsyncDisposable 接口，同样利用 using 关键字，但是要在 using 前面添加 await 关键字，如代码 15-19 所示。

代码15-19

```
class Program
```

```csharp
{
    static async Task Main(string[] args)
    {
        await using (var fileManager = new FileManager())
        {
        }
    }
}
```

当然,可以同时继承 IDisposable 接口,让其支持两种处理方式,如代码 15-20 所示。

代码15-20

```csharp
public class FileManager : IDisposable, IAsyncDisposable
{
    public async ValueTask DisposeAsync()
    {
        if (fileStream != null)
        {
            await fileStream.DisposeAsync();
        }
    }

    public void Dispose()
    {
        if (fileStream != null)
        {
            fileStream.Dispose();
        }
    }
}
```

IAsyncDisposable 接口的定义如代码 15-21 所示,定义了一个 DisposeAsync 方法,并且该方法返回一个 ValueTask 类型。

代码15-21

```csharp
public interface IAsyncDisposable
{
    ValueTask DisposeAsync();
}
```

15.4 垃圾回收器的设置

对性能调优而言，分为多个层次，通常来说架构调优优先，而垃圾回收器调优则按需来定，毕竟大多数的.NET 应用不需要使用垃圾回收器调优，因为默认的配置已经可以满足大多数的应用场景，而这些调整的参数也值得开发人员关注，可以由此了解垃圾回收器对哪些内容产生了限制。使用垃圾回收器可以帮助开发人员更好地管理内存，在大多数情况下不需要关注垃圾回收器调优。但是有时候开发人员可能会对程序的一些机制产生疑惑，所以了解垃圾回收器调优还是有必要的，这样就可以知道需要做哪些调整，以及存在哪些优化空间。

15.4.1 工作站模式与服务器模式的垃圾回收

垃圾回收器有两种不同的工作模式，分别为工作站模式（Workstation Mode）和服务器模式（Server Mode）。通常来说，工作站模式适用于桌面应用程序，需要注意的是，该模式适用于内存占用量较小的程序，可以将服务器模式应用于内存占用量较大的应用程序。

工作站模式的垃圾回收

- 垃圾处理会在触发垃圾回收的同一个线程中，并且保留相同的优先级，对于单核处理器的计算机来说，这是比较好的选择。
- 执行垃圾处理的频率比较高。
- 工作站模式的垃圾回收只能由一个线程处理一个托管堆。

服务器模式的垃圾回收

- 执行垃圾处理的频率比较低。
- 服务器模式的垃圾回收可以并行通过多个线程处理多个托管堆，默认线程为逻辑核心数。

在默认情况下，Console（控制台）和 WPF 为 false，采用工作站模式。ASP.NET Core 应用程序默认为 true，采用服务器模式。如代码 15-22 所示，可以修改模式。.NET Framework 通过 web.config/app.config 文件或"程序名称.config"文件设置模式。

代码15-22

```
<configuration>
    <runtime>
        <gcServer enabled="true"/>
```

```
        </runtime>
</configuration>
```

如代码 15-23 所示,在 .NET Core 之后,可以通过 .csproj 文件设置,将 ServerGarbageCollection 属性设置为 true 表示采用服务器模式,否则采用工作站模式。

代码15-23

```
<Project Sdk="Microsoft.NET.Sdk">

 <PropertyGroup>
    <!--指示运行时是否启用服务器模式的垃圾回收-->
   <ServerGarbageCollection>true</ServerGarbageCollection>
 </PropertyGroup>

</Project>
```

如代码 15-24 所示,也可以在发布文件"程序名称.runtimeconfig.json"中配置。

代码15-24

```
{
  "runtimeOptions": {
    "tfm": "net6.0",
    "frameworks": [
      {
        "name": "Microsoft.NETCore.App",
        "version": "6.0.0"
      },
      {
        "name": "Microsoft.AspNetCore.App",
        "version": "6.0.0"
      }
    ],
    "configProperties": {
      "System.GC.Server": true
    }
  }
}
```

在运行应用程序时,可以通过 GCSettings.IsServerGC 属性查看当前程序使用的是否是服务器模式的垃圾回收，该属性返回一个 bool 类型，如代码 15-25 所示。

代码15-25

```
var isServerGC = GCSettings.IsServerGC;
```

注意： 在一台只有一个 CPU 逻辑内核的计算机上，无论 gcServer 设置成什么，都将始终使用工作站模式的垃圾回收。

15.4.2 普通垃圾回收和后台垃圾回收

在.NET 中有两种垃圾回收的处理方式，分别为普通（Blocking）垃圾回收和后台（Background）垃圾回收。后台垃圾回收是并发垃圾回收的演进，所以类似于并发垃圾回收，但是在后台垃圾回收中，由单独的垃圾回收器执行，并且在第 2 代回收期间不会挂起托管线程，但是在第 0 代和第 1 代回收期间，必须暂停托管线程和第 2 代垃圾回收，在第 0 代和第 1 代垃圾回收完成后，第 2 代才会继续执行。

大多数垃圾回收都是 STW（Stop-The-World）的，它们会暂停所有应用线程，也就是垃圾回收停顿。实际上，后台垃圾回收也会暂停所有应用线程，但是都非常短暂。

在默认情况下会启用后台垃圾回收。在.NET Framework 程序中可以通过 gcConcurrent 配置，.NET Core 之后的版本可以使用 System.GC.Concurrent 来启用或禁用后台垃圾回收，如代码 15-26 所示。

代码15-26

```
<configuration>
    <runtime>
        <gcServer enabled="true"/>
        <gcConcurrent enabled="true"/>
    </runtime>
</configuration>
```

如代码 15-27 所示，在.NET Core 之后可以通过.csproj 文件配置，即通过 ConcurrentGarbageCollection 属性配置。

代码15-27

```
<Project Sdk="Microsoft.NET.Sdk">

  <PropertyGroup>
    <ConcurrentGarbageCollection>false</ConcurrentGarbageCollection>
```

```
    </PropertyGroup>
</Project>
```

如代码15-28所示，也可以在发布文件"程序名称.runtimeconfig.json"中配置。

代码15-28

```
{
  "runtimeOptions": {
    "configProperties": {
      "System.GC.Concurrent": false
    }
  }
}
```

15.4.3 设置延迟模式

垃圾回收有多种延迟模式可供开发人员选择（见表15-2），其中大部分是通过GCSettings.LatencyMode属性访问的，通常来说不需要更改这些模式，但是在一些情况下，这些模式会起到一定的作用。

表 15-2　GCLatencyMode 类属性/值说明

属性	值	说明
Batch	0	禁用后台垃圾回收，通过执行普通垃圾回收，所有非垃圾回收的线程会被挂起，直到垃圾回收结束
Interactive	1	启用后台垃圾回收和普通垃圾回收，允许回收第0代和第1代，并且允许后台线程回收第2代
LowLatency	2	禁止回收第2代，该模式仅在工作站模式下可用，禁用了第2代垃圾回收
SustainedLowLatency	3	禁用非并发完全垃圾回收，工作站模式和服务器模式都可使用
NoGCRegion	4	根据预分配对象的总大小决定何时启用垃圾回收，该属性不允许直接设置，而是借助 TryStartNoGCRegion 方法设置

如代码15-29所示，通过GCSettings的LatencyMode属性可以改变延迟模式，但是建议在执行完指定操作后再切换回原模式。

代码15-29

```
GCSettings.LatencyMode = GCLatencyMode.LowLatency;
```

如代码 15-30 所示，对于 NoGCRegion（无垃圾回收器区域模式）来说，可以通过如下语句进行设置（对于该内容，为了安全操作，建议通过如下类似的代码进行设置）。要切记在执行完指定操作后，调用 GC.EndNoGCRegion 方法结束。

代码15-30

```
if (GC.TryStartNoGCRegion(1024,true))
{
    try
    {
        //TODO
    }
    finally
    {
        try
        {
            GC.EndNoGCRegion();
        }
        catch (Exception e)
        {

        }
    }
}
```

15.4.4 垃圾回收器参数调优

关于垃圾回收器参数调优，可以通过如下方式进行处理，但是在修改时应该谨慎，这些设置推荐使用 DOTNET_环境变量。当然，也可以使用其他方式，如通过 runtimeconfig.json 文件定义，但是这种方式不适用于一些选项，所以推荐使用环境变量，这样就可以在运行程序之前轻松设置。

- **GCLatencyLevel**：设置垃圾回收器延迟级别，该属性可以限制垃圾回收器代数的大小，使垃圾回收器触发得更频繁，默认的延迟级别为 1。
- **GCHeapHardlimitPercent**：指定进程可以使用的物理内存（内存百分比形式）。通常，可以对服务器上的应用程序分配固定大小的内存，从而限制应用程序的内存使用量。
- **GCHighMemPercent**：设置内存总占用的大小，使垃圾回收器触发得更加频繁，使操作系统释放出更多的空闲内存。通过该属性可以调整内存最大的占用率（百分比），默认值为 90%，调整得过低和过高都不好，如果过低则会造成垃圾回收器触发得频繁，从而影响性能，如果过高则会影响服务器的整体资源占用，建议保持为 80%~97%。

- **gcTrimCommitOnLowMemory**：启用该属性后，如果内存使用量超过一定的阈值（90%），就会进行调整并减少分配过多的内存，直到降低到 85% 以下。
- **gcServer**：设置属性选择工作站模式或服务器模式，在多核心 CPU 计算机中，每个 CPU 核心都有一个专用的垃圾回收器线程，这样可以提高应用程序执行大型任务时的性能。如果计算机有 3 个或更多个 CPU 核心，则建议采用服务器模式；如果计算机只有 1 个 CPU 核心，则自动强制采用工作站模式；如果计算机有两个核心，则两者都可以考虑。
- **DOTNET_TieredPGO**：默认为禁用，用于设置在 .NET 6 及更高版本中启用动态或分层按配置优化（PGO）。
- **DOTNET_ReadyToRun**：配置 .NET 运行时是否要为具有可用 ReadyToRun 数据的映像使用预编译代码，如果禁用此选项，则强制运行时对框架代码进行 JIT 编译。默认为启用，如果禁用此选项启用 DOTNET_TieredPGO，则可以将分层按配置优化应用到整个 .NET 平台，而不仅仅是应用程序的代码。
- **DOTNET_TC_QuickJitForLoops**：配置 JIT 编译器是否对包含循环的方法使用快速 JIT，设置快速 JIT 可以提高启动的性能。

可以通过环境变量调整垃圾回收器的参数，在 Linux 操作系统中可以使用如代码 15-31 所示的代码。

代码15-31

```
export DOTNET_GCHeapHardLimitPercent=0x3C
export DOTNET_GCHighMemPercent=0x32
export DOTNET_GCLatencyLevel=0
export DOTNET_gcTrimCommitOnLowMemory=1
export DOTNET_gcServer=1
export DOTNET_TieredPGO=1
export DOTNET_ReadyToRun=0
export DOTNET_TC_QuickJitForLooks=1
```

如代码 15-32 所示，在 Windows 操作系统中可以使用 Powershell。

代码15-32

```
$Env: DOTNET_GCHeapHardLimitPercent=0x3C
$Env: DOTNET_GCHighMemPercent=0x32
$Env: DOTNET_GCLatencyLevel=0
```

```
$Env: DOTNET_gcTrimCommitOnLowMemory=1
$Env: DOTNET_gcServer=1
$Env: DOTNET_TieredPGO=1
$Env: DOTNET_ReadyToRun=0
$Env: DOTNET_TC_QuickJitForLoops=1
```

如果采用服务器模式的垃圾回收的内存开销比较高，则可以考虑调整 GCLatencyLevel 属性和 GCHeapHardLimitPercent 属性，但是建议保持默认状态，因为垃圾回收会在运行时进行自我优化，并且在操作系统真正需要时使用更少的内存。

15.5 小结

创建一个对象即代表需要开辟一些资源，所以需要为这些资源分配内存，在 C#中通常使用 new 关键字来完成。而在程序中调用 new 关键字创建对象时，如果没有足够的地址空间来分配该对象，CLR 就会执行垃圾回收。在托管应用中又分为本机堆和托管堆，CLR 在托管堆上为对象分配内存。而堆又分为两种，即大对象堆和小对象堆，通常将小于 85 000 字节的对象分配到小对象堆中，而大对象堆则用来分配超过 85 000 字节的对象，它们直接由第 2 代来跟踪。

第 16 章

诊断和调试

对开发人员而言，性能诊断是日常工作中经常面对和解决的问题，具备调试技能可以非常方便地应对应用程序所带来的问题。本章将探讨诊断和调试，16.1 节介绍了性能诊断工具，16.2 节介绍了不同环境中的调试技巧，如在 WSL2 中调试、使用 Visual Studio（Code）远程调试 Linux 部署的应用和使用容器工具调试。

16.1 性能诊断工具

在.NET 中，有非常多的诊断工具可供选择，如 Visual Studio、PerfView，也可以借助 ILSpy 查看 IL 代码，利用 WinDbg 分析转储文件，除此之外，还有基准测试工具 BenchmarkDotNet，以及 dotnet 一系列的 CLI 工具。

16.1.1 Visual Studio

Visual Studio 是一个 IDE。本节以 Visual Studio Professional（专业版）为例，对该 IDE 中性能分析工具的功能进行讲解，相关工具可以用于解决实际开发中的问题，以及提高应用程序的性能。

启动 Visual Studio，选择"调试"→"性能探查器"命令，打开如图 16-1 所示的"分析目标"界面。

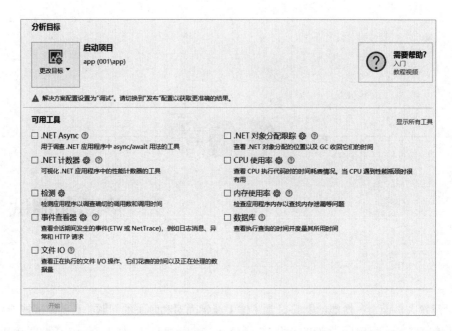

图 16-1 "分析目标"界面

内存使用率

可以通过"内存使用情况"工具分析应用程序的内存使用情况，在 Visual Studio 中可以利用该工具查看内存的实时使用情况，如图 16-2 所示。另外，也可以对应用程序中内存状态的详细信息拍摄快照，并对比不同时间段的内存使用情况。对内存进行分析的主要目的是通过收集内存使用情况方面的数据来分析内存是否健康，确定是否存在内存泄漏问题，从而对优化内存的使用提供引导。

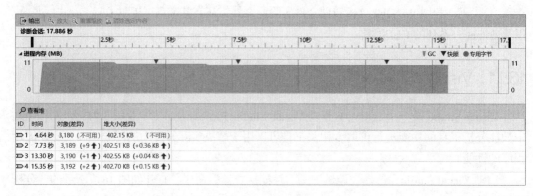

图 16-2 内存的实时使用情况

根据如图 16-2 所示的应用程序在运行时的内存波动情况，可以对比内存使用的走势。另外，对内存进行分析还可以利用内存快照的方式查看详细信息，如单击快照的条目查看内存的使用情况，当然，也可以根据相应的字节大小进行排序。图 16-3 所示为内存快照的详细信息。

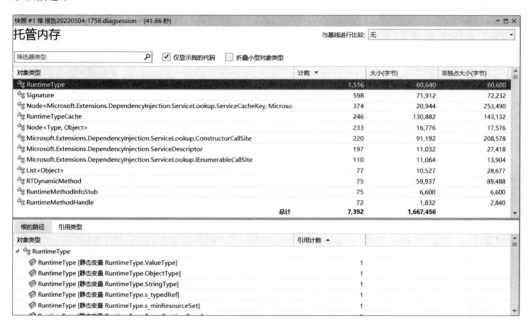

图 16-3　内存快照的详细信息

CPU 使用率

可以通过"CPU 使用情况"工具查看 CPU 使用率。随着时间的推移，CPU 使用率的百分比也会发生变化。图 16-4 所示为 CPU 使用率的数据报告。

如图 16-4 所示，可以通过"排名靠前的函数"区域和"热路径"区域了解 CPU 资源的占用情况，"热路径"区域中显示了占用 CPU 资源的代码。通过这些报告信息，开发人员可以清晰地看到 CPU 资源的占用情况，进而可以针对某个区域中的代码进行优化。

单击"查看详细信息"链接，在弹出的界面的"当前视图"下拉列表中选择"调用树"选项，如图 16-5 所示。选择工具栏中的"显示热路径"命令，展示如图 16-5 所示的数据，诊断报表按照"CPU 总计"的值从高到低进行排序，通过单击列表的标题可以更改排序依据。

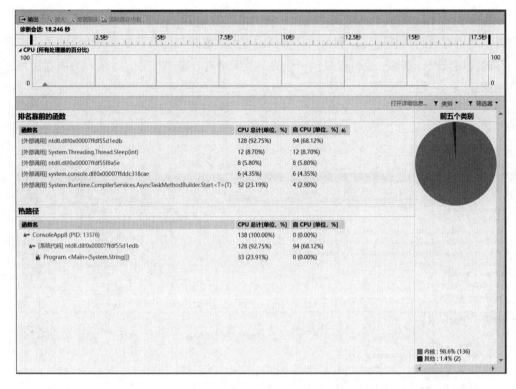

图 16-4　CPU 使用率的数据报告

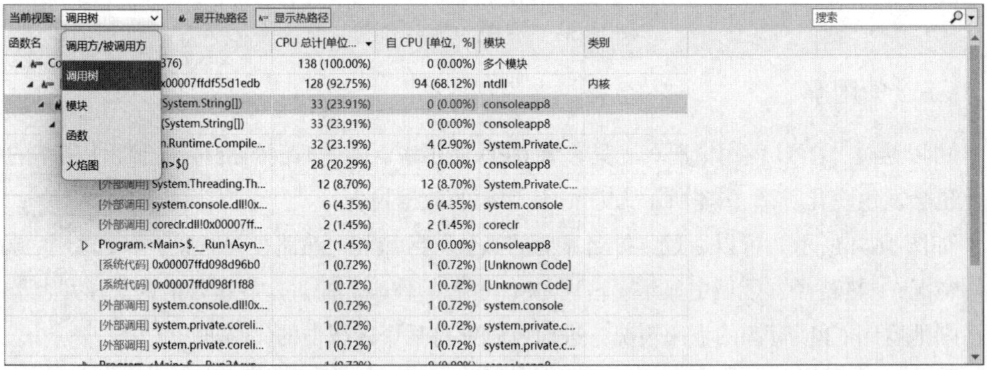

图 16-5　选择"调用树"选项

如图 16-5 所示，列表中展示了每个函数的路径，也可以展开界面左上角的"当前视图"下拉列表，选择其他的视图展示选项。

当然，还可以在 Visual Studio 性能分析器中使用其他的分析工具，如.NET 对象分配跟踪、.NET 计数器、检测、事件查看器、数据库等。

图 16-6 所示为 Visual Studio 实时使用报告，通过该工具开发人员可以对内存使用率和 CPU 使用率进行分析，同时可以了解 Visual Studio 在调试代码期间的性能报告，以及垃圾回收器的触发情况。

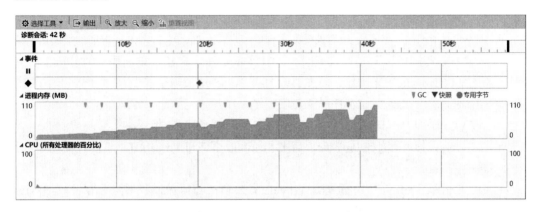

图 16-6　Visual Studio 实时使用报告

除此之外，还可以拍摄快照，这样就可以深入分析某个时间段的托管对象。

16.1.2　PerfView

PerfView 是免费的性能分析工具，体积小，免安装，并且功能强大。它是由.NET Runtime Performance 架构师 Vance Morrison 编写的，可以帮助开发人员分析与 CPU 和内存相关的性能问题。它是一个 Windows 工具，可以用于分析在 Linux 机器上收集的数据。利用 PerfView 可以对 ETW（Event Tracing for Windows）数据进行性能分析，并且它是一个集抓取和分析于一身的性能分析工具。PerfView 是一个可执行文件，所以只需要下载可执行文件即可。可以通过以下两种方式下载 PerfView。

- 访问 Microsoft 官网，搜索并下载 PerfView。
- 从 GitHub 官网的 microsoft/perfview 仓库中下载 PerfView。

启动程序后，可以看到一个包含大量帮助内容的窗口。选择 Collect 选项后，将打开如图 16-7 所示的窗口。

如图 16-7 所示，Advanced Options 区域中有许多选项，这些选项分别对应不同的.NET 信息。

- .NET：启用来自.NET Provider 的默认事件。
- .NET Stress：获取与压力测试相关的事件。

- .NET Alloc：在垃圾回收器堆上分配对象时，记录堆栈触发的事件，它是非常昂贵的开销，对性能会产生一定的影响。当然，应用程序分配对象的频率越高，触发事件的速度越快。
- .NET SampAlloc：在垃圾回收器堆上分配 10KB 对象的事件。
- ETW .NET Alloc：用于记录对象分配采样的事件，与.NET SampAlloc 选项的数据从本质上来看是相同的，只是方式不同。.NET SampAlloc 选项可以利用.NET Profiler Dll（ETWClrProvider）来工作，而 ETW .NET Alloc 选项则是基于.NET 4.5.3 中添加的 GCSampledObjectAllocationHigh 关键字工作。
- .NET Calls：启用该选项，当每次调用方法时，都会记录一个事件。
- GC Only：禁用所有的 Provider，启用 MemInfo 事件和 VirtualAlloc 事件，可以用于分析内存问题。
- GC Collect Only：禁用所有的 Provider，仅启用与垃圾回收器进程相关联的.NET Provider 事件，这种模式记录的数据比 GC Only 还要少。

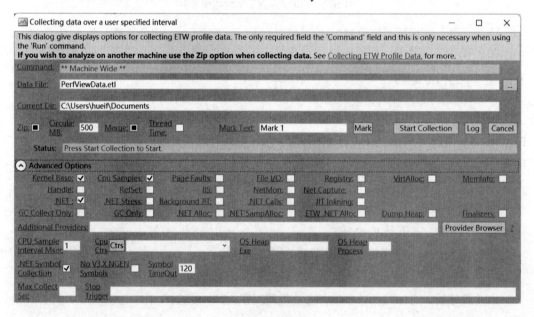

图 16-7 收集数据

图 16-7 中还包括以下几项。

- ZIP：将文件打包到一个存档文件中，通常这样更方便将数据复制到另一台机器上，以便日后通过该文件进行分析。
- Merge：当数据被收集在多个文件中时，可以将其合并为一个文件。

单击如图 16-7 所示的 Start Collection 按钮，PerfView 开始工作，日志输出将显示在右下角。在一定时间内，可以单击 Cancel 按钮关闭应用程序，此时 PerfView 已经创建了一个名为 PerfViewData.etl 的文件，如图 16-8 所示，可以打开该文件。

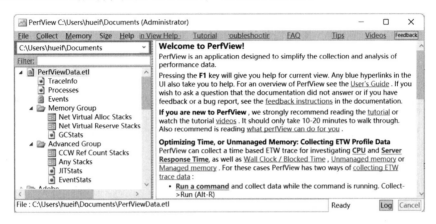

图 16-8　在 PerfView 中打开 ETL 文件

如图 16-9 所示，可以通过双击左窗格中的节点查看详情，详情包含所有已经记录的事件。另外，可以利用 Filter 功能，在输入 GC 时，筛选有关垃圾回收器的 DotNetRuntime 事件。图 16-9 所示为 PerfView Events GC 事件。

图 16-9　PerfView Events GC 事件

- Start：指定需要筛选的起始时间（单位为毫秒）。
- End：指定需要筛选的终止时间（单位为毫秒）。

- MaxRet：查询结果中每页的条数，默认为 10 000 条记录。
- Find：需要查找的文本。
- Process Filter：筛选要查看的应用程序。
- Text Filter：按文本过滤。

单击 GCStats 节点查看垃圾回收器的统计信息，该窗口中包括垃圾回收器的综合信息，如图 16-10 所示。

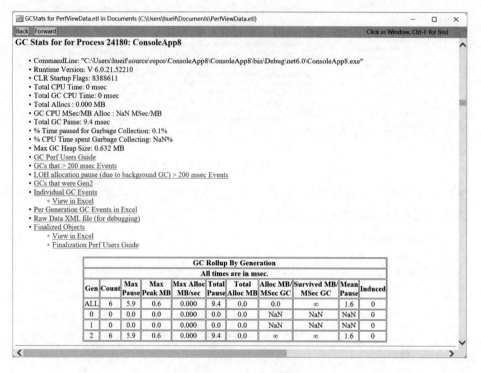

图 16-10　垃圾回收器的综合信息

除此之外，还有一些其他的视图。图 16-11 所示为 Any Stacks 视图。

- GroupPats：分组模板，多个模板之间以 ";" 分隔。
- Fold%：如果小于设置的值，那么该调用栈将折叠显示。
- FoldPats：折叠模板，多个模板之间以 ";" 分隔。
- IncPats：匹配函数名称，通常来说匹配进程名称，显示在列表中。
- ExcPats：匹配函数名称，在列表中排除。

如图 16-12 所示，在 PerfView 应用程序的菜单栏中展开 Memory 菜单，选择 Take Heap Snapshot 命令，打开 Collecting Memory Data 窗口。

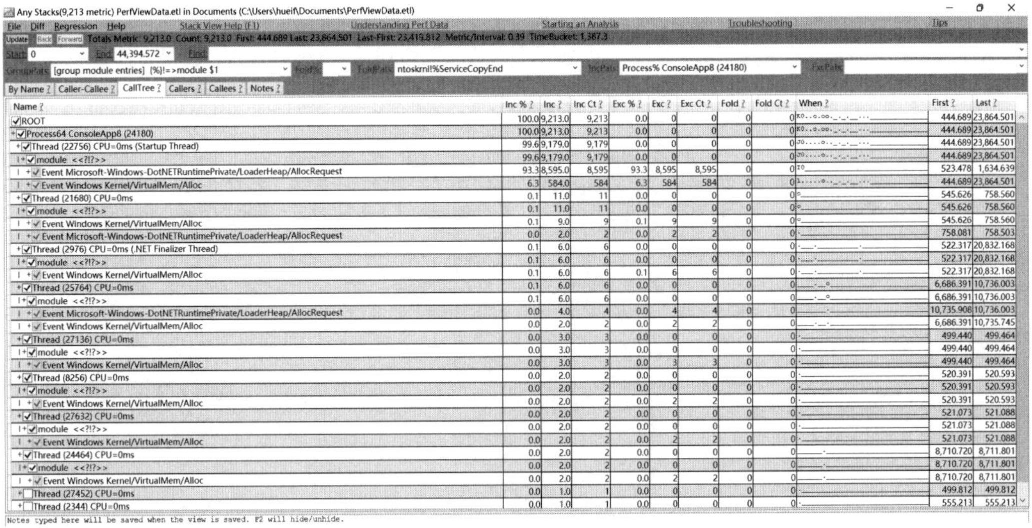

第 16 章 诊断和调试

图 16-11 Any Stacks 视图

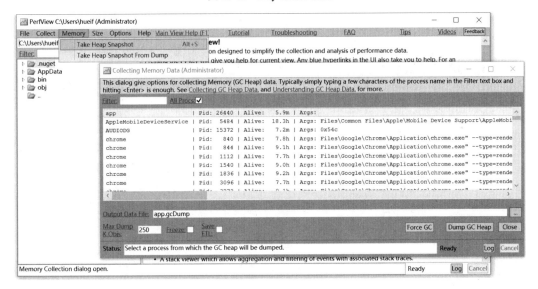

图 16-12 Collecting Memory Data 窗口

在 Collecting Memory Data 窗口中，先单击进程 app，再单击 Dump GC Heap 按钮，输出一个名为 app.gcdump 的文件，如图 16-13 所示，通过 PerfView 的左窗格，单击 Heap Stacks 节点，打开 Heap Stacks 窗口。

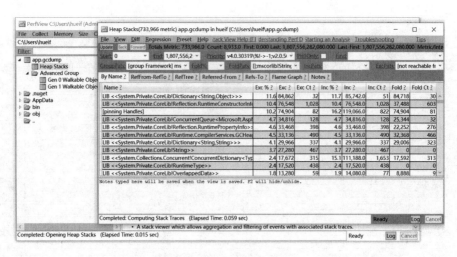

图 16-13　Heap Stacks 窗口

如图 16-14 所示，右击列表，在弹出的快捷菜单中选择 Memory→View Objects 命令，打开 ObjectViewer 窗口，如图 16-15 所示。

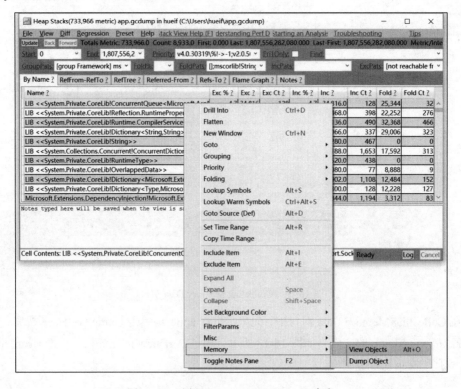

图 16-14　选择 Memory→View Objects 命令

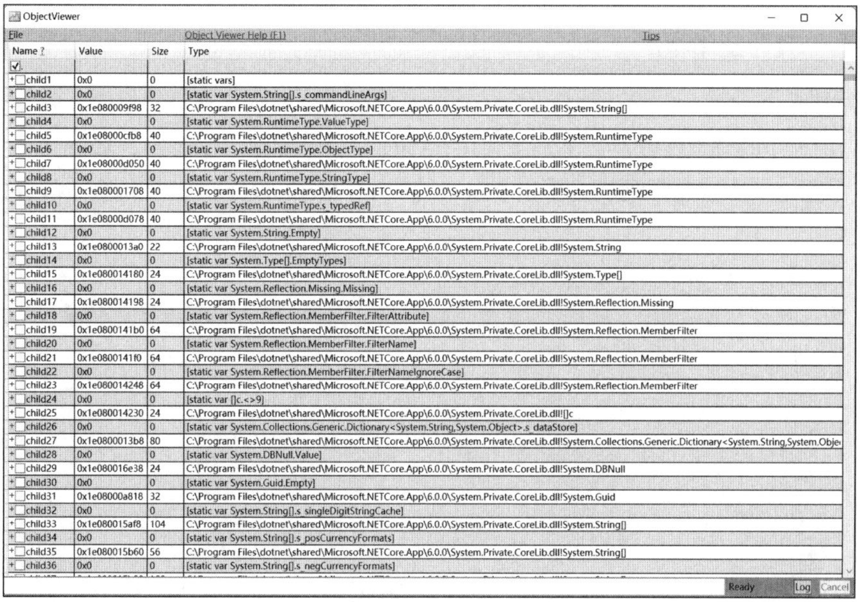

图 16-15　ObjectViewer 窗口

内存快照也是 PerfView 的重要功能，通过内存快照可以比较两个状态的内存差异性。在菜单栏中先选择 Diff 菜单，再选择要进行对比的文件，内存快照差异如图 16-16 所示。

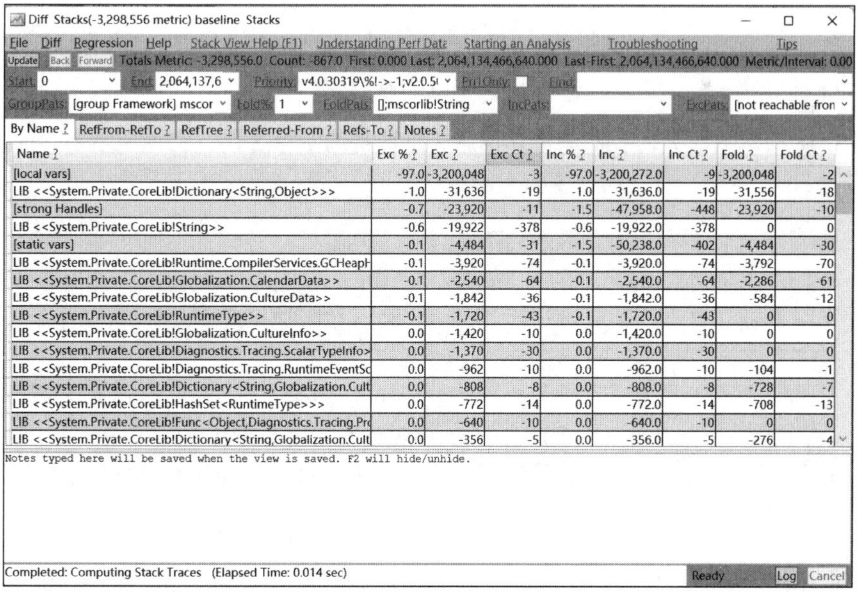

图 16-16　内存快照差异

场景：内存使用率高

首先打开 PerfView，然后选择 Collect 选项收集诊断信息。几秒钟后停止收集。首先了解当前应用程序的垃圾回收器，然后双击 GCStats 报告，弹出一个弹窗，如图 16-17 所示，GCStats 界面中展示了进程列表，单击 Process 17848: MemoryLeak 链接，查看垃圾回收器的汇总信息。

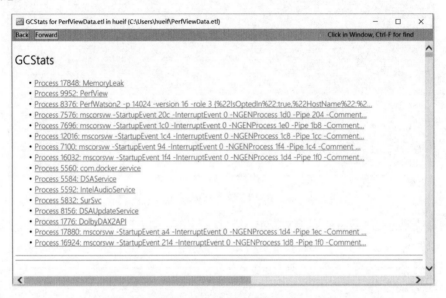

图 16-17　GCStats 界面

图 16-18 所示为垃圾回收器的汇总信息，这些都属于第 2 代，是预期结果，因为这些对象都大于 85 000 字节，所以均被视为大对象。

GC Rollup By Generation									
All times are in msec.									
Gen	Count	Max Pause	Max Peak MB	Max Alloc MB/sec	Total Pause	Total Alloc MB	Alloc MB/ MSec GC	Survived MB/ MSec GC	Mean Pause	Induced
ALL	4	32.8	4,801.8	1,762.531	51.2	8,000.0	156.1	∞	12.8	0
0	0	0.0	0.0	0.000	0.0	0.0	0.0	NaN	NaN	0
1	0	0.0	0.0	0.000	0.0	0.0	0.0	NaN	NaN	0
2	4	32.8	4,801.8	1,762.531	51.2	8,000.0	0.0	∞	12.8	0

图 16-18　垃圾回收器的汇总信息

图 16-19 所示为垃圾回收器的触发情况，所有的垃圾回收器都是第 2 代，并且都由 AllocLarge 触发，也就是分配了大对象，所以触发了该垃圾回收器。

第 16 章 诊断和调试

GC Index	Pause Start	Trigger Reason	Gen	Suspend Msec	Pause MSec	% Pause Time	% GC	Gen0 Alloc Rate MB/sec	Peak MB	After MB	Ratio Peak/After	Promoted MB	Gen0 MB	Gen0 Survival Rate %	Gen0 Frag %	Gen1 MB
145	383.123	AllocLarge	2B	15.076	16.388	23.3	NaN	0.000	0.00	1,601.806	1,601.846	1.00	1,600.103	1.483	0	64.01	0.000
146	2,846.497	AllocLarge	2B	0.019	0.915	0.0	NaN	0.000	0.00	4,801.806	801.854	5.99	800.108	1.491	0	63.74	0.000
147	3,620.033	AllocLarge	2B	31.302	32.831	5.4	NaN	0.000	0.00	1,601.807	1,601.855	1.00	1,600.108	1.492	0	63.70	0.000
148	6,129.069	AllocLarge	2B	0.021	1.116	0.0	NaN	0.000	0.00	4,801.808	801.848	5.99	800.108	1.485	0	63.93	0.000

图 16-19　垃圾回收器的触发情况

在 PerfView 窗口中，双击 GC Heap Alloc Ignore Free（Coarse Sampling）Stacks 节点后，如图 16-20 所示，在 Select Process Window 窗口中找到 MemoryLeak 进程。

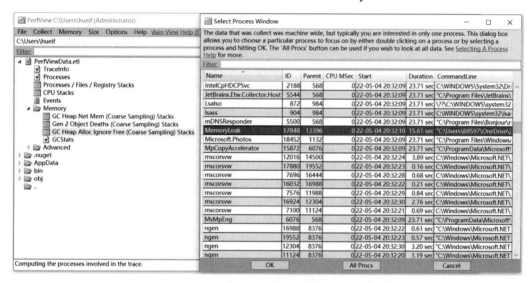

图 16-20　Select Process Window 窗口

在列表中找到相应的进程，双击进程名称 MemoryLeak，打开如图 16-21 所示的窗口，在该窗口中会显示分配对象的列表，在对象列表中可以看到引发的大对象 Type System.Double[]。其中，Inc%表示对象分配的字节与总字节数相比的百分比，Inc 是分配的字节数，Inc Ct 是分配的对象数。

代码 16-1 展示了引发该问题的代码，由于不断地创建数组，因此出现内存泄漏。

代码16-1

```
class Program
{
    static void Main(string[] args)
    {
```

```
        Console.WriteLine("Press any key to exit...");
        while (!Console.KeyAvailable)
        {
            var array = new double[100000000];
            for (int i = 0; i < array.Length; i++)
            {
                array[i] = i;
            }
            System.Threading.Thread.Sleep(10);
        }
        Console.WriteLine("Done");
    }
}
```

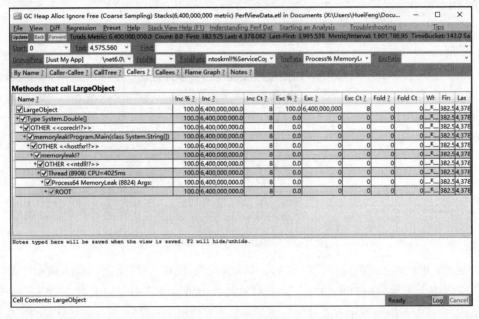

图 16-21　分配视图

16.1.3　ILSpy

ILSpy 是一个开源的工具,提供了.NET 程序集浏览和反编译功能。对.NET 应用程序而言,开发人员可以利用 ILSpy 将编译过的程序集反编译成可读的 IL 代码、C#源代码和 ReadyToRun 源代码,如图 16-22 所示。

第 16 章 诊断和调试

除了可以查看 C#源代码，还可以查看 IL 代码，如图 16-23 所示。

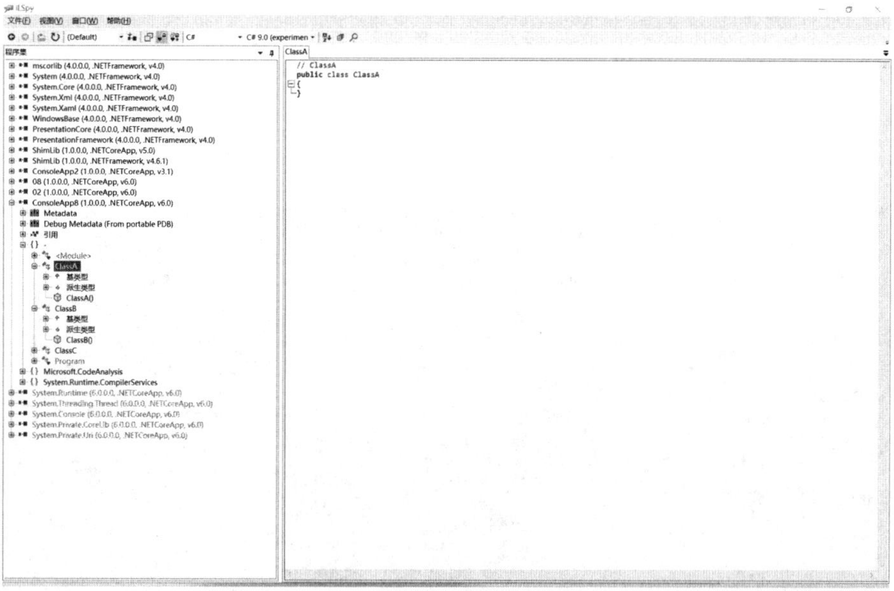

图 16-22　查看 C#源代码

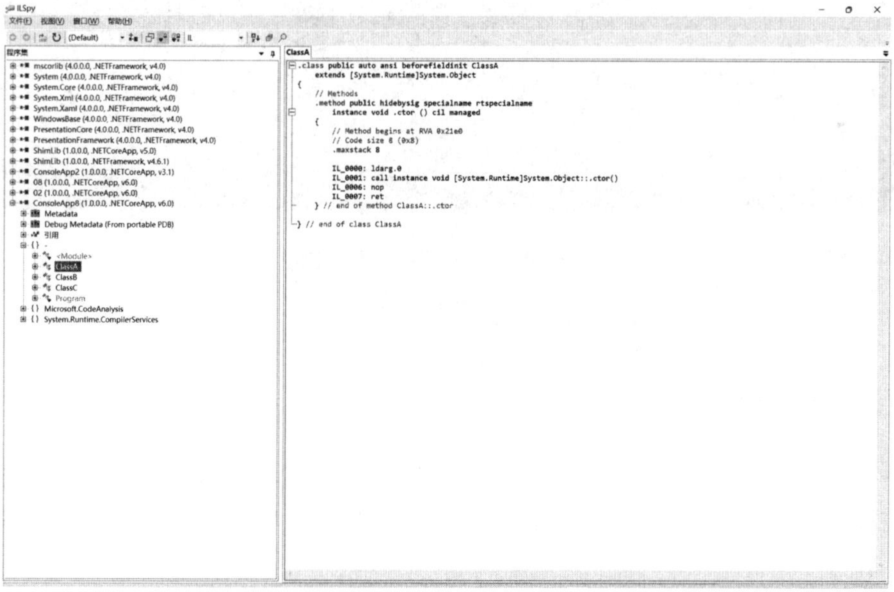

图 16-23　查看 IL 代码

关于 ILSpy 的获取，可以访问 GitHub 官网，在 icsharpcode/ILSpy 仓库中下载。

16.1.4　dnSpy

dnSpy 是 .NET 应用程序的逆向工程工具（如图 16-24 所示，可以使用 dnSpy 查看源代码，显示该工具的主界面）。dnSpy 不仅可以用于反编译 .dll 文件，还可以用于调试和编辑文件。也就是说，可以利用 dnSpy 修改 .dll 文件。

图 16-24　使用 dnSpy 查看源代码

访问 GitHub 官网，在 dnSpy/dnSpy 仓库中下载 dnSpy。

16.1.5　WinDbg

WinDbg 是 Windows 平台下一个强大且轻量级的调试工具，开发人员可以通过它附加到托管进程中调试，除此之外，还可以分析生成的内存转储，如调试内存泄漏、CPU 消耗、死锁等问题。

安装 WinDbg 的方式

- Windows 的调试工具包含在 Windows 驱动程序工具包（Windows Driver Kit，WDK）中。

- 使用 Windows 应用商店。

运行 WinDbg 后，将看到如图 16-25 所示的 WinDbg 主窗口，该窗口中提供了多种选项。

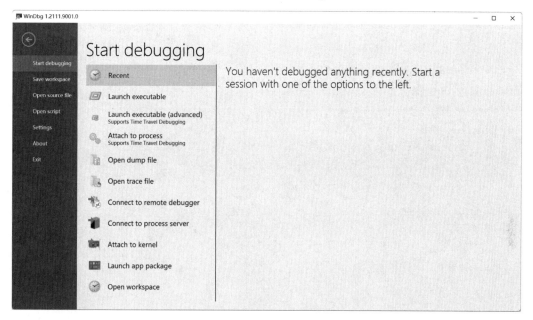

图 16-25　WinDbg 主窗口

最简单的方式是通过 WinDbg 附加进程，选择 Attach to process 命令，展开正在运行的进程列表，选择指定的进程即可。当然，也可以直接在 WinDbg 中打开转储文件。

WinDbg 同样支持插件。例如，使用.load <full to file>命令可以加载插件，如.load d:/Sosex/64bit/sosex.dll，利用该命令可以根据指定的路径加载插件。除此之外，还可以利用.loadby 命令，如使用.loadby sos clr 命令可以自动定位路径。

在进程中内存占用一直居高不下，有很多原因会引发这种现象。下面笔者以对象释放慢为例来介绍 WinDbg 的使用。如代码 16-2 所示，笔者以控制台实例为例展开介绍，在 Main 方法中，通过 while 语句循环创建自定义的 MyClass 对象，随后利用声明的 Random 对象获取随机数，该随机数通过 MyClass 对象的构造方法传入该实例中，用于缓解释放资源的时间。该应用程序会造成持续地创建对象实例，但内存得不到快速回收，导致内存一直增长。

代码16-2

```
public class MyClass
{
    private readonly int _delay;
```

```
    public MyClass(int delay) => _delay = delay;
    ~MyClass() => Thread.Sleep(_delay);
}

class Program
{
    static void Main(string[] args)
    {
        Console.WriteLine("Press any key to exit"...);
        var random = new Random();
        while (!Console.KeyAvailable)
        {
            var _ = new MyClass(random.Next(100, 5000));
        }
        Console.WriteLi"e("D"ne");
    }
}
```

如图 16-26 所示，在 Visual Studio 中利用诊断工具查看资源的表现情况。

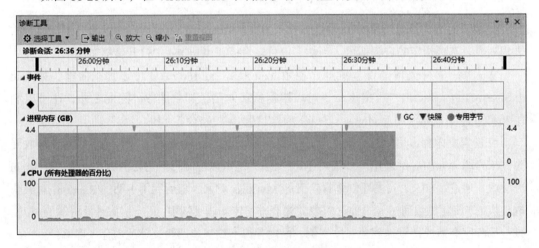

图 16-26　资源的表现情况

接下来利用 WinDbg 进行排查。在保存转储文件后，通过 WinDbg 打开转储文件，利用!address -summary 指令查看进程和目标计算机的内存信息，如代码 16-3 所示。

代码16-3

```
0:000> !address -summary
Mapping file section regions...
Mapping module regions...
Mapping PEB regions...
Mapping TEB and stack regions...
Mapping heap regions...
Mapping page heap regions...
Mapping other regions...
Mapping stack trace database regions...
Mapping activation context regions...

- Usage Summary ---- RgnCount ---- Total Size ----------%ofBusy %ofTotal
Free                      71       7dfe`7e72b000 ( 125.994 TB)          98.43%
<unknown>                136        201`54cf3000 (   2.005 TB) 99.97%    1.57%
Heap                      22          0`295c1000 ( 661.754 MB)  0.03%    0.00%
Image                    206          0`02e5c000 (  46.359 MB)  0.00%    0.00%
Stack                     12          0`00600000 (   6.000 MB)  0.00%    0.00%
Other                      7          0`001ac000 (   1.672 MB)  0.00%    0.00%
TEB                        4          0`00008000 (  32.000 kB)  0.00%    0.00%
PEB                        1          0`00001000 (   4.000 kB)  0.00%    0.00%


-Type Summary (for busy) --- RgnCount --- Total Size -------- %ofBusy %ofTotal
MEM_MAPPED                     72       200`01a1b000 (   2.000 TB) 99.71%   1.56%
MEM_PRIVATE                   110         1`7d04e000 (   5.953 GB)  0.29%   0.00%
MEM_IMAGE                     206         0`02e5c000 (  46.359 MB)  0.00%   0.00%


-State Summary --- RgnCount --- Total Size ----- %ofBusy %ofTotal
MEM_FREE                71       7dfe`7e72b000 ( 125.994 TB)         98.43%
MEM_RESERVE             75        201`1a67b000 (   2.004 TB) 99.92%   1.57%
MEM_COMMIT             313          0`6724a000 (   1.612 GB)  0.08%   0.00%


-Protect Summary (for commi) - RgnCount --- Total Size --- %ofBusy %ofTotal
PAGE_READWRITE                    91        0`61c73000 (   1.528 GB)  0.07%   0.00%
PAGE_NOACCESS                     19        0`02320000 (  35.125 MB)  0.00%   0.00%
```

PAGE_EXECUTE_READ	31	0`0219a000 (33.602 MB)	0.00%	0.00%
PAGE_READONLY	144	0`010a3000 (16.637 MB)	0.00%	0.00%
PAGE_WRITECOPY	14	0`00046000 (280.000 kB)	0.00%	0.00%
PAGE_EXECUTE_READWRITE	10	0`00026000 (152.000 kB)	0.00%	0.00%
PAGE_READWRITE | PAGE_GUARD	4	0`0000e000 (56.000 kB)	0.00%	0.00%

```
-Largest Region by Usage ---- Base Address --- Region Size -------
Free                          1a0`a334c000      7c53`bd2c4000 ( 124.327 TB)
<unknown>                     7dfb`e3ef0000     1f9`57de9000 (   1.974 TB)
Heap                          1a0`44c55000      0`15601000 ( 342.004 MB)
Image                         7ffa`0d8e1000     0`009ba000 (   9.727 MB)
Stack                         10`39c80000       0`0017b000 (   1.480 MB)
Other                         1a0`5d140000      0`00181000 (   1.504 MB)
TEB                           10`39777000       0`00002000 (   8.000 kB)
PEB                           10`39776000       0`00001000 (   4.000 kB)
```

如代码 16-3 所示，进程指标 MEM_COMMIT 为 1.612GB，接下来通过!dumpheap -stat 指令检查当前托管类型的统计信息，如代码 16-4 所示。

代码16-4

```
0:000> !dumpheap -stat
Statistics:
            MT    Count    TotalSize Class Name
00007ff9c2d63898     2          564 System.Char[]
00007ff9c2d07f48    33         1352 System.SByte[]
00007ff9c2d08410     4         3596 System.Int32[]
00007ff9c2c5b578     6        17896 System.Object[]
00007ff9c2d0d698    76        39100 System.String
000001a05cd96920 108654      2614440 Free
00007ff9c2d439e8 38913485  933923640 MyClass
Total 39022331 objects
```

代码 16-4 所示为输出的统计信息，其中的类型 MyClass 有 900 多 MB，而 MyClass 的数量也非常惊人，当然，这也是符合预期的。

为什么有这么多的对象没有被回收？是因为它们都有引用吗？其实应归根于应用程序在析构函数中通过随机数设置的堵塞时长。所以，接下来笔者利用!finalizequeue 指令查看终结者队列（Finalization Queue），在终结者队列中保存了处于存活状态的可终结对象，如代码 16-5 所示。

代码16-5

```
0:000> !finalizequeue
SyncBlocks to be cleaned up: 0
Free-Threaded Interfaces to be released: 0
MTA Interfaces to be released: 0
STA Interfaces to be released: 0
--------------------------------
generation 0 has 2 finalizable objects (000001A08F83B080-> 000001A08F83B090)
generation 1 has 8 finalizable objects (000001A08F83B040-> 000001A08F83B080)
generation 2 has 0 finalizable objects (000001A08F83B040->000001A08F83B040)
Ready for finalization 58973 objects (000001A08F83B090-> 000001A08F8AE378)
Statistics for all finalizable objects (including all objects ready for
finalization):
            MT    Count    TotalSize Class Name
00007ff9c2d439e8  58975    1415400   MyClass
Total 58983 objects
```

终结是一种用于回收对象时执行的某些操作，通常用于释放对象持有的资源。当然，在垃圾回收器中会维护终结者队列，这个队列持有对"预终结"对象的引用，目前还存在 58975 个 MyClass 实例。

其实，即使对象已经被标记为死亡，它们也不会立即被删除，因为它们的终结器尚未执行，所以这些对象会被移到 fReachable（finalization reachable）queue 中，在指令后面添加 -allReady 参数可以筛选出 fReachable queue 的内容，如代码 16-6 所示。

代码16-6

```
0:000> !finalizequeue -allReady
SyncBlocks to be cleaned up: 0
Free-Threaded Interfaces to be released: 0
MTA Interfaces to be released: 0
STA Interfaces to be released: 0
--------------------------------
generation 0 has 2 finalizable objects (000001A08F83B080-> 000001A08F83B090)
generation 1 has 8 finalizable objects (000001A08F83B040-> 000001A08F83B080)
generation 2 has 0 finalizable objects (000001A08F83B040-> 000001A08F83B040)
```

```
Finalizable but not rooted: 000001a0285472a8 000001a0289472c8
Ready for finalization 58973 objects (000001A08F83B090-> 000001A08F8AE378)
Statistics for all finalizable objects that are no longer rooted:
              MT    Count    TotalSize Class Name
00007ff9c2d439e8 58975      1415400    MyClass
Total 58975 objects
```

16.1.6　BenchmarkDotNet

BenchmarkDotNet 是一个用于.NET 基准测试的类库，是一个简单且易于使用开源的工具包。创建一个控制台项目，如代码 16-7 所示。

代码16-7

```
D:\>dotnet new console
已成功创建模板"控制台应用"。

正在处理创建后的操作...
在 D:\D.csproj 上运行 "dotnet restore"...
  正在确定要还原的项目…
  已还原 D:\D.csproj (用时 77 ms)。
已成功还原。
```

如代码 16-8 所示，安装"BenchmarkDotNet"NuGet 包。

代码16-8

```
D:\>dotnet add package BenchmarkDotNet
```

安装 NuGet 包之后，如代码 16-9 所示，创建一个需要测试的实例，并对需要做性能测试的方法添加 Benchmark 特性。定义一个名为 Md5VsSha256 的类，分别添加 Sha256 方法和 Md5 方法，用于性能测试。

代码16-9

```
[SimpleJob(RuntimeMoniker.Net472)]
[SimpleJob(RuntimeMoniker.Net60)]
[SimpleJob(RuntimeMoniker.NetCoreApp31)]
public class Md5VsSha256
{
```

```csharp
    private SHA256 sha256 = SHA256.Create();
    private MD5 md5 = MD5.Create();
    private byte[] data;

    [Params(1000, 10000)]
    public int N;

    [GlobalSetup]
    public void Setup()
    {
        data = new byte[N];
        new Random(42).NextBytes(data);
    }

    [Benchmark]
    public byte[] Sha256() => sha256.ComputeHash(data);

    [Benchmark]
    public byte[] Md5() => md5.ComputeHash(data);
}
```

通过 SimpleJob 特性指定测试执行的环境，如.NET Framework 472、.NET 6 等。定义 GlobalSetup 特性用于标记为参数初始方法，并在基准测试方法调用之前执行，每个基准测试方法仅执行一次。Params 特性用于标记类中的一个或多个字段或属性，在该特性中，可以指定一组值，每个值都必须是编译时常量，因此，可以获得每个参数值组合的结果。

如代码 16-10 所示，在 Program 类的 Main 方法中调用 BenchmarkRunner.Run<Md5VsSha256> 方法。

代码16-10

```csharp
public class Program
{
    public static void Main(string[] args)
    {
        BenchmarkRunner.Run<Md5VsSha256>();
        Console.ReadLine();
```

        }
}

运行应用程序，如图 16-27 所示，执行并展示相应的测试结果。

```
BenchmarkDotNet=v0.13.1, OS=Windows 10.0.19044.1586 (21H2)
Intel Core i7-1065G7 CPU 1.30GHz, 1 CPU, 8 logical and 4 physical cores
.NET SDK=7.0.100-preview.2.22153.17
  [Host]             : .NET 6.0.3 (6.0.322.12309), X64 RyuJIT
  .NET 6.0           : .NET 6.0.3 (6.0.322.12309), X64 RyuJIT
  .NET Core 3.1      : .NET Core 3.1.23 (CoreCLR 4.700.22.11601, CoreFX 4.700.22.12208), X64 RyuJIT
  .NET Framework 4.7.2 : .NET Framework 4.8 (4.8.4470.0), X64 RyuJIT
```

Method	Job	Runtime	N	Mean	Error	StdDev	Median
Sha256	.NET 6.0	.NET 6.0	1000	1.313 μs	0.0202 μs	0.0188 μs	1.3168 μs
Md5	.NET 6.0	.NET 6.0	1000	2.707 μs	0.0473 μs	0.0507 μs	2.7118 μs
Sha256	.NET Core 3.1	.NET Core 3.1	1000	1.043 μs	0.0324 μs	0.0914 μs	0.9999 μs
Md5	.NET Core 3.1	.NET Core 3.1	1000	2.411 μs	0.0478 μs	0.0550 μs	2.4101 μs
Sha256	.NET Framework 4.7.2	.NET Framework 4.7.2	1000	10.267 μs	0.5199 μs	1.4406 μs	9.9530 μs
Md5	.NET Framework 4.7.2	.NET Framework 4.7.2	1000	3.661 μs	0.0437 μs	0.0409 μs	3.6603 μs
Sha256	.NET 6.0	.NET 6.0	10000	8.740 μs	0.1723 μs	0.1527 μs	8.7271 μs
Md5	.NET 6.0	.NET 6.0	10000	22.506 μs	0.3653 μs	0.3417 μs	22.6690 μs
Sha256	.NET Core 3.1	.NET Core 3.1	10000	8.630 μs	0.1710 μs	0.3717 μs	8.5698 μs
Md5	.NET Core 3.1	.NET Core 3.1	10000	22.940 μs	0.4493 μs	0.6994 μs	23.0216 μs
Sha256	.NET Framework 4.7.2	.NET Framework 4.7.2	10000	85.111 μs	1.6259 μs	1.6697 μs	85.1095 μs
Md5	.NET Framework 4.7.2	.NET Framework 4.7.2	10000	24.330 μs	0.4797 μs	0.5332 μs	24.2369 μs

图 16-27　执行并展示相应的测试结果

16.1.7　LLDB

LLDB 是一个软件调试器，支持 C/C++的调试和 Linux 平台对核心转储（Core Dump）文件的分析。LLDB 和 WinDbg 之间的主要区别在于，LLDB 是命令行调试器，WinDbg 是基于 GUI 的调试器。

另外，LLDB 支持插件，可以通过 SOS 插件调试托管代码。利用 dotnet-sos 命令行工具可以安装 SOS 插件，该插件提供了许多用于调试托管代码的命令，下面展开介绍。

如代码 16-11 所示，安装 LLDB 软件包。

代码16-11

```
sudo apt-get update
sudo apt-get install lldb
```

> **注意**：SOS 插件需要 LLDB 3.9 及更高版本。在默认情况下，某些发行版只有旧版本可用，因此为这些平台提供了构建 LLDB 3.9 的说明和脚本，详情请参考 GitHub 官网中的 diagnostics 项目。

如代码 16-12 所示，要安装 SOS 插件需要先安装 dotnet-sos 命令行工具。

代码 16-12

```
$ dotnet tool install -g dotnet-sos
You can invoke the tool using the following command: dotnet-sos
Tool 'dotnet-sos' (version '6.0.257301') was successfully installed.
```

如代码 16-13 所示，通过 dotnet-sos 命令行工具安装 SOS 插件。

代码 16-13

```
$ dotnet-sos install
Installing SOS to /home/fh/.dotnet/sos
Installing over existing installation...
Creating installation directory...
Copying files from /home/fh/.dotnet/tools/.store/dotnet-sos/6.0.257301/dotnet-sos/6.0.257301/tools/netcoreapp3.1/any/linux-x64
Copying files from /home/fh/.dotnet/tools/.store/dotnet-sos/6.0.257301/dotnet-sos/6.0.257301/tools/netcoreapp3.1/any/lib
Updating existing /home/fh/.lldbinit file - LLDB will load SOS automatically at startup
Cleaning up...
SOS install succeeded
```

接下来就运行 LLDB。当输入 lldb 命令之后会自动加载 SOS 插件并启用符号下载，如代码 16-14 所示，随后输入 soshelp 命令，返回并展开可用的命令列表。

代码 16-14

```
$ lldb
Added Microsoft public symbol server
(lldb) soshelp
-------------------------------------------------------------------------
-----
SOS is a debugger extension DLL designed to aid in the debugging of managed
```

programs. Functions are listed by category, then roughly in order of
importance. Shortcut names for popular functions are listed in parenthesis.
Type "soshelp <functionname>" for detailed info on that function.

```
Object Inspection                  Examining code and stacks
-----------------------------      -----------------------------
DumpObj (dumpobj)                  Threads (clrthreads)
DumpALC (dumpalc)                  ThreadState
DumpArray                          IP2MD (ip2md)
DumpAsync (dumpasync)               u (clru)
DumpDelegate (dumpdelegate)         DumpStack (dumpstack)
DumpStackObjects (dso)             EEStack (eestack)
DumpHeap (dumpheap)                ClrStack (clrstack)
DumpVC                             GCInfo
FinalizeQueue (finalizequeue)       EHInfo
GCRoot (gcroot)                    bpmd (bpmd)
PrintException (pe)

Examining CLR data structures      Diagnostic Utilities
-----------------------------      -----------------------------
DumpDomain (dumpdomain)            VerifyHeap
EEHeap (eeheap)                    FindAppDomain
Name2EE (name2ee)                  DumpLog (dumplog)
SyncBlk (syncblk)                  SuppressJitOptimization
DumpMT (dumpmt)                    ThreadPool (threadpool)
DumpClass (dumpclass)
DumpMD (dumpmd)
Token2EE
DumpModule (dumpmodule)
DumpAssembly (dumpassembly)
DumpRuntimeTypes
DumpIL (dumpil)
DumpSig
DumpSigElem
```

```
Examining the GC history          Other
--------------------------        --------------------------
HistInit (histinit)               SetHostRuntime (sethostruntime)
HistRoot (histroot)               SetSymbolServer (setsymbolserver, loadsymbols)
HistObj  (histobj)                SetClrPath (setclrpath)
HistObjFind (histobjfind)         SOSFlush (sosflush)
HistClear (histclear)             SOSStatus (sosstatus)
                                  FAQ
                                  Help (soshelp)

Usage:
 > [command]

Commands:
 logging     Enable/disable internal logging
(lldb)
```

如代码 16-15 所示，附加调试的进程。

代码16-15

```
(lldb) process attach --pid 7738
```

使用 finalizequeue 命令可以查看终结者队列的内容，如代码 16-16 所示。

代码16-16

```
(lldb) finalizequeue
SyncBlocks to be cleaned up: 0
----------------------------------
generation 0 has 0 finalizable objects (00007F953C7E4070-> 00007F953C7E4070)
generation 1 has 12 finalizable objects (00007F953C7E4010-> 00007F953C7E4070)
generation 2 has 0 finalizable objects (00007F953C7E4010-> 00007F953C7E4010)
Ready for finalization 5 objects (00007F953C7E4070->00007F953C7E4098)
Statistics for all finalizable objects (including all objects ready for
finalization):
            MT    Count    TotalSize Class Name
00007f963a626ea0    1         24     MyClass
00007f963a6229f8    2         48
```

```
        System.WeakReference`1 [[System. Diagnostics.Tracing.EventSource,
System.Private.CoreLib]]
00007f963a6251b0       1         184
        System.Diagnostics.Tracing. NativeRuntimeEventSource
00007f963a60b280       1         384
                 System.Diagnostics.Tracing. RuntimeEventSource
00007f963a621f28       4         448
              System.Diagnostics.Tracing. EventSource+OverrideEventProvider
00007f963a62a6f8       8         512 Microsoft.Win32.SafeHandles.SafeFileHandle
Total 17 objects
```

如代码 16-17 所示，更新和删除 LLDB 配置及卸载 SOS 插件可以使用如下命令。

代码16-17

```
$ dotnet tool update -g dotnet-sos
$ dotnet-sos uninstall
Uninstalling SOS from /home/fh/.dotnet/sososhelp)
Reverting /home/fh/.lldbinit file - LLDB will no longer load SOS at startup
SOS uninstall succeeded
$ dotnet tool uninstall -g dotnet-sos
Tool 'dotnet-sos' (version '6.0.257301') was successfully uninstalled.
```

16.1.8　dotnet-dump

　　dotnet-dump 是一个集跨平台转储收集和分析托管代码于一身的命令行工具，并不涉及本机调试，但可以为 .NET 应用程序快速创建转储，并且可以在不适合 LLDB 的平台上使用。

　　dotnet-dump 只能在 .NET Core 3.1 及更高版本的 SDK 中才可以安装，安装 dotnet-dump 如代码 16-18 所示。

代码16-18

```
$ dotnet tool install -g dotnet-dump
You can invoke the tool using the following command: dotnet-dump
Tool 'dotnet-dump' (version '6.0.257301') was successfully installed.
```

　　如代码 16-19 所示，创建转储。

代码16-19

```
$ dotnet-dump collect --process-id 630

Writing full to /root/core_20220103_230816
Complete
```

随后可以利用以下命令进行转储分析，如代码 16-20 所示。

代码16-20

```
$ dotnet-dump analyze /root/core_20220103_230816
Loading core dump: /root/core_20220103_230816 ...
Ready to process analysis commands. Type 'help' to list available commands or
'help [command]' to get detailed help on a command.
Type 'quit' or 'exit' to exit the session.
>
```

如代码 16-21 所示，可以利用 clrstack 命令查看调用栈。

代码16-21

```
$ > clrstack
OS Thread Id: 0x276 (0)
        Child SP               IP Call Site
00007FFD16B21778 00007f34e597caed [HelperMethodFrame: 00007ffd16b21778]
00007FFD16B218A0        00007F346C5B3155        Program.Main(System.String[])
[/mnt/c/Users/hueif/OneDrive/.NET/dotnet-booklab/Part15/02/Program.cs @ 17]
>
```

dotnet-dump 还支持更多的指令，可以通过 help 命令来查看，如代码 16-22 所示。

代码16-22

```
> help
Usage:
  > [command]

Commands:
  d, readmemory <address>              Dump memory contents.
  db <address>                         Dump memory as bytes.
  dc <address>                         Dump memory as chars.
  da <address>                         Dump memory as zero-terminated byte strings.
```

```
du <address>                    Dump memory as zero-terminated char strings.
dw <address>                    Dump memory as words (ushort).
dd <address>                    Dump memory as dwords (uint).
dp <address>                    Dump memory as pointers.
dq <address>                    Dump memory as qwords (ulong).
//...
```

如代码 16-23 所示，卸载 dotnet-dump。

代码 16-23

```
$ dotnet tool uninstall -g dotnet-dump
Tool 'dotnet-dump' (version '6.0.257301') was successfully uninstalled.
```

16.1.9 dotnet-gcdump

dotnet-gcdump 是一个跨平台的命令行工具，通过它可以收集有关 .NET 进程的垃圾回收信息。

如代码 16-24 所示，安装 dotnet-gcdump。

代码 16-24

```
$ dotnet tool install --global dotnet-gcdump
You can invoke the tool using the following command: dotnet-gcdump
Tool 'dotnet-gcdump' (version '6.0.257301') was successfully installed.
```

如代码 16-25 所示，使用 dotnet gcdump ps 命令找出正在运行的 .NET 进程及进程 ID。

代码 16-25

```
$ dotnet gcdump ps
    1486  02               /mnt/c/Users/hueif/dotnet-booklab/Part15/02/bin/Debug/net6.0/02
     536  dotnet           /usr/share/dotnet/dotnet
     793  dotnet           /usr/share/dotnet/dotnet
     850  dotnet           /usr/share/dotnet/dotnet
     725  dotnet-dump      /root/.dotnet/tools/dotnet-dump
     810  dotnet-gcdump    /root/.dotnet/tools/dotnet-gcdump
```

如代码 16-26 所示，通过 dotnet gcdump collect 命令获取 gcdump 文件。

代码 16-26

```
$ dotnet gcdump collect -p 1486
```

```
Writing gcdump to '/root/20220103_232633_1486.gcdump'...
    Finished writing 192655912 bytes.
```

与 dotnet-dump 不同，使用 dotnet-gcdump 创建的转储可以用于 PerfView 和 Visual Studio（见图 16-28），这意味着无论进程是在 Linux 平台、macOS 平台还是 Windows 平台上运行，当使用 dotnet-gcdump 转储 gcdump 文件时，必须通过安装 Windows 机器来分析转储。

卸载 dotnet-gcdump 命令行工具，如代码 16-27 所示。

代码16-27

```
$ dotnet tool uninstall -g dotnet-gcdump
Tool 'dotnet-gcdump' (version '6.0.257301') was successfully uninstalled.
```

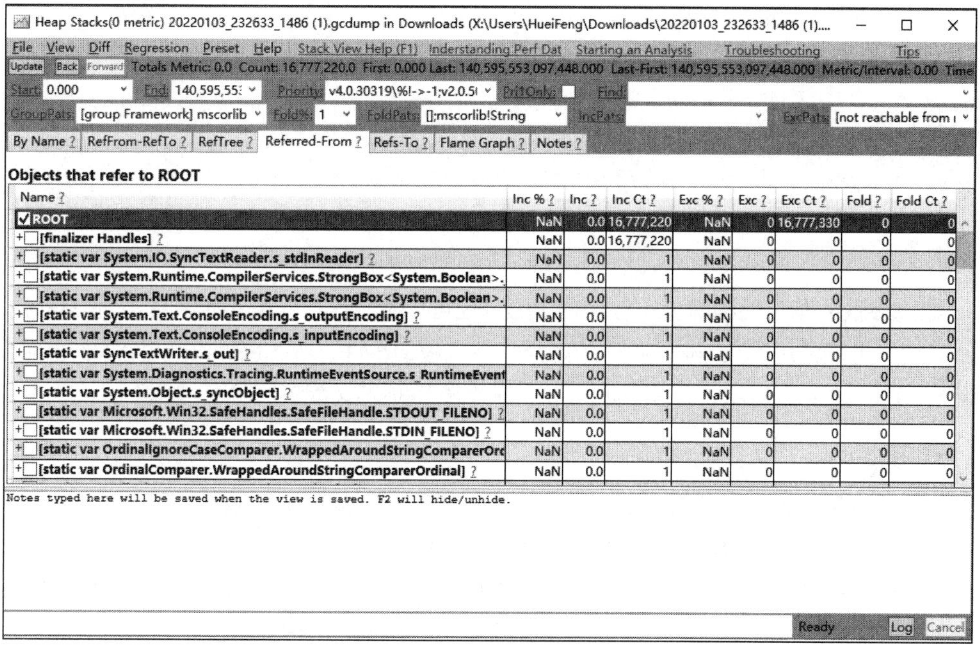

图 16-28　打开 gcdump 文件

16.1.10　dotnet-trace

dotnet-trace 是一个跨平台进程收集工具，并且是基于 EventPipe 构建的。使用 dotnet-trace 可以从正在运行的进程中收集诊断跟踪信息。

如代码 16-28 所示，安装 dotnet-trace。

代码 16-28

```
$ dotnet tool install --global dotnet-trace
You can invoke the tool using the following command: dotnet-trace
Tool 'dotnet-trace' (version '6.0.257301') was successfully installed.
```

如代码 16-29 所示，使用 dotnet trace ps 命令找出正在运行的.NET 进程及进程 ID。

代码 16-29

```
$ dotnet trace ps
943 02            /mnt/x/users/hueifeng/Part15/02/bin/Debug/net6.0/02
692 dotnet        /usr/share/dotnet/dotnet
1705 dotnet       /usr/share/dotnet/dotnet
```

如代码 16-30 所示，创建跟踪。

代码 16-30

```
$ dotnet trace collect -p <pid> [--buffersize 256] [-o <outputpath>]
[--providers <providerlist>]
[--profile <profilename>] [--format NetTrace|Speedscope]
```

通过参数 p 指定要跟踪应用程序的 PID，也可以指定缓存区大小（默认值为 256MB）。除此之外，可以利用参数--providers 指定添加感兴趣的事件，当然也可以通过参数--profile 来指定。如代码 16-31 所示，使用 list-profiles 命令查看预定义配置文件。

代码 16-31

```
$ dotnet trace list-profiles
cpu-sampling     - Useful for tracking CPU usage and general .NET runtime
information. This is the default option if no profile or providers are specified.
    gc-verbose   - Tracks GC collections and samples object allocations.
    gc-collect   - Tracks GC collections only at very low overhead.
    database     - Captures ADO.NET and Entity Framework database commands
```

如果未指定参数--providers 和--profile，则默认使用 cpu-sampling 配置文件。另外，可以利用","指定多个--provider 分隔它们，在默认情况下，允许捕获所有的关键字和级别事件。开发人员可以使用[<keyword_hex_nr>]:<loglevel_nr>在参数--providers 后面指定。例如，要跟踪 Microsoft-System-Net-Http 事件的 EventLevel.Informational，也可以指定--providers Microsoft-System-Net-Http::4，"*"可以用于跟踪所有事件（--providers '*'）。

如代码 16-32 所示，使用 dotnet trace collect 命令开始收集。

代码16-32

```
$ dotnet trace collect -p 943
No profile or providers specified, defaulting to trace profile 'cpu-sampling'

Provider Name                          Keywords              Level              Enabled By
Microsoft-DotNETCore-SampleProfiler    0x0000F00000000000    Informational(4)   --profile
Microsoft-Windows-DotNETRuntime        0x00000014C14FCCBD    Informational(4)   --profile


Process        : /mnt/x/users/hueifeng/Part15/02/bin/Debug/net6.0/02
Output File    : /root/02_20220104_130703.nettrace
    [00:00:12:47]   Recording trace 80.8607   (MB)
Press <Enter> or <Ctrl+C> to exit...167   (MB)
Stopping the trace. This may take several minutes depending on the application
being traced.


    Trace completed.
```

至此，收集已经结束，下面开始分析，在 PerfView 中打开 trace.nettrace 文件，如图 16-29 所示。

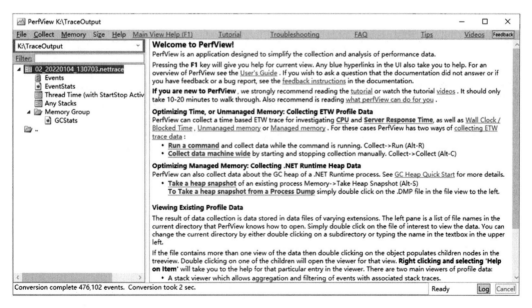

图 16-29　打开 trace.nettrace 文件

在图 16-29 左窗格展示的列表中通过双击 Thread Time，打开 Thread Time 窗口，如图 16-30 所示。在 Thread Time 窗口中可以看到一张表和一些选项卡，下面先介绍它们的含义。

- By Name：方法的名称，由 GroupPats 策略可知，它也可以是一组功能。
- Inc%：整个收集样本中的执行时间百分比，包括调用的方法。
- Inc：与 Inc%相同，是指标的标本数，如果每毫秒（默认）采样一次，那么代表在此方法中花费的毫秒数。
- Exc%：采集占据整个收集样本中的执行时间百分比，不包括被调用者。
- Exc：与 Exc%相同，不同之处在于指标是样本数。
- When：表示方法调用随时间分布，从左（开始时间）到右（结束时间）。
- First：第一次采集样本的时间点。
- Last：最后一次采集样本的时间点。

如图 16-31 所示，通过单击如图 16-29 所示的左窗格中的 Eventstats，可以显示事件计数器。

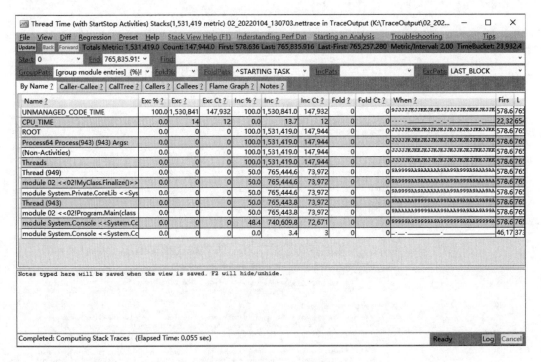

图 16-30　Thread Time 窗口

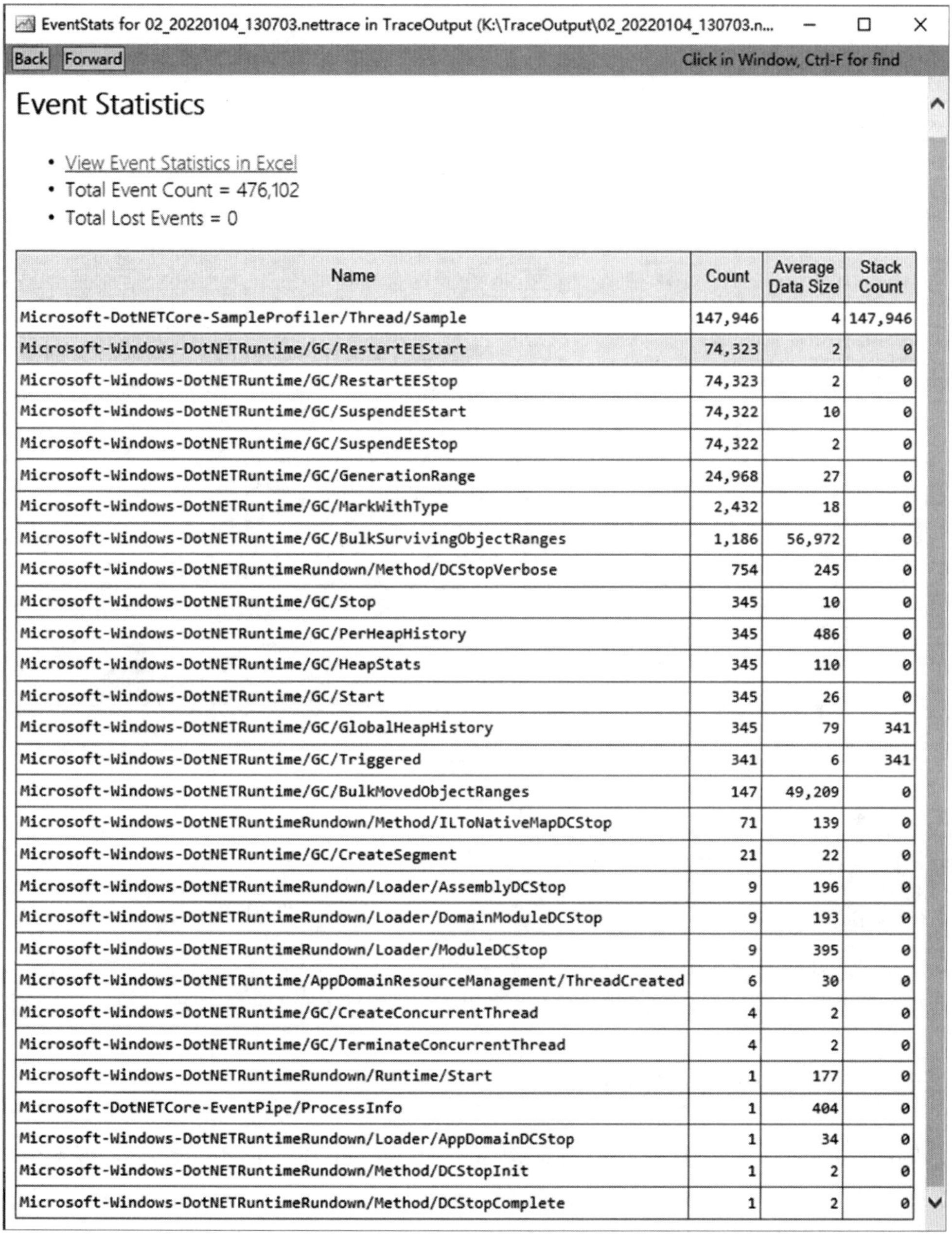

Name	Count	Average Data Size	Stack Count
Microsoft-DotNETCore-SampleProfiler/Thread/Sample	147,946	4	147,946
Microsoft-Windows-DotNETRuntime/GC/RestartEEStart	74,323	2	0
Microsoft-Windows-DotNETRuntime/GC/RestartEEStop	74,323	2	0
Microsoft-Windows-DotNETRuntime/GC/SuspendEEStart	74,322	10	0
Microsoft-Windows-DotNETRuntime/GC/SuspendEEStop	74,322	2	0
Microsoft-Windows-DotNETRuntime/GC/GenerationRange	24,968	27	0
Microsoft-Windows-DotNETRuntime/GC/MarkWithType	2,432	18	0
Microsoft-Windows-DotNETRuntime/GC/BulkSurvivingObjectRanges	1,186	56,972	0
Microsoft-Windows-DotNETRuntimeRundown/Method/DCStopVerbose	754	245	0
Microsoft-Windows-DotNETRuntime/GC/Stop	345	10	0
Microsoft-Windows-DotNETRuntime/GC/PerHeapHistory	345	486	0
Microsoft-Windows-DotNETRuntime/GC/HeapStats	345	110	0
Microsoft-Windows-DotNETRuntime/GC/Start	345	26	0
Microsoft-Windows-DotNETRuntime/GC/GlobalHeapHistory	345	79	341
Microsoft-Windows-DotNETRuntime/GC/Triggered	341	6	341
Microsoft-Windows-DotNETRuntime/GC/BulkMovedObjectRanges	147	49,209	0
Microsoft-Windows-DotNETRuntimeRundown/Method/ILToNativeMapDCStop	71	139	0
Microsoft-Windows-DotNETRuntime/GC/CreateSegment	21	22	0
Microsoft-Windows-DotNETRuntimeRundown/Loader/AssemblyDCStop	9	196	0
Microsoft-Windows-DotNETRuntimeRundown/Loader/DomainModuleDCStop	9	193	0
Microsoft-Windows-DotNETRuntimeRundown/Loader/ModuleDCStop	9	395	0
Microsoft-Windows-DotNETRuntime/AppDomainResourceManagement/ThreadCreated	6	30	0
Microsoft-Windows-DotNETRuntime/GC/CreateConcurrentThread	4	2	0
Microsoft-Windows-DotNETRuntime/GC/TerminateConcurrentThread	4	2	0
Microsoft-Windows-DotNETRuntimeRundown/Runtime/Start	1	177	0
Microsoft-DotNETCore-EventPipe/ProcessInfo	1	404	0
Microsoft-Windows-DotNETRuntimeRundown/Loader/AppDomainDCStop	1	34	0
Microsoft-Windows-DotNETRuntimeRundown/Method/DCStopInit	1	2	0
Microsoft-Windows-DotNETRuntimeRundown/Method/DCStopComplete	1	2	0

图 16-31　事件计数器

通过单击如图 16-29 所示的 Events 节点，可以展开事件的详情，如图 16-32 所示。

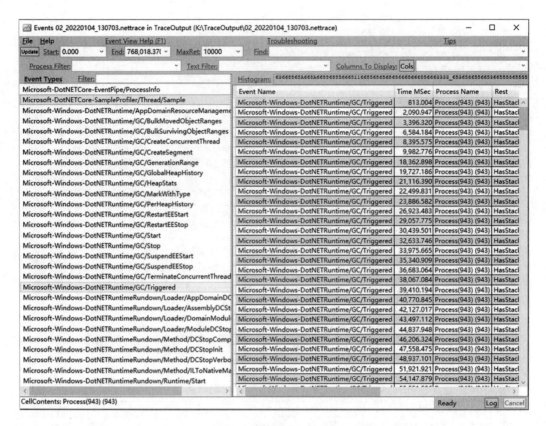

图 16-32　事件的详情

如代码 16-33 所示,卸载 dotnet-trace。

代码16-33

```
$ dotnet tool uninstall -g dotnet-trace
Tool 'dotnet-trace' (version '6.0.257301') was successfully uninstalled.
```

16.1.11　dotnet-counters

dotnet-counters 是一个性能监视工具,是基于 EventCounter API 发布的指标监视器,可以用于收集 CPU、内存、垃圾回收器、线程等指标信息。

如代码 16-34 所示,安装 dotnet-counters。

代码16-34

```
$ dotnet tool install --global dotnet-counters
You can invoke the tool using the following command: dotnet-counters
```

```
Tool 'dotnet-counters' (version '6.0.257301') was successfully installed.
```

如代码 16-35 所示，使用 dotnet-counters ps 命令找出正在运行的.NET 进程及进程 ID。

代码16-35

```
$ dotnet-counters ps
707 02         /mnt/x/Users/HueiFeng/Part15/02/bin/Debug/net6.0/02
391 dotnet     /usr/share/dotnet/dotnet
653 dotnet     /usr/share/dotnet/dotnet
```

如代码 16-36 所示，查看收集的计数器信息，目前这些计数器信息由 System.Runtime 和 Microsoft.AspNetCore.Hosting 提供程序公开。

代码16-36

```
$ dotnet-counters list
Showing well-known counters for .NET (Core) version 3.1 only. Specific
processes may support additional counters.
System.Runtime
    cpu-usage                       The percent of process' CPU usage relative to
all of the system CPU resources [0-100]
    working-set                     Amount of working set used by the process (MB)
    gc-heap-size                    Total heap size reported by the GC (MB)
    gen-0-gc-count                  Number of Gen 0 GCs between update intervals
    gen-1-gc-count                  Number of Gen 1 GCs between update intervals
    gen-2-gc-count                  Number of Gen 2 GCs between update intervals
    time-in-gc                      % time in GC since the last GC
    gen-0-size                      Gen 0 Heap Size
    gen-1-size                      Gen 1 Heap Size
    gen-2-size                      Gen 2 Heap Size
    loh-size                        LOH Size
    alloc-rate                      Number of bytes allocated in the managed heap
between update intervals
    assembly-count                  Number of Assemblies Loaded
    exception-count                 Number of Exceptions / sec
    threadpool-thread-count         Number of ThreadPool Threads
    monitor-lock-contention-count   Number of times there were contention
when trying to take the monitor lock between update intervals
    threadpool-queue-length         ThreadPool Work Items Queue Length
```

```
    threadpool-completed-items-count     ThreadPool Completed Work Items Count
    active-timer-count                Number of timers that are currently active

Microsoft.AspNetCore.Hosting
    requests-per-second              Number of requests between update intervals
    total-requests                   Total number of requests
    current-requests                 Current number of requests
    failed-requests                  Failed number of requests
```

如代码 16-37 所示，收集监控指标，并将结果默认保存到 counter.csv 文件中。

代码16-37

```
$ dotnet-counters collect -p 707
    --counters is unspecified. Monitoring System.Runtime counters by default.
    Starting a counter session. Press Q to quit.
    File saved to counter.csv
```

导出的 counter.csv 文件如图 16-33 所示。

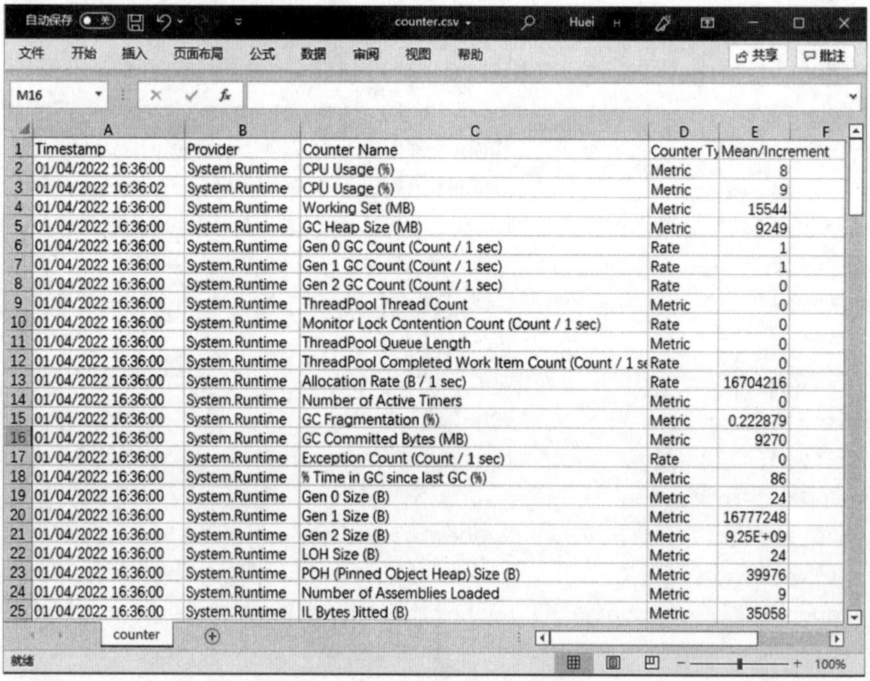

	A	B	C	D	E	F
1	Timestamp	Provider	Counter Name	Counter Ty	Mean/Increment	
2	01/04/2022 16:36:00	System.Runtime	CPU Usage (%)	Metric	8	
3	01/04/2022 16:36:02	System.Runtime	CPU Usage (%)	Metric	9	
4	01/04/2022 16:36:00	System.Runtime	Working Set (MB)	Metric	15544	
5	01/04/2022 16:36:00	System.Runtime	GC Heap Size (MB)	Metric	9249	
6	01/04/2022 16:36:00	System.Runtime	Gen 0 GC Count (Count / 1 sec)	Rate	1	
7	01/04/2022 16:36:00	System.Runtime	Gen 1 GC Count (Count / 1 sec)	Rate	1	
8	01/04/2022 16:36:00	System.Runtime	Gen 2 GC Count (Count / 1 sec)	Rate	0	
9	01/04/2022 16:36:00	System.Runtime	ThreadPool Thread Count	Metric	0	
10	01/04/2022 16:36:00	System.Runtime	Monitor Lock Contention Count (Count / 1 sec)	Rate	0	
11	01/04/2022 16:36:00	System.Runtime	ThreadPool Queue Length	Metric	0	
12	01/04/2022 16:36:00	System.Runtime	ThreadPool Completed Work Item Count (Count / 1 se	Rate	0	
13	01/04/2022 16:36:00	System.Runtime	Allocation Rate (B / 1 sec)	Rate	16704216	
14	01/04/2022 16:36:00	System.Runtime	Number of Active Timers	Metric	0	
15	01/04/2022 16:36:00	System.Runtime	GC Fragmentation (%)	Metric	0.222879	
16	01/04/2022 16:36:00	System.Runtime	GC Committed Bytes (MB)	Metric	9270	
17	01/04/2022 16:36:00	System.Runtime	Exception Count (Count / 1 sec)	Rate	0	
18	01/04/2022 16:36:00	System.Runtime	% Time in GC since last GC (%)	Metric	86	
19	01/04/2022 16:36:00	System.Runtime	Gen 0 Size (B)	Metric	24	
20	01/04/2022 16:36:00	System.Runtime	Gen 1 Size (B)	Metric	16777248	
21	01/04/2022 16:36:00	System.Runtime	Gen 2 Size (B)	Metric	9.25E+09	
22	01/04/2022 16:36:00	System.Runtime	LOH Size (B)	Metric	24	
23	01/04/2022 16:36:00	System.Runtime	POH (Pinned Object Heap) Size (B)	Metric	39976	
24	01/04/2022 16:36:00	System.Runtime	Number of Assemblies Loaded	Metric	9	
25	01/04/2022 16:36:00	System.Runtime	IL Bytes Jitted (B)	Metric	35058	

图 16-33　导出的 counter.csv 文件

如代码 16-38 所示，以 10 秒的刷新间隔收集运行时性能计数器，并将其导出到名为 test 的 JSON 文件中。

代码16-38

```
$ dotnet-counters collect --process-id 707 --refresh-interval 10 --output test --format json
```

除此之外，还可以指定要收集的类型指标，如代码 16-39 所示。

代码16-39

```
//以 3 秒的刷新间隔监视 System.Runtime 运行时信息
dotnet-counters monitor --process-id 707  --refresh-interval 3 System.Runtime

//以 3 秒的刷新间隔监视 Microsoft.AspNetCore.Hosting 运行信息
dotnet-counters monitor --process-id 707
                        --refresh-interval 3 Microsoft.AspNetCore.Hosting
```

收集实时监控指标，以 3 秒的刷新间隔收集运行时性能计数器，该结果会直接输出到控制台中，如代码 16-40 所示。

代码16-40

```
dotnet-counters monitor -p 707 --refresh-interval 3
Press p to pause, r to resume, q to quit.
Status: Running

[System.Runtime]
% Time in GC since last GC (%)                           83
Allocation Rate (B / 3 sec)                       33,456,104
CPU Usage (%)                                             7
Exception Count (Count / 3 sec)                           0
GC Committed Bytes (MB)                               6,119
GC Fragmentation (%)                                  0.226
GC Heap Size (MB)                                     6,105
Gen 0 GC Count (Count / 3 sec)                            2
Gen 0 Size (B)                                           24
Gen 1 GC Count (Count / 3 sec)                            1
Gen 1 Size (B)                                   16,777,248
Gen 2 GC Count (Count / 3 sec)                            0
```

```
Gen 2 Size (B)                                           6.1023e+09
IL Bytes Jitted (B)                                      28,420
LOH Size (B)                                             24
Monitor Lock Contention Count (Count / 3 sec)            0
Number of Active Timers                                  0
Number of Assemblies Loaded                              9
Number of Methods Jitted                                 211
POH (Pinned Object Heap) Size (B)                        39,976
ThreadPool Completed Work Item Count (Count / 3 sec)     0
ThreadPool Queue Length                                  0
ThreadPool Thread Count                                  0
Time spent in JIT (ms / 3 sec)                           0
Working Set (MB)                                         10,220
```

如代码 16-41 所示，卸载 dotnet-counters。

代码16-41

```
$ dotnet tool uninstall -g dotnet-counters
Tool 'dotnet-counters' (version '6.0.257301') was successfully uninstalled.
```

16.1.12　dotnet-symbol

dotnet-symbol 可以用于下载符号文件（Symbol Files）、模块文件等。符号文件包含应用程序二进制文件的调试信息，以 .pdb 为扩展名。

如代码 16-42 所示，安装 dotnet-symbol。

代码16-42

```
$ dotnet tool install -g dotnet-symbol
You can invoke the tool using the following command: dotnet-symbol
Tool 'dotnet-symbol' (version '1.0.252801') was successfully installed.
```

如代码 16-43 所示，使用 ps 命令获取进程信息，并筛选有关 dotnet 的进程。

代码16-43

```
$ ps -ef|grep dotnet
root        77    62 90 11:50 pts/1    00:00:02 dotnet 02.dll
root        87    29  0 11:50 pts/0    00:00:00 grep --color=auto dotnet
```

接下来通过 createdump 生成转储文件，如代码 16-44 所示，createdump 在安装 .NET SDK

时下载下来,开发人员可以使用/usr/share/dotnet/shared/Microsoft.NETCore.App/<version>/createdump PID 创建转储文件。

代码16-44

```
$ /usr/share/dotnet/shared/Microsoft.NETCore.App/6.0.1/createdump 77
Gathering state for process 77 dotnet
Writing minidump with heap to file /tmp/coredump.77
Written 4779175936 bytes (1166791 pages) to core file
Dump successfully written
```

获取到转储文件之后,在 tmp 文件夹下创建 dump 目录,并将转储文件复制到 dump 目录下,如代码 16-45 所示。

代码16-45

```
$ mkdir /tmp/dump
$ cp /tmp/coredump.77 /tmp/dump
```

如代码 16-46 所示,使用 dotnet-symbol 命令下载模块和符号。

代码16-46

```
$ dotnet-symbol /tmp/dump/coredump.77
```

启动 LLDB,并且通过参数--core 指定转储文件的地址,如代码 16-47 所示。

代码16-47

```
$ lldb --core /tmp/dump/coredump.77
Added Microsoft public symbol server
(lldb)
```

如代码 16-48 所示,将符号目录通过 setsymbolserver 命令来指定路径。

代码16-48

```
(lldb) setsymbolserver -directory /tmp/dump
 Added symbol directory path: /tmp/dump
(lldb)
```

如代码 16-49 所示,可以通过 clrstack 命令来获取相应的信息。

代码16-49

```
(lldb) clrstack
OS Thread Id: 0x4d (1)
```

```
         Child SP               IP Call Site
00007FFFD4B01418 00007f01204c4bdd [HelperMethodFrame: 00007fffd4b01418]
00007FFFD4B01540      00007F00A70E3128      Program.Main(System.String[])
[X:\Users\HueiFeng\Part15\02\Program.cs @ 17]
(lldb) clrthreads
ThreadCount:      3
UnstartedThread:  0
BackgroundThread: 2
PendingThread:    0
DeadThread:       0
Hosted Runtime:   no

Lock
 DBG   ID     OSID ThreadOBJ            State GC Mode     GC Alloc Context
Domain         Count Apt Exception
   1    1       4d 0000564722C4BD90     20020 Cooperative 0000000000000000:
0000000000000000 0000564722C336E0 -00001 Ukn (GC)
   6    2       52 0000564722C593B0     21220 Preemptive  0000000000000000:
0000000000000000 0000564722C336E0 -00001 Ukn (Finalizer)
   8    3       59 0000564722C77B80     21220 Preemptive  0000000000000000:
0000000000000000 0000564722C336E0 -00001 Ukn
(lldb)
```

如代码 16-50 所示，卸载 dotnet-symbol。

代码16-50

```
$ dotnet tool uninstall -g dotnet-symbol
Tool 'dotnet-symbol' (version '1.0.252801') was successfully uninstalled.
```

16.2 Linux 调试

在 Windows 环境下可以利用 Visual Studio 等进行代码调试，但是在 Linux 环境下应该如何调试呢？其实方法也不少，可以利用 WSL2 在 Visual Studio 中调试 Linux 环境下的代码，方便模拟和测试，除此之外，还可以利用远程调试，或者利用容器工具对 Linux 环境下的代码进行调试。

16.2.1 在 WSL2 中调试

Microsoft 提供了 Windows Subsystem for Linux（WSL），在 Windows 环境下可以通过 Visual Studio 调试 Linux/WSL2 上正在运行的应用程序。即使开发人员使用的是 Windows 操作系统，也可以通过它面向 Linux 环境进行调试。

如图 16-34 所示，Visual Studio 会自动识别 WSL 的安装，当然，也可以选择手动添加 WSL 配置来运行和调试。

在菜单中选择 WSL 命令时，Visual Studio 会在"解决方案资源管理器"面板中添加一个 Properties\launchSettings.json 文件，如图 16-35 所示。

图 16-34　WSL 调试器

图 16-35　文件结构

launchSettings.json 文件如代码 16-51 所示。

代码16-51

```
{
  "profiles": {
    "ConsoleApp2": {
      "commandName": "Project"
    },
    "WSL": {
      "commandName": "WSL2",
      "environmentVariables": {},
      "distributionName": ""
    }
  }
}
```

在 launchSettings.json 文件中可以自定义多个 WSL2 启动配置，如分别添加 Ubuntu 和 Debian 的配置项，如代码 16-52 所示。

代码16-52

```json
{
  "profiles": {
    "ConsoleApp2": {
      "commandName": "Project"
    },
    "WSL": {
      "commandName": "WSL2",
      "environmentVariables": {},
      "distributionName": ""
    },
    "WSL 2 : Ubuntu 20.04": {
      "commandName": "WSL2",
      "distributionName": "Ubuntu-20.04"
    },
    "WSL 2 : Debian": {
      "commandName": "WSL2",
      "distributionName": "Debian"
    }
  }
}
```

图 16-36 所示为多种可选的调试器。

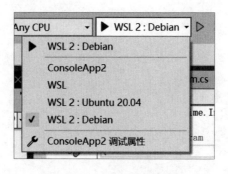

图 16-36　多种可选的调试器

创建一个控制台实例，如代码 16-53 所示，打印计算机相关的信息。

代码16-53

```
using System.Runtime.InteropServices;
public class Program
{
    public static void Main(string[] args)
    {
        Console.WriteLine(
                $"系统架构：{RuntimeInformation.OSArchitecture}");
        Console.WriteLine(
                $"系统名称：{RuntimeInformation.OSDescription}");
        Console.WriteLine(
                $"进程架构：{RuntimeInformation.ProcessArchitecture}");
        Console.WriteLine(
            $"是否为64位操作系统：{Environment.Is64BitOperatingSystem}");
        Console.Read();
    }
}
```

WSL 环境下的输出结果如图 16-37 所示，Windows 环境下的输出结果如图 16-38 所示。

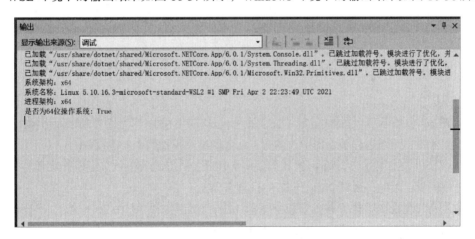

图 16-37　WSL 环境下的输出结果

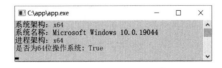

图 16-38　Windows 环境下的输出结果

16.2.2　使用 Visual Studio（Code）远程调试 Linux 部署的应用

Visual Studio 和 Visual Studio Code 支持远程调试，使开发人员可以在本地轻松地调试线上的应用进程。Visual Studio 和 Visual Studio Code 的 Remote 功能都可以支持 3 种不同场景的远程调试，Remote-SSH 利用 SSH 连接远程主机进行调试，Remote-Container 通过连接容器进行调试，Remote-WSL 通过连接子系统（Windows Subsystem for Linux）进行调试。

图 16-39 所示为 Visual Studio Code 远程调试进程的过程，通过 Visual Studio Code 的 SSH 模式调试远程的应用进程。

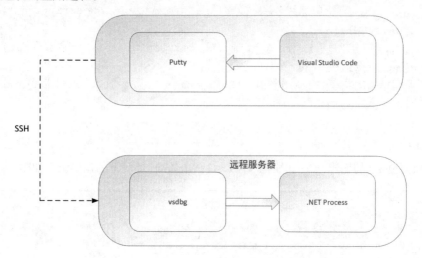

图 16-39　Visual Studio Code 远程调试进程的过程

如图 16-39 所示，Visual Studio Code 通过 Putty 利用 SSH 作为远程调试的传输协议，连接远程服务器，而在远程服务器中，通过 vsdbg 调试器附加到.NET 进程中。

接下来以 SSH 模式一步步配置和调试应用程序。Visual Studio Code 中无法弹出 UI 交互窗口进行授权，因此，通过 Putty 连接 SSH 服务器。

先进行环境配置，再下载并安装 Putty（Putty 安装列表如图 16-40 所示）。Putty 是一个支持 Telnet、SSH、Rlogin 的纯 TCP 协议的串口连接工具，这里需要使用它的 SSH 功能。

如代码 16-54 所示，在目标服务器上安装 vsdbg 调试器。

代码16-54

```
curl -sSL https://******.ms/getvsdbgsh | bash /dev/stdin -v latest -l ~/vsdbg
```

如图 16-41 所示，打开 Visual Studio Code，先单击左侧栏中的 Debugger 图标，打开 Debug 视图，再单击"运行和调试"按钮，弹出下拉列表，选择.NET Auto Attach 选项（需要具备 Visual

Studio Code C#扩展）即可在项目目录中创建一个.vscode 目录和一个 launch.json 文件。

图 16-40　Putty 安装列表

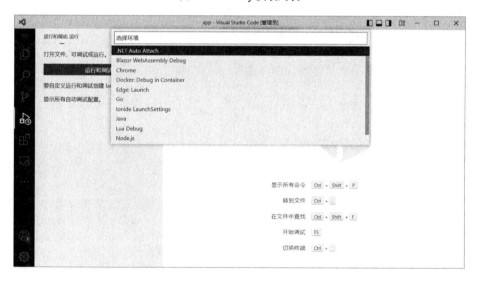

图 16-41　打开 Visual Studio Code

launch.json 文件默认的配置内容如代码 16-55 所示。

代码16-55

```
{
    "version": "0.2.0",
    "configurations": [
```

```json
{
    "name": ".NET Core Launch (web)",
    "type": "coreclr",
    "request": "launch",
    "preLaunchTask": "build",
    "program": "${workspaceFolder}/bin/Debug/net7.0/app.dll",
    "args": [],
    "cwd": "${workspaceFolder}",
    "stopAtEntry": false,
    "serverReadyAction": {
        "action": "openExternally",
        "pattern": "\\bNow listening on:\\s+(https?://\\S+)"
    },
    "env": {
        "ASPNETCORE_ENVIRONMENT": "Development"
    },
    //"envFile": "${workspaceFolder}/.env",
    "sourceFileMap": {
        "/Views": "${workspaceFolder}/Views"
    }
},
{
    "name": ".NET Core Attach",
    "type": "coreclr",
    "request": "attach"
}
]
}
```

launch.json 文件中常见的属性如表 16-1 所示。

表 16-1　launch.json 文件中常见的属性

属性	说明
name	配置名称，显示在启动配置调试界面下拉菜单中的名称
type	配置类型，设置为 coreclr，Visual Studio Code 调试代码时指定的扩展类型
program	调试程序的路径（绝对路径）
env	调试时使用的环境变量
envFile	环境变量文件的绝对路径，在 env 中设置的属性会覆盖 envFile 中的配置
args	为需要调试的程序指定相关的命令行参数

如代码 16-56 所示，添加远程调试的相关配置信息。

代码16-56

```json
{
    "name": ".NET Core Remote Attach",
    "type": "coreclr",
    "request": "attach",
    "processId": "${command:pickRemoteProcess}",
    "pipeTransport":
    {
        "debuggerPath": "~/vsdbg/vsdbg",
        "pipeCwd": "${workspaceFolder}",
        "windows": {
            "pipeProgram": "plink.exe",
            "pipeArgs": [
                "-l",
                "root",
                "-pw",
                "password",
                "localhost",
                "-P",
                "22",
                "-T" ],
            "quoteArgs": true
        }
    },
    "sourceFileMap": {
        "/src": "${workspaceFolder}",
    }
}
```

接下来选择新增的.NET Core Remote Attach 选项，如图 16-42 所示，建立连接后会显示该远程服务器的进程列表。

这样就可以在本地调试远程服务器的.NET 进程，如图 16-43 所示。

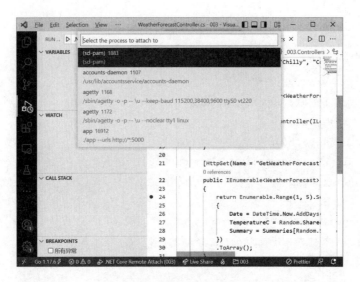

图 16-42　远程服务器的进程列表

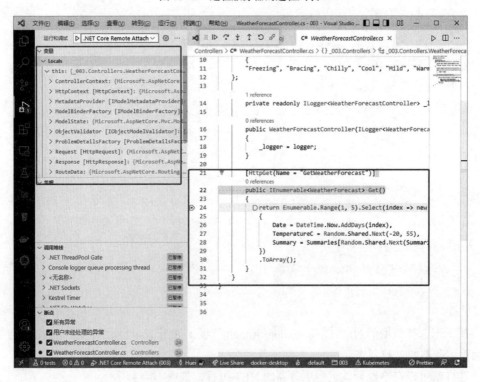

图 16-43　调试远程服务器的.NET 进程

除了可以在本地通过 Visual Studio Code 对远程的.NET 进程进行调试，还可以通过 Visual

Studio 进行调试，并且在 Visual Studio 中调试会变得更加简单。图 16-44 所示为 Visual Studio 远程调试基本的连接过程。

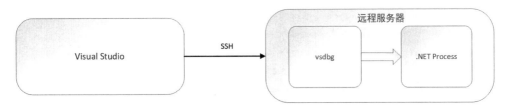

图 16-44　Visual Studio 远程调试基本的连接过程

如图 16-44 所示，当使用 Visual Studio 进行远程调试时，需要利用 SSH 连接到远程服务器的进程后再进行调试。当然，万变不离其宗，远程服务器将 vsdbg 作为调试器进行交互处理。

在 Visual Studio 工具栏中选择"调试"→"附加进程选项"命令，打开如图 16-45 所示的"附加到进程"对话框，在该对话框中将"连接类型"设置为 SSH，同时设置连接目标，输入远程服务器的连接信息，显示上面运行的进程，最后只需要选择需要调试的 dotnet 进程即可。

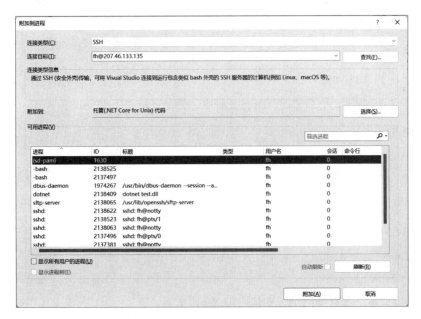

图 16-45　"附加到进程"对话框

综上所述，在 Visual Studio 中通过 SSH 模式调试的步骤很简单，这是因为该 IDE 会在远程服务器建立连接成功后，自动将 vsdbg 下载到.vs-debugger 目录下，根据当前 IDE 的版本在该目录下创建指定版本的文件夹。图 16-46 所示为在 Linux 中搜索 vsdbg 文件。

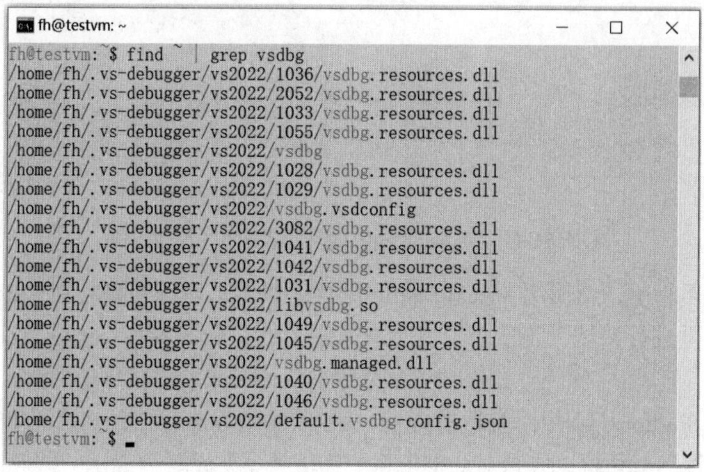

图 16-46 在 Linux 中搜索 vsdbg 文件

除此之外，还可以选择手动安装 vsdbg，但是需要注意的是目录，如代码 16-57 所示，在.vs-debugger 的主目录下创建一个名为 vs2022（根据 Visual Studio 的版本来定义）的文件夹。

代码16-57

```
mkdir -p ~/.vs-debugger/vs2022
```

如代码 16-58 所示，下载 GetVsDbg.sh 脚本，另外还需要下载 vsdbg。

代码16-58

```
wget https://******.ms/getvsdbgsh -O ~/.vs-debugger/GetVsDbg.sh
wget https://vsdebugger.***.net/vsdbg-17-0-10712-2/vsdbg-linux-x64.zip
```

在目标服务器下载完成后，先解压缩到目录下，再删除，如代码 16-59 所示。

代码16-59

```
unzip vsdbg-linux-x64.zip -d ~/.vs-debugger/vs2022
rm vsdbg-linux-x64.zip
```

修改环境变量，加权并测试，如代码 16-60 所示。

代码16-60

```
cd ~/.vs-debugger/vs2022
chmod +x vsdbg
```

```
export DOTNET_SYSTEM_GLOBALIZATION_INVARIANT=true
export COMPlus_LTTng=0
./vsdbg
```

在测试过程中没有出现异常，创建 success.txt 文件，并填入版本号，如代码 16-61 所示。

代码16-61

```
cd ~/.vs-debugger/vs2022
touch success.txt
echo "17.0.10712.2" >> success.txt
```

如代码 16-62 所示，执行如下操作并输出结果。

代码16-62

```
$ ~/.vs-debugger
$ sh GetVsDbg.sh -v vs2022 -l ~/.vs-debugger/vs2022 -d vscode
Info: Last installed version of vsdbg is '17.0.10712.2'
Info: VsDbg is up-to-date
Info: Using vsdbg version '17.0.10712.2'
Using arguments
    Version                     : 'vs2022'
    Location                    : '/root/.vs-debugger/vs2022'
    SkipDownloads               : 'true'
    LaunchVsDbgAfter            : 'true'
        VsDbgMode               : 'vscode'
    RemoveExistingOnUpgrade     : 'false'
Info: Skipping downloads
Info: Launching vsdbg
```

16.2.3 使用容器工具调试

Visual Studio 中的 Docker 容器工具非常易用，并且大大简化了生成、调试和部署容器化应用程序的过程。下面从调试的角度来介绍 Docker 容器工具。

在创建新项目时可以通过勾选"启用 Docker"复选框来启用 Docker 支持，如图 16-47 所示。

注意：对于 .NET Framework 项目来说，只有 Windows 容器可用。

通过在"解决方案资源管理器"面板中选择 Add→"Docker 支持"命令来向现有项目添加 Docker 支持，如图 16-48 所示。

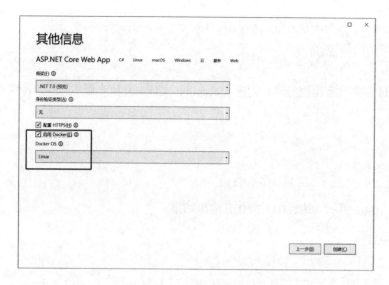

图 16-47　启用 Docker 支持

图 16-48　添加 Docker 支持

当启用或添加 Docker 支持后，Visual Studio 会向项目中添加如下文件和引用。
- Dockerfile 文件。
- .dockerignore 文件。
- Microsoft.VisualStudio.Azure.Containers.Tools.Targets 的 NuGet 包引用。

Dockerfile 文件如代码 16-63 所示。

代码16-63

```
FROM mcr.microsoft.com/dotnet/aspnet:6.0 AS base
WORKDIR /app
EXPOSE 80
EXPOSE 443

FROM mcr.microsoft.com/dotnet/sdk:6.0 AS build
WORKDIR /src
COPY ["test.csproj", "."]
RUN dotnet restore "./test.csproj"
COPY . .
WORKDIR "/src/."
RUN dotnet build "test.csproj" -c Release -o /app/build

FROM build AS publish
RUN dotnet publish "test.csproj" -c Release -o /app/publish

FROM base AS final
WORKDIR /app
COPY --from=publish /app/publish .
ENTRYPOINT ["dotnet", "test.dll"]
```

接下来就可以利用 Docker 调试器对应用程序进行调试，如图 16-49 所示。

图 16-49　Docker 调试器

16.3　小结

本章主要介绍 .NET 技术的调试工具，读者可以利用这些工具对应用程序进行调试。对于调试来说，我们无须专注于某一种调试工具，可以将多种调试工具结合使用。

第 17 章 编译技术精讲

对于有志成为.NET 专家的开发人员来说，学习和了解编译技术尤为重要。通过学习编译技术，读者可以深入了解.NET 平台。本章主要介绍 IL 解析、JIT 编译、AOT 编译等内容。

17.1 IL 解析

IL（Intermediate Language，中间语言）是一种编程语言，并且具备可读性。通过编译器将源代码编译成 IL 代码后，保存到.dll 文件中。通常，使用反编译工具可以查看 IL 代码，以及编译器对它做了哪些优化，或者一些语法糖背后到底做了什么，这是一件很有趣的事情，并且对深入理解.NET 尤为重要。

17.1.1 IL 简介

IL 也称为 CIL 和 MSIL，是一种编程语言。其目标是通过编译器先将源代码编译成 IL 代码，再汇编成字节码。IL 类似于面向对象的汇编语言，等执行的时候会被加载到运行时库，由运行时的 Just-In-Time（JIT）编译器编译成机器代码再执行。

17.1.2 从.NET 代码到 IL 代码

源代码被编译成 IL 代码后，并不是特定于平台或处理器的目标代码。IL 代码是独立于

CPU 和平台指令集的，可以在.NET 运行时支持的任何环境中运行，如可以在 Linux、Windows 等环境中运行同一份代码。

执行过程如下。

（1）将源代码编译为 IL 代码并创建程序集。

（2）在执行 IL 程序集时，代码通过运行时的 JIT 编译器编译为机器代码。

（3）计算机处理器执行机器代码。

如代码 17-1 所示，在 x86 汇编语言中将两个数相加，其中的 eax 和 edx 用于指定不同的通用寄存器。

代码17-1

```
add eax, edx
```

如代码 17-2 所示，在 IL 程序集中，0 表示 eax，1 表示 edx。

代码17-2

```
//把方法的第一个参数压入栈中
ldarg.0
//把方法的第二个参数压入栈中
ldarg.1
//执行 add 操作后，将结果保存到栈顶
add
//把栈顶的值存储到本地或临时变量 0 中
stloc.0
```

寄存器 eax 和 edx 的值先推送到栈中，再调用 add 指令将结果保存到栈顶，最后出栈的结果值将保存在 eax 寄存器中。

类和静态方法如代码 17-3 所示。

代码17-3

```
.class public Foo
{
  .method private static int32
    Add(
      int32 a,
      int32 b
    ) cil managed
  {
    .maxstack 8
```

```
    //加载第一个参数
    IL_0000: ldarg.0      //a
    //加载第二个参数
    IL_0001: ldarg.1      //b
    //执行 add 操作，将栈中最上面的两个数相加，并弹出结果
    IL_0002: add
    //返回结果
    IL_0003: ret

  }
}
```

如代码 17-4 所示，在 C#代码中调用 Add 方法。

代码17-4

```
Foo.Add(3, 2); //5
```

如代码 17-5 所示，在 IL 代码中调用 Add 方法。

代码17-5

```
ldc.i4.3
ldc.i4.2
call        int32 Foo::Add(int32, int32)
pop
ret
```

17.1.3　IL 语法的格式

为了使读者可以更好地了解 IL 语法，下面以 IL 项目为例创建一个简单的实例，但是该实例只是一个 DEMO，在实际的编程过程中，建议直接采用 C#编写代码，因为编译器会为代码做出更多的优化，所以直接采用 IL 编写代码会对性能产生一定的影响。

如代码 17-6 所示，创建一个类库。

代码17-6

```
mkdir MyILProject\MyILlib
cd MyILProject\MyILlib
dotnet new classlib
```

首先创建文件夹目录，在 MyILProject 目录下创建 MyILlib 文件夹，然后通过 cd 命令切

换到 MyILlib 目录下，最后通过 dotnet 命令行工具创建一个类库项目。

如代码 17-7 所示，创建并定义 global.json 文件中的内容。

代码17-7

```
{
    "msbuild-sdks": {
      "Microsoft.NET.Sdk.IL": "6.0.0"
    }
}
```

将 MyILlib.csproj 文件重命名为 MyILlib.ilproj，将项目修改为 IL 类库项目。

接下来打开 MyILlib.ilproj 文件，修改项目属性，如代码 17-8 所示。

代码17-8

```
<Project Sdk="Microsoft.NET.Sdk.IL">
  <PropertyGroup>
    <TargetFramework>net6.0</TargetFramework>
    <ProduceReferenceAssembly>false</ProduceReferenceAssembly>
  </PropertyGroup>
</Project>
```

在 MyILlib.ilproj 文件中，将 SDK 更改为 Microsoft.NET.Sdk.IL。

如代码 17-9 所示，新建一个 helloworld.il 文件，并添加如下 IL 代码。

代码17-9

```
.assembly extern System.Runtime
{
    .publickeytoken = (B0 3F 5F 7F 11 D5 0A 3A )
    .ver 4:0:0:0
}

.assembly MyILlib
{
  .ver 1:0:0:0
}

.module MyILlib.dll

.class public auto ansi beforefieldinit Hello
  extends [System.Runtime]System.Object
```

```
{
  .method public hidebysig static string World() cil managed
  {
//评价堆栈的最大深度
.maxstack 8
//把字符串"Hello World!"存入评价堆栈
ldstr    "Hello World!"
//return, 返回值
    ret
  }
}
```

上述代码中包含程序集运行所需的外部程序集（System.Runtime.dll），通过 extern 修饰的 .assembly 表示外部程序集。

.publickeytoken 表示程序集的公开键；.ver 用于标识程序集的版本；.assembly 用于标识当前程序集的名称，通常还可以定义程序集的版本等信息；.module 用于标识模块的名称；public 关键字定义了类型的可访问性。

auto 关键字定义了类的布局风格（自动布局，默认值）。其他可选关键字有 sequential（保存字段的指定顺序）和 explicit（显示通过 FieldOffset 特性设置每个成员的位置），在 C# 中可以通过 StructLayoutAttribute 特性设置类或结构的字段在托管内存中的物理布局。

ansi 关键字定义了类与其他非托管代码互操作时的字符串转换模式，这个关键字默认指定字符串与标准 C 语言风格的字节字符串进行转换，其他可选关键字还有 unicode（字符串和 UTF-16 格式的 Unicode 编码相互转换）和 autochar（由底层平台决定字符串转换的模式）。

static 关键字标识了静态方法；hidebysig 关键字表示隐藏通过 hidebysig 关键字标识的方法签名；string 关键字表示方法的返回类型；cil managed 是方法的实现标志，表示这个方法是在 IL 中实现的。除此之外，还有 native（机器代码实现）属性、optil（优化 IL 代码时实现）属性和 runtime（.NET 运行时内部实现）属性。

如代码 17-10 所示，创建一个控制台项目并引入该 IL 类库。

代码17-10

```
cd ../
mkdir MyApp
cd MyApp
dotnet new console
dotnet add reference ../MyILlib
```

如代码 17-11 所示，修改 Program.cs 文件。

代码17-11

```
Console.WriteLine(Hello.World());
```

如代码 17-12 所示，运行控制台项目。

代码17-12

```
$ dotnet run
Hello World!
```

17.2 JIT 简介

JIT 是 Just-In-Time 的缩写形式，也就是即时编译。使用 JIT 技术可以避免硬件环境的差异性，并且可以根据不同的平台给出最优的机器代码。

17.2.1 JIT 编译器

在.NET 中，编译分为两部分，首先编译器将代码编译为 IL 代码，然后.NET 运行时会通过 JIT 编译器将 IL 代码实时编译为机器代码，这种方式给开发人员带来了便捷性，可以做到"一次编译，处处运行"，开发人员无须关注平台指令的差异性（见图 17-1）。另外，JIT 编译器可以根据当前硬件环境的实际情况编译出最优的机器代码，这也是 JIT 最明显的特征。

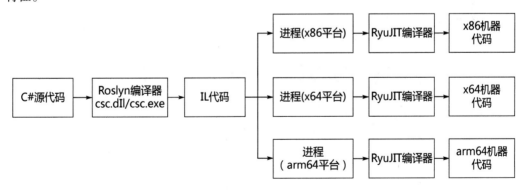

图 17-1 编译过程示意图

JIT 编译器在应用程序跨平台的实现中具有重要作用，可以避免编译器直接生成本机机器代码，而通过 JIT 进行实时编译，在 JIT 编译器的帮助下开发人员无须分别维护多个平台的代码。

17.2.2 在 Visual Studio 中查看汇编代码

Visual Studio 支持在调试时查看 .NET 应用程序生成的汇编代码。在调试时，可以选择"调试"→"窗口"→"反汇编"命令，如图 17-2 所示。

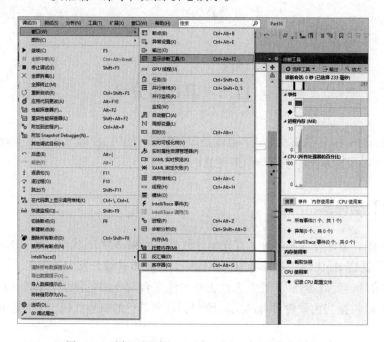

图 17-2 选择"调试"→"窗口"→"反汇编"语言

如图 17-3 所示，查看汇编代码。

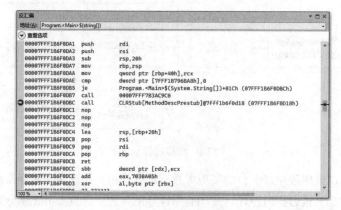

图 17-3 查看汇编代码

17.3 JIT 编译

在.NET 中，对 JIT 的优化也提出了不少功能，本节以分层编译（Tiered Compilation）和配置文件引导优化（Profile-Guided Optimization，PGO）两个功能为例展开介绍，探讨如何优化 JIT 可以使其达到更优（合适）的模式。

17.3.1 分层编译

从 IL 代码到机器代码的过程中，需要进行一次 JIT 编译，这会影响应用程序第一次运行的速度。为了解决这个问题，可以加入分层编译的特性，在启动应用程序时，JIT 先快速生成具有一定的优化功能的 tier0 代码，由于生成 tier0 代码时做的优化程度不深，因此 JIT 生成速度快，吞吐量也高，可以改善一定的延时。但是，随着应用程序的运行，对于频繁调用的方法，根据计数器测量相应的"热点代码"，再次生成具有 JIT 优化的 tier1 代码，以提升程序的执行效率，这样可以平衡编译速度和性能。

注意：在.NET Core 3.0 中默认启用分层编译，在.NET Core 2.1 和.NET Core 2.2 中默认关闭分层编译。

如代码 17-13 所示，在.csproj 文件中添加 TieredCompilation 属性可以启用或关闭分层编译。

代码17-13

```
<Project Sdk="Microsoft.NET.Sdk">
  <PropertyGroup>
    <TieredCompilation>false</TieredCompilation>
  </PropertyGroup>
</Project>
```

代码 17-14 展示了如何在 runtimeconfig.json 文件中开启分层编译。

代码17-14

```
{
  "runtimeOptions": {
    "configProperties": {
      "System.Runtime.TieredCompilation": false
    }
  }
}
```

}
```

启用 Quick（快速）JIT 可以更快速地完成编译，虽然不会进行优化，但是可以缩短启动的时间。如果禁用了 TieredCompilationQuickJit 属性，启用了分层编译，则只有预编译的代码会参与分层编译。

如代码 17-15 所示，在 .csproj 文件中添加 TieredCompilationQuickJit 属性。

**代码 17-15**

```xml
<Project Sdk="Microsoft.NET.Sdk">
  <PropertyGroup>
    <TieredCompilationQuickJit>false</TieredCompilationQuickJit>
  </PropertyGroup>
</Project>
```

如代码 17-16 所示，在 runtimeconfig.json 文件中添加 System.Runtime.TieredCompilation.QuickJit 属性。

代码 17-16

```json
{
  "runtimeOptions": {
    "configProperties": {
      "System.Runtime.TieredCompilation.QuickJit": false
    }
  }
}
```

为循环方法启用 Quick JIT，可以提高应用程序的性能。如果禁用 Quick JIT，那么 TieredCompilationQuickJitForLoops 属性就起不到优化的作用。

如代码 17-17 所示，在 .csproj 文件中添加 TieredCompilationQuickJitForLoops 属性。

代码 17-17

```xml
<Project Sdk="Microsoft.NET.Sdk">
  <PropertyGroup>
    <TieredCompilationQuickJitForLoops>true</TieredCompilationQuickJitForLoops>
  </PropertyGroup>
</Project>
```

代码 17-18 展示了如何在 runtimeconfig.json 文件中添加 System.Runtime.TieredCompilation.QuickJitForLoops 属性。
```

代码17-18

```
{
  "runtimeOptions": {
    "configProperties": {
      "System.Runtime.TieredCompilation.QuickJitForLoops": false
    }
  }
}
```

17.3.2 使用 PGO 优化程序编译（动态 PGO）

PGO 是一种 JIT 编译器优化技术，允许 JIT 编译器在 tier0 代码生成中收集运行时信息，以便后期从 tier0 升级到 tier1 的过程中依赖 PGO 来获取"热点代码"，从而提升执行代码的效率。.NET 提供了动态（Dynamic）PGO 和静态（Static）PGO。静态 PGO 通过工具收集 profile 数据，应用程序会在下一次编译时引导编译器对代码进行优化；动态 PGO 则是在运行时一边收集 profile 数据一边进行优化。

如代码 17-19 方式，可以使用环境变量启用动态 PGO。

代码17-19

```
#启用动态 PGO
export DOTNET_TieredPGO=1
#禁用 AOT
export DOTNET_ReadyToRun=0
#为循环启用 Quick JIT
export DOTNET_TC_QuickJitForLoops=1
```

如代码 17-20 所示，在 Windows 操作系统中通过 PowerShell 添加环境变量。

代码17-20

```
$env:DOTNET_TieredPGO=1
$env:DOTNET_ReadyToRun=0
$env:DOTNET_TC_QuickJitForLoops=1
```

如代码 17-21 所示，添加一个基准测试项目，用于测试动态 PGO 和默认模式的性能的差异性。

代码17-21

```
//运行基准测试项目
```

```csharp
BenchmarkRunner.Run<PgoBenchmarks>();

[Config(typeof(MyEnvVars))]
public class PgoBenchmarks
{
    //自定义配置环境 Default 和 DPGO
    class MyEnvVars : ManualConfig
    {
        public MyEnvVars()
        {
            //使用默认模式
            AddJob(Job.Default.WithId("Default mode"));
            //使用动态 PGO
            AddJob(Job.Default.WithId("Dynamic PGO")
                .WithEnvironmentVariables(
                    new EnvironmentVariable("DOTNET_TieredPGO", "1"),
                    new EnvironmentVariable("DOTNET_TC_QuickJitForLoops", "1"),
                    new EnvironmentVariable("DOTNET_ReadyToRun", "0")));

        }
    }

    [Benchmark]
    [Arguments(6, 100)]
    public int HotColdBasicBlockReorder(int key, int data)
    {
        if (key == 1)
            return data - 5;
        if (key == 2)
            return data += 4;
        if (key == 3)
            return data >> 3;
        if (key == 4)
            return data * 2;
        if (key == 5)
            return data / 1;
        return data; //默认 key 为 6,所以会使用返回的 data
    }
}
```

表 17-1 所示为性能测试的输出结果。表 17-1 简化了输出条目,通过默认模式和动态 PGO 配置的优化可以看出,两种模式存在一个明显的差距。

表 17-1 性能测试的输出结果

方法	模式	环境变量	均值	误差	标准误差
HotColdBasicBlockReorder	默认模式	Empty	1.8672 纳秒	0.0412 纳秒	0.0386 纳秒
HotColdBasicBlockReorder	动态 PGO	DOTNET_TieredPGO=1, DOTNET_TC_QuickJitForLoops=1, DOTNET_ReadyToRun=0	0.0370 纳秒	0.0445 纳秒	0.0416 纳秒

通常,编译器会将应用程序进行冷热分区,但不够智能化,在动态 PGO 模式下,可以采集到在"实际"业务运行时代码的调用信息,使用该信息可以引导应用程序的编译,以达到优化的目的。利用记录的 Basic Block(BB)块调用频率,可以更加准确地划分出冷热块,做到冷热分离,并利用 BB 块进行优化,其中包括循环展开、函数内联等。在划分完代码中的冷区和热区后,编译器还会权衡性能,在指令重排时,将它们合理地放到相应的位置(见图 17-4)。

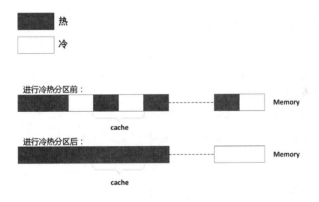

图 17-4 冷热分区

冷代码是指不频繁执行的代码,热代码是指频繁执行的代码。使用动态 PGO 对冷热区域进行划分,有利于提高指令缓存的利用率。

17.3.3 使用 PGO 优化程序编译(静态 PGO)

静态 PGO 允许开发人员提前收集 profile 数据,将收集的 profile 数据在应用程序启动时引导并优化。

在编写本书期间，正式版 PGO 工具还未发布，所以笔者在项目目录下新建一个 NuGet.config 文件，并添加 NuGet 源，如代码 17-22 所示。

代码17-22

```xml
<configuration>
  <packageSources>
    <add key="dotnet-public" value="https://pkgs.dev.***.com/dnceng/public/_packaging/dotnet-public/nuget/v3/index.json" />
    <add key="dotnet-tools" value="https://pkgs.dev.***.com/dnceng/public/_packaging/dotnet-tools/nuget/v3/index.json" />
    <add key="dotnet-eng" value="https://pkgs.dev.***.com/dnceng/public/_packaging/dotnet-eng/nuget/v3/index.json" />
    <add key="dotnet6" value="https://pkgs.dev.***.com/dnceng/public/_packaging/dotnet6/nuget/v3/index.json" />
    <add key="dotnet6-transport" value="https://pkgs.dev.***.com/dnceng/public/_packaging/dotnet6-transport/nuget/v3/index.json" />
  </packageSources>
</configuration>
```

如代码 17-23 所示，安装 dotnet-pgo，该工具用于生成静态的 PGO 数据。

代码17-23

```
$ dotnet tool install dotnet-pgo --version 6.0.0-* -g
可以使用以下命令调用工具：dotnet-pgo
已成功安装工具"dotnet-pgo"（版本"6.0.0-rtm.21522.10"）。
```

如代码 17-24 所示，在 Windows 操作系统中添加环境变量。

代码17-24

```
SETX DOTNET_ReadyToRun 0              #禁用 AOT
SETX DOTNET_TieredPGO 1               #启用分层 PGO
SETX DOTNET_TC_CallCounting 0         #永远不产生 tier1 代码
SETX DOTNET_TC_QuickJitForLoops 1
SETX DOTNET_JitCollect64BitCounts 1
SETX DOTNET_JitEdgeProfiling 1
```

如代码 17-25 所示，在 Linux 操作系统中添加环境变量。

代码17-25

```
export DOTNET_ReadyToRun 0
```

```
export DOTNET_TieredPGO 1
export DOTNET_TC_CallCounting 0
export DOTNET_TC_QuickJitForLoops 1
export DOTNET_JitCollect64BitCounts 1
export DOTNET_JitEdgeProfiling 1
```

新建一个控制台项目,如代码 17-26 所示。

代码17-26

```
public interface IFoo
{
    int Value { get; }
}

public interface IFooFactory
{
    IFoo Foo { get; }
}

public class FooImpl : IFoo
{
    public int Value => 42;
}

public class FooFactoryImpl : IFooFactory
{
    public IFoo Foo { get; } = new FooImpl();
}
```

如代码 17-27 所示,编写测试代码。

代码17-27

```
using System;
using System.Diagnostics;
using System.Runtime.CompilerServices;
using System.Threading;

public static class Tests
{
    static void Main()
```

```csharp
{
    Console.WriteLine("Running...");
    var sw = Stopwatch.StartNew();
    var factory = new FooFactoryImpl();

    for (int iteration = 0; iteration < 10; iteration++)
    {
        sw.Restart();
        for (int i = 0; i < 10000000; i++)
            Test(factory);
        sw.Stop();
        Console.WriteLine(
                $"[{iteration}/9]: {sw.ElapsedMilliseconds} ms.");
        Thread.Sleep(20);
    }
}

[MethodImpl(MethodImplOptions.NoInlining)]
static int Test(IFooFactory factory)
{
    IFoo? foo = factory?.Foo;
    return foo?.Value ?? 0;
}
```

如代码 17-28 所示，通过 dotnet trace 收集性能文件，会得到一个 trace.nettrace 文件，里面包含追踪的数据，同时利用 dotnet-pgo 生成 PGO 数据。

代码17-28

```
$ dotnet trace collect
  --providers Microsoft-Windows-DotNETRuntime:0x1F000080018:5
-o trace.nettrace -- dotnet 04.dll
```

注意：参数 04.dll 为具体的文件名称，当运行 dotnet trace collect 命令时，文件一定要指定正确，否则对当前应用程序起不到优化作用。

如代码 17-29 所示，通过 dotnet-pgo 将 trace.nettrace 文件生成出 PGO 数据。

代码17-29

```
$ dotnet-pgo create-mibc -t trace.nettrace -o pgo.mibc
```

如代码 17-30 所示，通过 PGO 数据引导对代码的编译。

代码17-30

```
$ dotnet publish -c Release -r win-x64 /p:PublishReadyToRun=true
/p:PublishReadyToRunComposite=true    /p:PublishReadyToRunCrossgen2ExtraArgs=
--embed-pgo-data%3b--mibc%3apgo.mibc
```

如代码 17-31 所示，在 PGO 优化后，运行应用程序。

代码17-31

```
$ cd bin/Release/net6.0/win-x64/publish
$ dotnet 04.dll
Running...
[0/9]: 17 ms.
[1/9]: 17 ms.
[2/9]: 17 ms.
[3/9]: 17 ms.
[4/9]: 17 ms.
[5/9]: 16 ms.
[6/9]: 17 ms.
[7/9]: 17 ms.
[8/9]: 17 ms.
[9/9]: 17 ms.
```

如代码 17-32 所示，查看默认模式下的输出。

代码17-32

```
$ dotnet run -c Release
Running...
[0/9]: 54 ms.
[1/9]: 54 ms.
[2/9]: 64 ms.
[3/9]: 51 ms.
[4/9]: 51 ms.
[5/9]: 52 ms.
[6/9]: 53 ms.
[7/9]: 53 ms.
[8/9]: 52 ms.
[9/9]: 52 ms.
```

17.4 AOT 编译

AOT 编译是一种优化手段，使应用程序在运行前编译，既可以避免 JIT 编译器带来的性能损耗，又可以显著加快应用程序的启动速度。

17.4.1 用 Crossgen2 进行 AOT 预编译

.NET 6 中引入了 Crossgen2，用来代替原有的 Crossgen。Crossgen 和 Crossgen2 是提供预编译的工具，可以改进应用程序的启动时间，并根据运行情况利用 JIT 编译器进行优化，这是混合 AOT 编译的一种策略，通过这种方式可以减少 JIT 编译器在执行期间的工作量。

Crossgen2 首先对 System.Private.Corelib 项目开展工作，至于为什么开展它，官方的解释是这个项目并不是为了提升性能而改进，而是为托管 RyuJIT 提供更好的体系结构。它可以在非运行期间以离线的方式生成代码，以便更好地支持交叉编译。

目前有两种方式可以将应用程序通过 Crossgen2 发布。

第一种方式是利用 dotnet publish 命令，如代码 17-33 所示。

代码 17-33

```
$ dotnet publish -c Release -r win-x64 /p:PublishReadyToRun=true
 /p:PublishReadyToRunUseCrossgen2=true
```

发布之后，在发布目录下可以看到有一个 [program].r2r.dll 文件，AOT 编译后的代码就在其中。

第二种方式是添加指定的项目属性，如代码 17-34 所示，修改 .csproj 文件。

代码 17-34

```xml
<PropertyGroup>
  <PublishReadyToRun>true</PublishReadyToRun>
  <PublishReadyToRunComposite>True</PublishReadyToRunComposite>
  <PublishReadyToRunUseCrossgen2>true</PublishReadyToRunUseCrossgen2>
</PropertyGroup>
```

如代码 17-35 所示，执行程序。

代码 17-35

```
$ dotnet publish -c Release -r win-x64
```

17.4.2　Source Generators

元编程（Metaprogramming）是一种编程范式，相当于以代码生成代码，在编译和运行时生成或更改代码。下面介绍 Source Generators 功能。利用 Source Generators 功能可以在编译时生成代码，也就是所谓的"编译时元编程"。图 17-5 所示为 Source Generators 编译阶段。

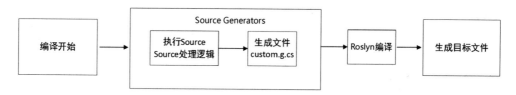

图 17-5　Source Generators 编译阶段

Source Generators 会在代码编译时生成代码，并将编译时生成的代码参与编译进行构建。

Source Generators 抽象了 Microsoft.CodeAnalysis.ISourceGenerator 接口，如代码 17-36 所示。

代码17-36

```
public interface ISourceGenerator
{
    void Initialize(GeneratorInitializationContext context);
    void Execute(GeneratorExecutionContext context);
}
```

ISourceGenerator 接口提供了两个方法：Initialize 方法接收一个 GeneratorInitializationContext 类型，用于初始化调用，在生成之前调用该方法，生成器可以利用该方法进行注册并执行所需的回调方法；Execute 方法通过 GeneratorExecutionContext 对象对具体的代码执行生成操作，并将代码进行织入。

接下来创建一个控制台项目，如代码 17-37 所示。

代码17-37

```
HelloFrom("HueiFeng");
partial class Program
{
    static partial void HelloFrom(string name);
}
```

创建一个类库,将 TargetFramework 属性修改为 netstandard2.0,如代码 17-38 所示。

代码17-38

```xml
<Project Sdk="Microsoft.NET.Sdk">
<Project Sdk="Microsoft.NET.Sdk">
    <PropertyGroup>
        <TargetFramework>netstandard2.0</TargetFramework>
    </PropertyGroup>
    <ItemGroup>
        <PackageReference Include="Microsoft.CodeAnalysis" Version="4.0.1"
                                                PrivateAssets="all" />
        <PackageReference
          Include="Microsoft.CodeAnalysis.Analyzers" Version="3.3.3"
                                                PrivateAssets="all"/>
    </ItemGroup>
</Project>
```

生成器实现 Microsoft.CodeAnalysis.ISourceGenerator 接口之后,还需要添加 Microsoft.CodeAnalysis.Generator 属性进行标记。如果需要调试 Source Generator,则可以添加 Debugger.Launch 方法,添加该方法后,可以在编译时调用调试器进行调试,如代码 17-39 所示。

代码17-39

```csharp
using System.Diagnostics;
using System.Text;
using Microsoft.CodeAnalysis;
using Microsoft.CodeAnalysis.Text;

namespace SourceGenerator
{
    [Generator]
    public class CustomGenerator : ISourceGenerator
    {
        public void Initialize(GeneratorInitializationContext context)
        {
            //如果需要调试 Source Generator,则可以设置如下代码
            Debugger.Launch();
        }

        public void Execute(GeneratorExecutionContext context)
```

```
        {
            var mainMethod = 
             context.Compilation.GetEntryPoint(context.CancellationToken);

            //设置需要织入的代码
            string source = $@" //Auto-generated code
using System;

  public static partial class {mainMethod.ContainingType.Name}
  {{
      static partial void HelloFrom(string name) =>
          Console.WriteLine($""Generator says: Hi from '{{name}}'"");
  }}
";
            SourceText sourceText = SourceText.From(source, Encoding.UTF8);
            var typeName = mainMethod.ContainingType.Name;

            //Add the source code to the compilation
            context.AddSource($"{typeName}.g.cs", sourceText);
        }
    }
}
```

首先定义一个 mainMethod 实例在编译时获取 main 入口，mainMethod 实例返回一个 IMethodSymbol 对象，用于获取命名空间、方法名称和类型名称。然后创建一个需要织入的代码片段，创建 SourceText 对象，通过 SourceText.From 静态方法添加需要插入的源代码，并设置目标编码格式。最后利用 context.AddSource 方法将源代码添加到编译中。

设置完之后，通过控制台项目引入类库，如代码 17-40 所示。

代码17-40

```
<ItemGroup>
  <ProjectReference Include="..\SourceGenerator\SourceGenerator.csproj"
                OutputItemType="Analyzer"
                ReferenceOutputAssembly="false"/>
</ItemGroup>
```

- ReferenceOutputAssembly：默认为 true，如果设置为 false，则不包括引用项目的输出作为此项目的引用，但仍可确保在此项目之前生成其他项目。

- OutputItemType：可选项，默认为空，设置要输出的类型。

输出结果如图 17-6 所示。

图 17-6　输出结果

17.5　小结

本章首先介绍了 .NET 的多种编译技术（如 IL），然后介绍了 JIT 编译器的工作方式、JIT 优化手段（如分层编译、PGO 编译），最后介绍了 AOT 编译。通过学习本章，读者基本上可以了解编译技术的相关知识。

第 18 章

部署

通常,应用程序在开发和测试后,就可以部署到生产环境中使用。本章对应用程序的部署展开介绍,18.1 节会通过一个实例将代码发布并上传到 Ubuntu、CentOS 及 Windows Server IIS 中进行部署。

本章以 Docker 为例探讨容器技术。除此之外,本章还介绍了 Docker 家族的 Docker Compose 和 Docker Swarm。18.7 节将探讨 Kubernetes 的安装与使用,以及如何轻松地将应用程序部署到容器云。

18.1 发布与部署

上面介绍了如何开发 ASP.NET Core 应用程序。当应用程序开发完成并且经过测试后,还需要部署到服务器中,对外提供服务访问。.NET 是跨平台的开发框架,能够在 Windows 操作系统、Linux 操作系统、macOS 操作系统中运行,基于.NET 开发的应用程序可以运行在主流的 CPU 和操作系统中,目前.NET 官方已经支持在 x86、x64、ARM32 和 ARM64 架构的硬件平台上运行。值得一提的是,国内的龙芯团队也将.NET 移植到龙芯 CPU 中。

18.1.1 部署至 Ubuntu 操作系统中

如代码 18-1 所示,通过 dotnet 命令行工具发布应用程序。

代码18-1

```
dotnet publish -c Release -r linux-x64
```

输出的发布内容保存在 bin\Release\net6.0\linux-x64\publish 目录下，接下来通过 Xftp 文件传输软件将发布的目录复制并传输到 Ubuntu 服务器上。

注意：安装运行环境可以参考官网，查看支持的具体的操作系统及相应版本的环境。

如代码 18-2 所示，安装环境。笔者的环境是 Ubuntu 18.04，通过执行如下命令，将 Microsoft 包签名密钥添加到受信任的密钥列表中，并添加包存储库。

代码18-2

```
wget https://packages.**.com/config/ubuntu/18.04/packages-microsoft-prod.deb
 -O package-microsoft-prod.deb
 sudo dpkg -I packages-microsoft-prod.deb
 rm package-microsoft-prod.deb
```

如代码 18-3 所示，安装运行时。

代码18-3

```
sudo apt-get update; \
 sudo apt-get install -y apt-transport-https && \
 sudo apt-get update && \
 sudo apt-get install -y aspnetcore-runtime-6.0
```

在环境安装完成后，可以直接运行应用程序，如代码 18-4 所示，MyProject 为发布后的可执行文件。

代码18-4

```
fh@vm:/$ cd wwwroot
fh@vm:/wwwroot$ ./MyProject
info: Microsoft.Hosting.Lifetime[14]
      Now listening on: http://localhost:5000
info: Microsoft.Hosting.Lifetime[0]
      Application started. Press Ctrl+C to shut down.
info: Microsoft.Hosting.Lifetime[0]
      Hosting environment: Production
info: Microsoft.Hosting.Lifetime[0]
      Content root path: /wwwroot/
```

如代码 18-4 所示，可以看到 Now listening on: http://localhost:5000，它监听 localhost 本地端口，端口号为 5000，但它仅能在本地访问，接下来进行修改，使其允许公网 IP 地址访问，如代码 18-5 所示。

代码18-5

```csharp
using System.Runtime.InteropServices;
using System.Text;

var builder = WebApplication.CreateBuilder(args);

var app = builder.Build();

app.MapGet("/", () =>
    {
        StringBuilder sb = new StringBuilder();
        sb.AppendLine($"系统架构: {RuntimeInformation.OSArchitecture}");
        sb.AppendLine($"系统名称: {RuntimeInformation.OSDescription}");
        sb.AppendLine($"进程架构: {RuntimeInformation.ProcessArchitecture}");
        sb.AppendLine($"是否为64位操作系统: {Environment.Is64BitOperatingSystem}");
        return sb.ToString();
    })
    .WithName("GetInfo");

app.Run("http://*:5000");
```

通过在 app.Run 方法中添加需要监听的 URL，就可以使用公网 IP 地址访问服务器上的应用程序。如代码 18-6 所示，运行应用程序。

代码18-6

```
root@vm:/wwwroot# ./MyProject
info: Microsoft.Hosting.Lifetime[14]
      Now listening on: http://[::]:5000
info: Microsoft.Hosting.Lifetime[0]
      Application started. Press Ctrl+C to shut down.
info: Microsoft.Hosting.Lifetime[0]
      Hosting environment: Production
info: Microsoft.Hosting.Lifetime[0]
      Content root path: /wwwroot/;
```

453

接下来在客户端浏览器中访问应用程序，如图18-1所示，可以看到在Ubuntu服务器中获取的详细信息。

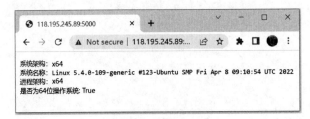

图18-1　详细信息

这还不算结束，在应用程序启动后，如果命令窗口（终端）关闭，则应用程序也会随着命令窗口的关闭而关闭，接下来利用Systemd添加一个守护进程。在/ect/systemd/system目录下创建一个myproject.service文件，如代码18-7所示。

代码18-7

```
[Unit]
Description=myproject

[Service]
WorkingDirectory=/wwwroot/
ExecStart=/wwwroot/MyProject
Restart=always

[Install]
WantedBy=multi-user.target
```

WorkingDirectory属性为工作目录，ExecStart属性用于定义执行启动的脚本，Restart=always表示如果此过程崩溃或停止，则始终进行重启。

如代码18-8所示，执行systemctl daemon-reload命令，以便Systemd加载新配置文件。

代码18-8

```
sudo systemctl daemon-reload
```

如代码18-9所示，启动和停止服务。

代码18-9

```
sudo systemctl start myproject.service
sudo systemctl stop myproject.service
sudo systemctl restart myproject.service
```

如代码 18-10 所示，通过 systemctl status 命令查看服务的状态。如图 18-2 所示，通过 systemctl 命令查看服务的状态。

代码18-10

```
sudo systemctl status myproject.service
```

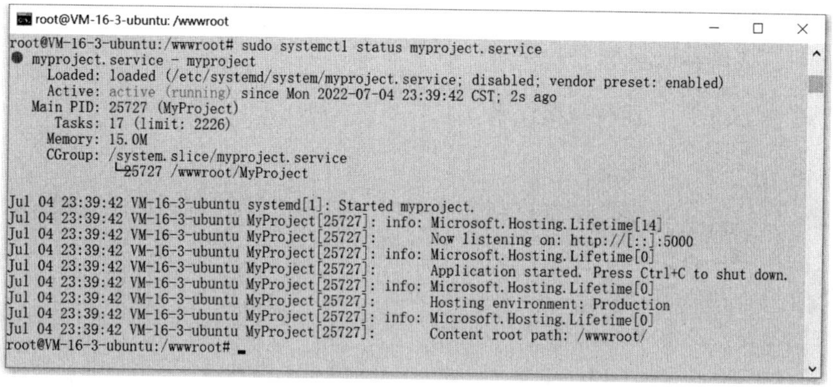

图 18-2　通过 systemctl 命令查看服务的状态

18.1.2　部署至 CentOS 操作系统中

如代码 18-11 所示，发布应用程序。

代码18-11

```
dotnet publish -c Release -r linux-x64
```

输出的发布文件默认在 bin\Release\net6.0\linux-x64\publish 目录下，随后，通过 Xftp 文件传输软件将发布的目录复制到 CentOS 服务器上。

如代码 18-12 所示，安装环境。笔者的环境是 CentOS 7.6，通过执行如下命令，将 Microsoft 包签名密钥添加受信任的密钥列表，并添加包存储库。

代码18-12

```
sudo rpm -Uvh https://packages.******.com/config/centos/7/packages-microsoft-prod.rpm
```

如代码 18-13 所示，安装 ASP.NET Core 运行时。

代码18-13

```
sudo yum install aspnetcore-runtime-6.0
```

注意：将生成的文件复制到 CentOS 操作系统中，在运行时需要设置执行权限，777 表示拥有完全控制的权限。

如代码 18-14 所示，设置执行权限。

代码18-14

```
chmod 777 ./MyProject
```

如代码 18-15 所示，运行应用程序。

代码18-15

```
[root@centosvm wwwroot]# ./MyProject
info: Microsoft.Hosting.Lifetime[14]
      Now listening on: http://[::]:5000
info: Microsoft.Hosting.Lifetime[0]
      Application started. Press Ctrl+C to shut down.
info: Microsoft.Hosting.Lifetime[0]
      Hosting environment: Production
info: Microsoft.Hosting.Lifetime[0]
      Content root path: /wwwroot/
```

图 18-3 所示为获取的详细信息。

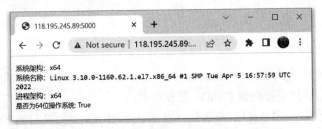

图 18-3　获取的详细信息

如代码 18-16 所示，添加守护进程，接下来利用 Systemd 添加一个守护进程服务，并且在 /etc/systemd/system 目录下创建一个 myproject.service 文件。

代码18-16

```
[Unit]
Description=myproject

[Service]
```

```
WorkingDirectory=/wwwroot/
ExecStart=/wwwroot/MyProject
Restart=always

[Install]
WantedBy=multi-user.target
```

WorkingDirectory 属性为工作目录，ExecStart 属性用于定义执行启动的脚本，Restart=always 表示如果应用程序崩溃或停止，则始终重启。

如代码 18-17 所示，执行 systemctl daemon-reload 命令，以便 Systemd 加载新配置文件。

代码18-17

```
sudo systemctl daemon-reload
```

如代码 18-18 所示，启动和停止服务。

代码18-18

```
sudo systemctl start myproject.service
sudo systemctl stop myproject.service
sudo systemctl restart myproject.service
```

如代码 18-19 所示，查看服务的状态。如图 18-4 所示，使用 systemctl 命令可以查看服务的状态。

代码18-19

```
sudo systemctl status myproject.service
```

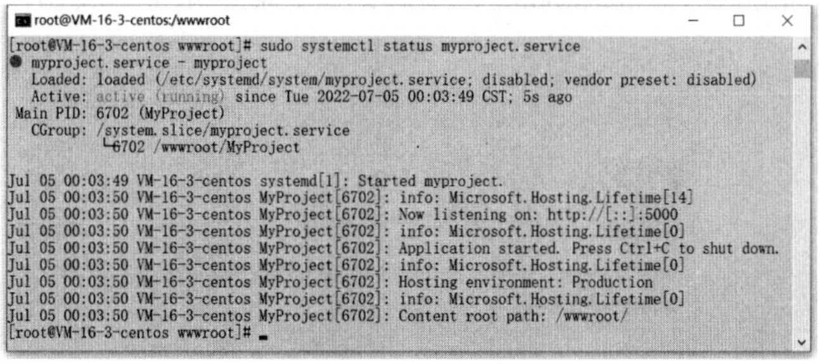

图 18-4 服务的状态

18.1.3 部署至 Windows Server IIS 中

下面以 Windows Server 2016 Datacenter 为例，安装 Windows 运行环境。

安装 IIS

首先安装 IIS（Internet Information Services），打开服务器管理器后选择"添加角色和功能"命令，然后在打开的"添加角色和功能向导"窗口的"选择服务器角色"界面中勾选"Web服务器（IIS）"复选框（见图 18-5）进行安装即可。

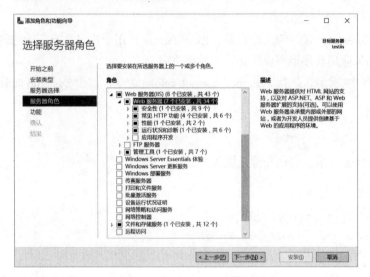

图 18-5　添加角色和功能向导

安装 .NET Windows Server Hosting

如图 18-6 所示，安装 .NET 6.0.1 Windows Server Hosting 软件包。该包属于捆绑包，包含 .NET 运行时和 ASP.NET Core Module 的捆绑，属于 ASP.NET Core 应用程序在 IIS 中运行的基础包。

图 18-6　安装 .NET 6.0.1 Windows Server Hosting 软件包

如图 18-7 所示，通过 Visual Studio 发布应用程序。

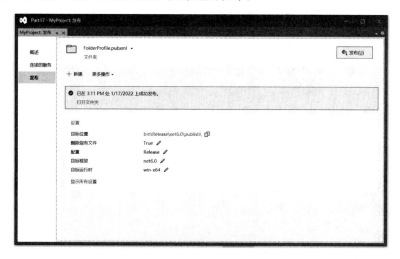

图 18-7　发布应用程序

应用程序被发布到指定的目录下，先将它上传到服务器中，并且创建一个 Web 站点，再进行调整，更改应用池的配置，将其改为"无托管代码"（见图 18-8）。设置为"无托管代码"，只需要利用它的转发请求功能即可，而不需要通过它进行代码托管。

如图 18-9 所示，通过浏览器访问站点并查看网站输出信息。

图 18-8　"编辑应用程序池"对话框　　　　图 18-9　网站输出信息

进程内托管和进程外托管

IIS 部署有两种模式可以选择，分别为进程内托管和进程外托管。进程内托管是指不使用 Kestrel 服务器，直接通过 IIS 托管应用程序；进程外托管则通过 Kestrel 服务器托管应用程序，并利用 IIS 转发到 Kestrel 服务器中。所以，二者的区别是，一个是独立的进程，另一个则是单进程。图 18-10 所示为进程内托管。

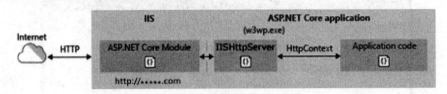

图 18-10　进程内托管

在默认情况下，在 IIS 中部署通常采用进程内托管。如代码 18-20 所示，如果需要显式配置进程内托管，则可以在.csproj 文件中通过 AspNetCoreHostingModel 属性设置进程内托管的 InProcess 属性值。

代码18-20

```
<PropertyGroup>
  <AspNetCoreHostingModel>InProcess</AspNetCoreHostingModel>
</PropertyGroup>
```

除了进程内托管，还可以修改为进程外托管，在.csproj 文件中通过 AspNetCoreHostingModel 属性设置进程外托管的 OutOfProcess 属性值。图 18-11 所示为进程外托管。

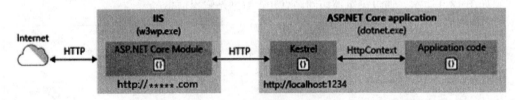

图 18-11　进程外托管

18.2　Docker

Docker 是一个开源项目，诞生于 2013 年年初，是一种容器技术。容器这个概念并不是 Docker 公司发明的，Docker 只是众多容器技术中的一种。容器类似于沙盒技术，具备沙盒的特性，使容器可以像集装箱一样，把应用装到集装箱内。通过这种天然的特性，容器与容器之间彼此隔离，互不干扰。镜像与容器均可通过 Docker 提供的命令进行创建，而在 Docker 中创建容器可以利用 Dockerfile 文件，该文件中包含 Docker 能够解析的指令，利用这些指令可以在 Docker 中构建一个镜像，使容器可以轻易"搬来搬去"。

18.2.1 容器技术解决了什么

容器技术有很多种，Docker 只是其中的一种。Docker 与虚拟化看起来类似。虚拟化的核心是为了解决资源调配问题，最大限度地解决机器的占用情况，有效地提高资源的使用率；而容器技术的核心是缩小粒度，提高程序的密度，解决应用环境依赖的问题，规避编译、打包与部署之间的问题，提高应用程序开发和运维的效率。

容器最直观的是环境的统一，容器技术提供了一个隔离的环境，使用容器技术开发人员可以快速创建和销毁应用程序及运行环境。另外，容器还具有启动快和易迁移等优势，可以将容器从本地迁移到云端，或者将一份环境复制到另一台开发机器上都是非常简单的事情。因为容器直接运行于宿主机内核，所以无须启动一个完整的系统环境，而它又具备比传统虚拟机启动快的优势，可以做到秒级启动。Docker 容器技术与传统虚拟机的对比如图 18-12 所示。

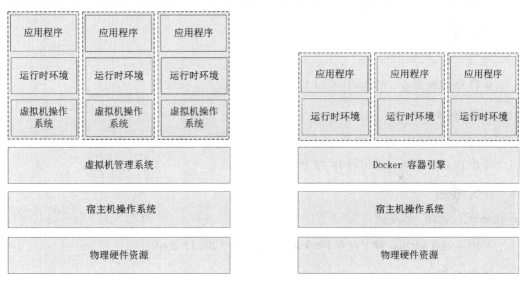

图 18-12　Docker 容器技术与传统虚拟机的对比

在应用开发中，利用容器技术不仅可以节省开发、测试、部署的时间，还可以轻松地复制一个相同的环境用来进行测试和部署。

在应用部署中，利用容器技术可以快速从一个平台复制到另一个平台上，而无须担心运行环境的差异性，所以它提供了一次构建、多处运行的方式。

与传统的虚拟化技术不同，利用容器技术不仅可以避免系统等额外的开销，还可以提高对资源的利用率。无论是启动速度，还是应用的执行速度，容器技术都比传统的虚拟化技术更高效。

18.2.2 安装 Docker

在 Linux 操作系统中简化了安装,可以在 CentOS 操作系统、Debian 操作系统、Ubuntu 操作系统中通过脚本一键安装,如代码 18-21 所示。

代码 18-21
```
curl -sSL https://get.******.com/ | sh
```

如代码 18-22 所示,启动 Docker 服务。

代码 18-22
```
systemctl start docker
```

设置开机自动启动 Docker 服务,如代码 18-23 所示。

代码 18-23
```
systemctl enable docker
```

重启 Docker 服务,如代码 18-24 所示。

代码 18-24
```
systemctl restart docker
```

停止 Docker 服务,如代码 18-25 所示。

代码 18-25
```
systemctl stop docker
```

利用 docker version 命令查看 Docker 的版本,如代码 18-26 所示。

代码 18-26
```
$ docker version
Client: Docker Engine - Community
 Version:           20.10.12
 API version:       1.41
 Go version:        go1.16.12
 Git commit:        e91ed57
 Built:             Mon Dec 13 11:45:41 2021
 OS/Arch:           linux/amd64
 Context:           default
```

第 18 章 部署

```
 Experimental:      true

Server: Docker Engine - Community
 Engine:
  Version:          20.10.12
  API version:      1.41 (minimum version 1.12)
  Go version:       go1.16.12
  Git commit:       459d0df
  Built:            Mon Dec 13 11:44:05 2021
  OS/Arch:          linux/amd64
  Experimental:     false
 containerd:
  Version:          1.4.12
  GitCommit:        7b11cfaabd73bb80907dd23182b9347b4245eb5d
 runc:
  Version:          1.0.2
  GitCommit:        v1.0.2-0-g52b36a2
 docker-init:
  Version:          0.19.0
  GitCommit:        de40ad0
```

如图 18-13 所示，可以通过 docker 命令搜索镜像。

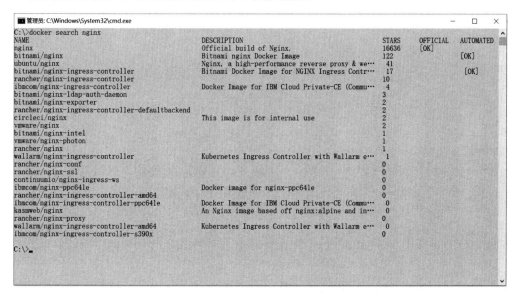

图 18-13　通过 docker 命令搜索镜像

docker search 命令常用的参数如表 18-1 所示。

表 18-1 docker search 命令常用的参数

参数	说明
--automated	只列出自动构建的镜像
--no-trunc	显示完整的镜像描述
-f <过滤条件>	根据提供的 filter 过滤镜像

如代码 18-27 所示，通过运行 docker run hello-world 命令可以验证 Docker 的安装是否正确。

代码18-27

```
$ docker run hello-world
#在本地找不到 hello-world 镜像
Unable to find image 'hello-world:latest' locally
#拉取最新版本的 hello-world 镜像
latest: Pulling from library/hello-world
2db29710123e: Pull complete
Digest: sha256:975f4b14f326b05db86e16de00144f9c12257553bba9484fed41f9b6f2257800
Status: Downloaded newer image for hello-world:latest

#表示已正常安装
Hello from Docker!
This message shows that your installation appears to be working correctly.

To generate this message, Docker took the following steps:
 1. The Docker client contacted the Docker daemon.
 2. The Docker daemon pulled the "hello-world" image from the Docker Hub.
    (amd64)
 3. The Docker daemon created a new container from that image which runs the
    executable that produces the output you are currently reading.
 4. The Docker daemon streamed that output to the Docker client, which sent it
    to your terminal.

To try something more ambitious, you can run an Ubuntu container with:
 $ docker run -it ubuntu bash

Share images, automate workflows, and more with a free Docker ID:
```

```
https://hub.******.com/

For more examples and ideas, visit:
 https://docs.******.com/get-started/
```

如代码 18-28 所示，拉取镜像。

代码 18-28

```
$ docker pull centos
Using default tag: latest
latest: Pulling from library/centos
a1d0c7532777: Pull complete
Digest: sha256:a27fd8080b517143cbbbab9dfb7c8571c40d67d534bbdee55bd6c473f432b177
Status: Downloaded newer image for centos:latest
docker.io/library/centos:latest
```

docker pull 命令的参数如表 18-2 所示。

表 18-2　docker pull 命令的参数

参数	说明
-a	拉取镜像库中所有标签名为 tag 的镜像
--disable-content-trust	忽略镜像校验，默认开启

如代码 18-29 所示，查看本地所有的镜像。

代码 18-29

```
$ docker images
REPOSITORY      TAG        IMAGE ID        CREATED         SIZE
hello-world     latest     feb5d9fea6a5    3 months ago    13.3kB
centos          latest     5d0da3dc9764    4 months ago    231MB
```

如代码 18-30 所示，运行容器。

代码 18-30

```
$ docker run centos echo "hello world"
hello world
```

如代码 18-31 所示，通过 systemctl status docker 命令查看 Docker 的状态。

代码18-31

```
$ systemctl status docker
● docker.service - Docker Application Container Engine
   Loaded: loaded (/usr/lib/systemd/system/docker.service; enabled; vendor preset: disabled)
   Active: active (running) since Tue 2022-01-18 09:25:24 UTC; 13min ago
     Docs: https://docs.******.com
 Main PID: 30960 (dockerd)
   CGroup: /system.slice/docker.service
           └─30960 /usr/bin/dockerd -H fd://--containerd=/run/containerd/containerd.sock

Jan 18 09:25:24 centosvm dockerd[30960]: time="2022-01-18T09:25:24.453604783Z" level=info msg="ccResolverWrapper:...=grpc
Jan 18 09:25:24 centosvm dockerd[30960]: time="2022-01-18T09:25:24.453616683Z" level=info msg="ClientConn switchi...=grpc
Jan 18 09:25:24 centosvm dockerd[30960]: time="2022-01-18T09:25:24.511816386Z" level=info msg="Loading containers...art."
Jan 18 09:25:24 centosvm dockerd[30960]: time="2022-01-18T09:25:24.731312613Z" level=info msg="Default bridge (do...ress"
Jan 18 09:25:24 centosvm dockerd[30960]: time="2022-01-18T09:25:24.826292635Z" level=info msg="Loading containers: done."
Jan 18 09:25:24 centosvm dockerd[30960]: time="2022-01-18T09:25:24.845654867Z" level=info msg="Docker daemon" com...10.12
Jan 18 09:25:24 centosvm dockerd[30960]: time="2022-01-18T09:25:24.846156604Z" level=info msg="Daemon has complet...tion"
Jan 18 09:25:24 centosvm systemd[1]: Started Docker Application Container Engine.
Jan 18 09:25:24 centosvm dockerd[30960]: time="2022-01-18T09:25:24.890245963Z" level=info msg="API listen on /var...sock"
Jan 18 09:37:47 centosvm dockerd[30960]: time="2022-01-18T09:37:47.765449596Z" level=info msg="ignoring event" co...lete"
Hint: Some lines were ellipsized, use -l to show in full.
```

查看使用yum命令安装的Docker文件包，如代码18-32所示。

代码18-32

```
$ yum list installed | grep docker
containerd.io.x86_64        1.4.12-3.1.el7              @docker-ce-stable
```

```
docker-ce.x86_64                        3:20.10.12-3.el7        @docker-ce-stable
docker-ce-cli.x86_64                    1:20.10.12-3.el7        @docker-ce-stable
docker-ce-rootless-extras.x86_64        20.10.12-3.el7          @docker-ce-stable
docker-scan-plugin.x86_64               0.12.0-3.el7            @docker-ce-stable
```

如代码 18-33 所示，卸载 Docker。

代码18-33

```
$ sudo yum remove docker-ce docker-ce-cli containerd.io
$ sudo rm -rf /var/lib/docker
```

如代码 18-34 所示，删除镜像或容器。

代码18-34

```
#删除镜像
$ docker rmi <your-image-id> <your-image-id> ...

#根据容器的名称或 ID 删除容器
$ docker rm [OPTIONS] CONTAINER [CONTAINER...]
```

如代码 18-35 所示，批量删除镜像和容器。

代码18-35

```
#删除所有镜像
$ docker rmi $(docker images -q)
#停止所有的容器
$ docker stop $(docker ps -a -q)
#删除所有停止的容器
$ docker rm $(docker ps -a -q)
```

Docker CLI 如上所示，在操作时可以进行条件筛选。Docker CLI 支持 shell 语法，可以通过 "$( )" 在括号内填写条件筛选命令。采用这种命令筛选的方式，可以快速停止和删除多个容器。

利用 docker ps 命令列出容器。

下面对 docker ps 命令的参数进行说明。

- -a：用于显示所有容器，包括已经停止运行的。
- -q：用于仅显示容器的 ID。

18.2.3 使用加速器

为了加快在仓库中下载镜像的速度，可以配置国内镜像源，从国内获取镜像，从而提高镜像的下载速度，还可以配置镜像加速器。

可以通过修改配置文件/etc/docker/daemon.json 来使用加速器，如果没有该文件则可以新建，如代码 18-36 所示。

代码18-36

```
$ vim /etc/docker/daemon.json
```

如代码 18-37 所示，添加加速器地址。

代码18-37

```
{
    "registry-mirrors": [加速器地址]
}
```

如代码 18-38 所示，可以使用中国科学技术大学（LUG@USTC）的开源镜像 docker.mirrors.ustc.edu.cn 和网易的开源镜像 hub-mirror.c.163.com。

代码18-38

```
{
    "registry-mirrors":[
        "http://hub-******.c.163.com",
        "https://docker.******.ustc.edu.cn"
    ]
}
```

如代码 18-39 所示，重载配置文件并重启 Docker 服务。

代码18-39

```
$ sudo systemctl daemon-reload
$ sudo systemctl restart docker
```

18.2.4　下载基础镜像

如代码 18-40 所示，下载 Ubuntu 镜像。

代码18-40

```
$ docker pull ubuntu
```

如代码 18-41 所示，查看已经下载的镜像。

代码18-41

```
$ docker images
REPOSITORY      TAG         IMAGE ID        CREATED         SIZE
ubuntu          latest      d13c942271d6    11 days ago     72.8MB
hello-world     latest      feb5d9fea6a5    3 months ago    13.3kB
centos          latest      5d0da3dc9764    4 months ago    231MB
```

运行容器，因为容器内部没有任何可以驻留的应用程序，所以需要加上参数-it，如代码 18-42 所示。

代码18-42

```
$ docker run -d -it d13c942271d6
241328b7c6213d0f5a9386be6d5477627090569db2f1e8433516fd1c7c9799c2
```

如代码 18-43 所示，查看运行中的容器。

代码18-43

```
$ docker ps
CONTAINER ID    IMAGE           COMMAND     CREATED         STATUS          PORTS       NAMES
241328b7c621    d13c942271d6    "bash"      7seconds ago    Up 5 seconds                sleepy_lehmann
```

18.3 编写 Dockerfile 文件

使用 docker build 命令可以在当前目录下找到 Dockerfile 文件进行构建，而命令后面的"."代表当前目录，如代码 18-44 所示。

代码18-44

```
$ docker build .
```

FROM 指令

FROM 指令用于指定基础镜像，允许开发人员基于基础镜像进行更改，基础镜像是必须指定的，如代码 18-45 所示。

代码18-45

```
FROM Ubuntu
```

RUN 指令

RUN 指令用于执行命令行命令，如安装依赖项等操作，是定制化镜像的常用指令之一，如代码 18-46 所示。

代码 18-46

```
RUN apt-get install -y vim
RUN ls
```

CMD 指令

CMD 指令用于指定默认容器主进程的启动命令，通常放在 Dockerfile 文件的末尾。例如，开发人员可以在容器启动时通过 docker run -it ubuntu 命令直接进入 bash，也可以在运行时指定命令，如创建一个名为 test 的镜像。可以通过 docker run 命令在启动容器时添加需要执行的指令，如代码 18-47 所示。

代码 18-47

```
docker run -d test_nginx /usr/sbin/nginx -g "daemon off;"
```

如代码 18-48 所示，通过 CMD 指令启动服务。

代码 18-48

```
CMD ["nginx", "-g", "daemon off;"]
```

注意：Dockerfile 文件只能指定一条 CMD 指令，如果存在多条也只执行最后一条，docker run 命令如果指定了指令，就会覆盖 Dockerfile 文件中的 CMD 指令。

ENTRYPOINT 指令

ENTRYPOINT 指令和 CMD 指令都可以用来执行命令，区别是 CMD 指令会因为 docker run 命令指定的参数被覆盖，而 ENTRYPOINT 指令可以接收 docker run 命令带来的参数并结合使用。

例如，在 Dockerfile 文件中指定 ENTRYPOINT ["nginx"]，并通过如代码 18-49 所示的命令执行。

代码 18-49

```
docker run -d test_nginx -g "daemon off;"
```

执行如代码 18-49 所示的命令，等同于执行如代码 18-50 所示的命令。

代码 18-50

```
docker run -d test_nginx /usr/sbin/nginx -g "daemon off;"
```

第 18 章　部署

如代码 18-51 所示，ENTRYPOINT 指令还可以结合 CMD 指令使用。

代码18-51

```
ENTRYPOINT ["nginx"]
CMD ["-h"]
```

代码 18-51 等同于代码 18-52。

代码18-52

```
docker run -d test__nginx nginx -h
```

VOLUME 指令

为了防止应用的输出文件写入容器存储层，如代码 18-53 所示，可以事先在 Dockerfile 文件中指定某些需要挂载的文件目录，即使用户不指定挂载目录，这些目录也会以匿名卷的方式存在。

代码18-53

```
VOLUME ["/var/www/html"]
```

这里的/var/www/html 目录会自动挂载为匿名卷，任何向/var/www/html 目录输出的文件都存储在容器中，从而保证容器存储层的无状态化。当然，在运行容器时可以覆盖这个挂载设置，如代码 18-54 所示。

代码18-54

```
$ docker run -p 8080:80 -d -v /d/mydata:/var/www/html test__nginx
```

上述命令使用了 mydata 这个命名卷，挂载到了/var/www/html 目录下，代替了 Dockerfile 文件中定义的匿名卷的挂载配置。

-v /d/mydata:/var/www/html 表示将本地文件夹挂载到 Docker 容器中，/d/mydata 对应的 Windows 操作系统中的文件夹路径为 D:\mydata，/var/www/html 为容器目录中的绝对路径。

COPY 指令与 ADD 指令

如代码 18-55 所示，将当前目录下的内容复制到容器的 app 目录下。

代码18-55

```
COPY . /app
```

ADD 指令和 COPY 指令的性质基本一致，但是 ADD 指令在 COPY 指令的基础上增加了一些功能，如<源路径>可以指定 URL，也就是说允许将文件下载到<目标路径>，如代码 18-56 所示。

代码18-56

```
ADD ["http://download.******.io/releases/redis-5.0.4.tar.gz",
"/www/redis/"]
```

EXPORT 指令

如代码 18-57 所示，在 Dockerfiler 文件中开放 80 端口。

代码18-57

```
EXPORT 80
```

如果要在本地主机访问，则可以通过参数 -p 映射到容器内的端口上，如代码 18-58 所示。

代码18-58

```
docker run -p 8080:80 -d --name nginxtest test_nginx
```

WORKDIR 指令

如代码 18-59 所示，WORKDIR 指令用于切换目录，类似于 cd 命令。

代码18-59

```
WORKDIR /wwwroot/web
RUN dotnet project.dll
```

18.4 构建.NET 应用镜像

Dockerfile 是一个用于承载 Docker 镜像而构建的脚本文件。开发人员可以先编写 Dockerfile 文件，再通过该文件执行 docker build 命令，生成一个镜像（Docker Image），最后通过 docker run 命令运行容器。产生容器的过程如图 18-14 所示。

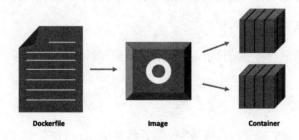

图 18-14 产生容器的过程

下面通过一个简单的实例来介绍如何运行一个容器。执行 docker build 命令在当前目录下找到 Dockerfile 文件并构建，命令后面的"."代表当前目录，如代码 18-60 所示。

代码18-60

```
docker build .
```

创建一个 Dockerfile 文件，如代码 18-61 所示。

代码18-61

```
FROM mcr.microsoft.com/dotnet/aspnet:6.0-alpine AS base
WORKDIR /app
EXPOSE 80

FROM mcr.microsoft.com/dotnet/sdk:6.0-alpine AS publish
WORKDIR /app
COPY . .

RUN dotnet publish -c Release -o output

FROM base AS final
WORKDIR /app
COPY --from=publish /app/output .
ENTRYPOINT ["dotnet", "MyProject.dll"]
```

在 base 阶段定义了容器内的 80 端口；在 publish 阶段将目录切换到/app 目录下，同时将当前目录的本地文件复制到容器内的/app 目录下，通过 Dockerfile 文件中的 RUN 指令定义 dotnet publish 命令，并将应用程序发布到 output 目录下；将 publish 阶段的结果复制到 final 阶段新的容器中，并利用 ENTRYPOINT 指令指定容器的启动命令。表 18-3 所示为 dotnet 镜像。

表 18-3　dotnet 镜像

镜像地址	镜像	镜像说明
mcr.microsoft.com/dotnet/runtime	.NET Runtime	.NET 运行时
mcr.microsoft.com/dotnet/runtime-deps	.NET Runtime Dependencies	用于独立发布的应用程序
mcr.microsoft.com/dotnet/sdk	.NET SDK	构建.NET 或 ASP.NET Core 应用程序
mcr.microsoft.com/dotnet/aspnet	ASP.NET Core Runtime	部署 ASP.NET Core 应用程序

如代码 18-62 所示，构建一个名为 aspnetapp 的镜像。

代码18-62

```
docker build -t aspnetapp .
```

运行 aspnetapp 镜像,并且定义该容器名为 myapp,同时对外开放 8000 端口,如代码 18-63 所示。

代码18-63

```
docker run --name myapp -p 8000:80 -d aspnetapp
```

如代码 18-64 所示,使用 docker images 命令可以查看镜像。

代码18-64

```
$ docker images
REPOSITORY                       TAG          IMAGE ID        CREATED          SIZE
<none>                           <none>       19d0af582978    3 minutes ago    592MB
aspnetapp                        latest       5f01befdd381    3 minutes ago    100MB
mcr.microsoft.com/dotnet/sdk     6.0-alpine   9d2c47a10a43    6 weeks ago      579MB
mcr.microsoft.com/dotnet/aspnet  6.0-alpine   043f270b380c    6 weeks ago      100MB
```

要把本地镜像推送到 Docker Hub 上,需要把镜像加上 tag。代码 18-65 展示了如何使用 docker tag 命令,以及 docker tag ${镜像名称} DockerHub 账号/镜像名。

代码18-65

```
$ docker tag aspnetapp hueifeng/aspnetapp
$ docker images
REPOSITORY                       TAG          IMAGE ID        CREATED           SIZE
hueifeng/aspnetapp               latest       5f01befdd381    22 minutes ago    100MB
aspnetapp                        latest       5f01befdd381    22 minutes ago    100MB
<none>                           <none>       19d0af582978    22 minutes ago    592MB
mcr.microsoft.com/dotnet/sdk     6.0-alpine   9d2c47a10a43    6 weeks ago       579MB
mcr.microsoft.com/dotnet/aspnet  6.0-alpine   043f270b380c    6 weeks ago       100MB
```

如代码 18-66 所示,使用 docker login 命令可以登录 Docker Hub 官网。

代码18-66

```
$ docker login
Login with your Docker ID to push and pull images from Docker Hub. If you don't
have a Docker ID, head over to https://hub.******.com to create one.
Username: hueifeng
Password:
WARNING! Your password will be stored unencrypted in /root/.docker/
```

```
config.json.
Configure a credential helper to remove this warning. See
https://docs.******.com/engine/reference/commandline/login/#credentials-
store


Login Succeeded
```

如代码 18-67 所示，使用 docker push 命令可以把镜像推送到 Docker Hub 中。图 18-15 所示为通过 Docker Hub 官网查到的镜像详情。

代码18-67

```
$ docker push hueifeng/aspnetapp
Using default tag: latest
The push refers to repository [docker.io/hueifeng/aspnetapp]
996e5254f895: Pushed
2fc5396cf731: Pushed
11c15b90ebce: Pushed
98bc857b3aed: Pushed
1a058d5342cc: Pushed
latest: digest: sha256:
5502bca044e8b3320564d9ac69cf56b7b429bc1d0a816b82386dbb7adccd7e92 size: 1371
```

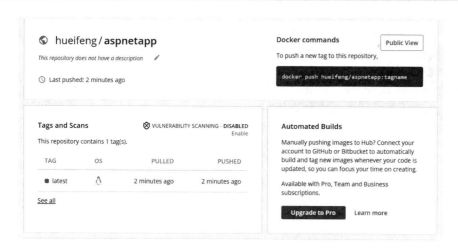

图 18-15　镜像详情

先使用 docker rmi 命令将本地镜像删除，再从 Docker Hub 中拉取镜像，如代码 18-68 所示。

代码18-68

```
$ docker rmi -f hueifeng/aspnetapp
Untagged: hueifeng/aspnetapp:latest
Untagged: hueifeng/aspnetapp@sha256:5502bca044e8b3320564d9ac69cf56b7b429bc1d0a816b82386dbb7adccd7e92
[root@centosvm ~]#docker rmi -f aspnetapp
Untagged: aspnetapp:latest
Deleted: sha256:5f01befdd38144a70a7f5c34665d4bc10ad3c59451a2d7ffc2aa4cbff9801b54
Deleted: sha256:5a391f6e00654414c4330c4e0c5b75aa0e1e5e29455f6659152f8ba61c252a1b
Deleted: sha256:720f806c485366c92103384f4512ac948fa7b17c7c7e38d6d5fac8d58a3de882
Deleted: sha256:cf0b2f3aee2fe68b71f588a928cf5bb9d60ab38fa1e9c5f10ab0cee83c258fd2
```

如代码 18-69 所示，使用 docker pull 命令可以拉取镜像。

代码18-69

```
$ docker pull hueifeng/aspnetapp
Using default tag: latest
latest: Pulling from hueifeng/aspnetapp
97518928ae5f: Already exists
e907ef816d7d: Already exists
b96604a8317d: Already exists
32010f400c96: Already exists
b4fca380e1f5: Pull complete
Digest: sha256:5502bca044e8b3320564d9ac69cf56b7b429bc1d0a816b82386dbb7adccd7e92
Status: Downloaded newer image for hueifeng/aspnetapp:latest
docker.io/hueifeng/aspnetapp:latest
```

如代码 18-70 所示，运行容器。

代码18-70

```
$ docker run -p 8000:80 hueifeng/aspnetapp
```

18.5　Docker Compose

Docker Compose 是 Docker 官方的开源项目，负责实现 Docker 容器编排（Container Orchestration），同时提供 docker-compose 命令行工具。类似于 Docker 的 Dockerfile 文件，docker-compose 命令行工具会通过 YAML 文件对容器进行管理。

18.5.1　安装 docker-compose 命令行工具

使用 pip 命令安装 docker-compose 命令行工具，如代码 18-71 所示。

代码 18-71

```
$ pip install docker-compose
```

通过访问 GitHub 官网的 docker/compose 仓库获取最新版本的 docker-compose（如 docker-compose 2.2.3），执行以下命令安装 docker-compose 命令行工具，如代码 18-72 所示。

代码 18-72

```
#将最新版本的 docker-compose 下载到/usr/bin 目录下
curl  -L  https://******.com/docker/compose/releases/download/2.2.3/docker-compose-`uname -s`-`uname -m` -o /usr/bin/docker-compose
#为 docker-compose 授权
chmod +x /usr/bin/docker-compose
```

18.5.2　命令行

build 命令

如代码 18-73 所示，构建或重新构建服务，执行该命令后，自定义 Dockerfile 文件并运行，先构建再运行，如果镜像不存在则从镜像仓库中拉取。

代码 18-73

```
docker-compose build
```

up 命令

使用 up 命令可以启动 docker-compose.yml 中的所有服务，如代码 18-74 所示，利用参数 -d 可以让服务在后台运行，也就是说日志不会输出到控制台中。

代码18-74

```
docker-compose up -d
```

logs 命令

如代码 18-75 所示，查看服务日志的输出。

代码18-75

```
docker-compose logs
```

down 命令

如代码 18-76 所示，停止并销毁所有通过 up 命令创建的容器。

代码18-76

```
docker-compose down
```

ps 命令

如代码 18-77 所示，列出运行的所有容器。

代码18-77

```
docker-compose ps
```

stop 命令

如代码 18-78 所示，停止正在运行的容器，也可以通过 docker-compose start 命令再次启动。

代码18-78

```
docker-compose stop
```

start 命令

如代码 18-79 所示，启动已经存在的容器。

代码18-79

```
docker-compose start [SERVICE...]
docker-compose start
```

restart 命令

如代码 18-80 所示，重启容器。

代码18-80

```
docker-compose restart
```

exec 命令

如代码 18-81 所示，进入容器。

代码18-81

```
docker-compose exec [serviceName] sh
```

18.5.3 部署.NET 实例

创建一个 MyApp 目录，同时利用 dotnet cli 命令创建一个 Web 应用程序，如代码 18-82 所示。

代码18-82

```
mkdir MyApp
cd MyApp
dotnet new webapp
```

在 MyApp 目录下创建一个 Dockerfile 文件，如代码 18-83 所示。

代码18-83

```
FROM mcr.microsoft.com/dotnet/aspnet:6.0-alpine AS base
WORKDIR /app
EXPOSE 80


FROM mcr.microsoft.com/dotnet/sdk:6.0-alpine AS publish
WORKDIR /app
COPY . .

RUN dotnet publish -c Release -o output

FROM base AS final
WORKDIR /app
COPY --from=publish /app/output .
ENTRYPOINT ["dotnet", "MyApp.dll"]
```

创建一个 docker-compose.yml 文件，如代码 18-84 所示。

代码18-84

```
version: '3.4'
```

```
services:
  myapp:
    image: ${DOCKER_REGISTRY-}myapp
    build:
      context: .
      dockerfile: MyApp/Dockerfile
    ports:
      - "8000:80"
```

version 描述当前的版本，而 services 则指明了提供服务的容器，myapp 是一个容器，image 表示当前镜像的名称，build 配置项定义了 context 属性和 dockerfile 属性，context 表示在当前目录下，dockerfile 表示 Dockerfile 文件的位置，ports 则定义了本机 8000 端口并且映射到容器内部的 80 端口。

如代码 18-85 所示，构建镜像。

代码18-85

```
docker-compose build
```

如代码 18-86 所示，启动容器。

代码18-86

```
docker-compose up
```

18.6 Docker Swarm

作为 Docker 的三大"剑客"之一，Docker Swarm 是 Docker 公司推出的用来管理 Docker 集群的平台。和 Docker Compose 一样，Docker Swarm 也是 Docker 容器的编排项目。

Docker Compose 可用于单台服务器，而 Docker Swarm 则可以在多个服务器容器集群中使用。Docker Swarm 提供了 Docker 容器集群的管理服务，是 Docker 对容器云管理的核心方案。使用 Docker Swarm，开发人员可以打通多台主机，从而进行统一化管理，快速打造一套容器云平台。

18.6.1 集群初始化

初始化主节点，如代码 18-87 所示。

代码18-87

```
docker swarm init --advertise-addr xxx.xxx.xxx.xxx
```

xxx.xxx.xxx.xxx 代表主节点的 IP 地址，docker swarm init 命令运行成功后，会给出一条 join 的命令，同时附带一个 Token。

连接主节点，如代码 18-88 所示。

代码18-88

```
docker swarm join --token xxxxx xxx.xxx.xxx.xxx:2377
```

根据上面给出的 Token 与主节点进行连接。

在集群中部署应用，如代码 18-89 所示。

代码18-89

```
docker stack deploy -c docker-compose.yml myapp
```

查看部署的应用，如代码 18-90 所示。

代码18-90

```
docker stack services myapp
```

删除应用，如代码 18-91 所示。

代码18-91

```
docker stack rm myapp
```

开发人员可以使用 docker stack 命令对服务进行编排。如代码 18-92 所示，定义一个 docker-compose.yml 配置文件。

代码18-92

```
version: '3.4'

services:
  redis:
    image: redis:alpine
    deploy:
      replicas: 6
      update_config:
        parallelism: 2
        delay: 10s
```

```
    restart_policy:
      condition: on-failure
```

由此可知，通过 replicas 属性指定 6 个容器；update_config 属性用于指定更新策略；parallelism 属性指定每组更新两个，完成之后延迟 10 秒再更新下一组；restart_policy 属性用于指定容器重启的策略；condition 是条件，表示失败后重启。

18.6.2　构建.NET 实例

如代码 18-93 所示，更改 Index.cshtml 文件，用于获取不同容器实例的 IP 地址。

代码18-93

```
@page
@model IndexModel
@{
    ViewData["Title"] = "Home page";
}

<div class="text-center">
<h1 class="display-4">
@(HttpContext.Connection.LocalIpAddress?.MapToIPv4().ToString())
</h1>
</div>
```

添加 docker-compose.yml 文件，如代码 18-94 所示。

代码18-94

```
version: '3.9'

services:
  web:
    image: myapp
    deploy:
      replicas: 6
      resources:
        limits:
          cpus: "0.1"
          memory: 50M
      restart_policy:
        condition: on-failure
```

```
      ports:
        - "8080:80"
```

集群模式不支持构建镜像，所以只能将构建任务分开，resources 的 limits 属性用于对系统资源进行限制，如代码 18-94 所示，限制容器使用不超过 0.1（10%）的 CPU 与 50MB 的内存。

如代码 18-95 所示，初始化节点。

代码18-95

```
docker swarm init
```

执行 docker swarm init 命令，如图 18-16 所示。

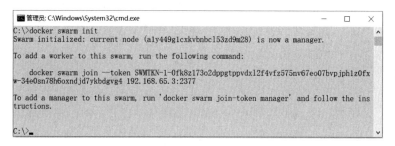

图 18-16　执行 docker swarm init 命令

添加 Dockerfile 文件，如代码 18-96 所示。

代码18-96

```
FROM mcr.microsoft.com/dotnet/aspnet:6.0-alpine AS base
WORKDIR /app
EXPOSE 80


FROM mcr.microsoft.com/dotnet/sdk:6.0-alpine AS publish
WORKDIR /app
COPY . .

RUN dotnet publish -c Release -o output

FROM base AS final
WORKDIR /app
COPY --from=publish /app/output .
ENTRYPOINT ["dotnet", "MyApp.dll"]
```

如代码 18-97 所示，构建镜像。

代码 18-97

```
docker build -t myapp .
```

构建镜像，如图 18-17 所示。

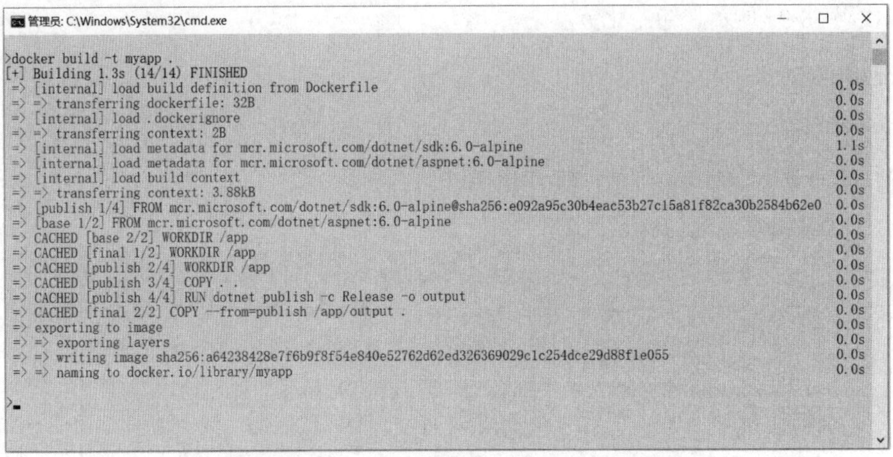

图 18-17 构建镜像

如代码 18-98 所示，运行服务。

代码 18-98

```
docker stack deploy -c docker-compose.yml myapp
Creating network myapp_default
Creating service myapp_web
```

myapp 是这个集群的名称。

如代码 18-99 所示，查看服务。

代码 18-99

```
docker stack services myapp
```

查看服务，如图 18-18 所示。

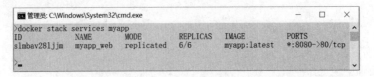

图 18-18 查看服务

访问 Web 应用程序，如图 18-19 所示。

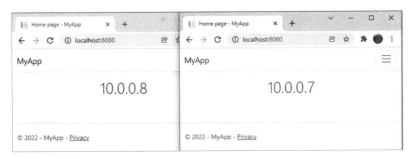

图 18-19　访问 Web 应用程序

18.7　Kubernetes

Kubernetes 简称 K8s，是一个容器集群管理平台，并且是开源的，可以实现容器集群自动化部署、自动扩容/缩容等功能。Kubernetes 是 CNCF（Cloud Native Computing Foundation）的项目，依托的是一个成熟的生态系统。

Kubernetes 是一个 PaaS（Platform as a Service，平台即服务）平台，满足了在生产环境中运行应用的一些常见需求，如滚动更新、弹性伸缩、负载均衡、资源监控等。

18.7.1　安装 Kubernetes

如代码 18-100 所示，关闭防火墙。

代码 18-100

```
systemctl stop firewalld
systemctl disable firewalld
```

如代码 18-101 所示，关闭 SELinux。

代码 18-101

```
setenforce 0
sed -i 's/^SELINUX=enforcing$/SELINUX=disabled/' /etc/selinux/config
```

自 Kubernetes 1.8 起，要求关闭 Swap（见代码 18-102），如果不关闭则默认导致 Kubelet 无法正常启动。

代码18-102

```
sudo sed -i '/swap/d' /etc/fstab
sudo swapoff -a
```

创建 k8s.conf 文件，调整内核参数，如代码 18-103 所示。

代码18-103

```
cat <<EOF > /etc/sysctl.d/k8s.conf
net.bridge.bridge-nf-call-ip6tables = 1
net.bridge.bridge-nf-call-iptables = 1
EOF
sysctl --system
```

如代码 18-104 所示，在节点上设置主机名。

代码18-104

```
hostnamectl set-hostname <hostname>
```

如代码 18-105 所示，添加 Kubernetes 的 yum 源。

代码18-105

```
cat <<EOF > /etc/yum.repos.d/kubernetes.repo
[kubernetes]
name=Kubernetes
baseurl=https://packages.cloud.******.com/yum/repos/kubernetes-el7-x86_64
enabled=1
gpgcheck=1
repo_gpgcheck=1
gpgkey=https://packages.cloud.******.com/yum/doc/yum-key.gpg
       https://packages.cloud.******.com/yum/doc/rpm-package-key.gpg
EOF
```

如代码 18-106 所示，安装 Kubelet、Kubeadm 和 Kubectl。

代码18-106

```
sudo yum install -y kubelet kubeadm kubectl
```

- Kubelet 分布在集群的每个节点上，是一个可以用于启动 Pod 和容器的工具。
- Kubeadm 是一个可以用于初始化集群和启动集群的命令行工具。
- Kubectl 是一个用来与集群通信的命令行工具。使用 Kubectl 不仅可以部署和管理应用，还可以查看各种资源，以及创建、删除和更新各种组件。

如代码 18-107 所示,启动 Kubelet 并设置开机启动。

代码18-107
```
systemctl enable kubelet
systemctl start kubelet
```

使用 systemd 来代替 cgroupfs,如代码 18-108 所示。

代码18-108
```
$ vim /etc/docker/daemon.json

{
  "exec-opts": ["native.cgroupdriver=systemd"]
}
```

修改 Kubelet,如代码 18-109 所示。

代码18-109
```
$ cat > /var/lib/kubelet/config.yaml <<EOF
apiVersion: kubelet.config.k8s.io/v1beta1
kind: KubeletConfiguration
cgroupDriver: systemd
EOF
```

如代码 18-110 所示,重启 Docker 与 Kubelet。

代码18-110
```
$ systemctl daemon-reload
$ systemctl restart docker
$ systemctl restart kubelet
```

如代码 18-111 所示,初始化集群。

代码18-111
```
$ kubeadm init --pod-network-cidr=10.244.0.0/16
```

Kubeadm 还会提示开发人员在第一次使用 Kubernetes 时需要配置的命令,如代码 18-112 所示,设置配置文件。

代码18-112
```
mkdir -p $HOME/.kube
sudo cp -i /etc/kubernetes/admin.conf $HOME/.kube/config
```

```
sudo chown $(id -u):$(id -g) $HOME/.kube/config
```

如代码 18-113 所示，部署网络插件，以 Flannel 插件为例，只需要执行一条 kubectl apply 命令即可。

代码 18-113

```
sudo kubectl apply -f      https://raw.*********.com/coreos
/flannel/master/Documentation/kube-flannel.yml
```

此时就可以使用 kubectl get 命令来查看当前节点的状态，如代码 18-114 所示。

代码 18-114

```
$ kubectl get nodes
NAME         STATUS     ROLES                  AGE     VERSION
centosvm     Ready      control-plane,master   108s    v1.23.3
```

向集群中添加新节点，执行在主节点上 kubeadm init 输出的 kubeadm join 命令，如代码 18-115 所示。

代码 18-115

```
$ kubeadm join 172.18.0.5:6443 --token k50281.npyirun0yey2kb1a \
    --discovery-token-ca-cert-hash
sha256:f59523b1a7611cfd0dfab04b1573f67f3be2016f9280b01e042b2c03de80882a
```

18.7.2 安装 Kubernetes Dashboard

下载 Kubernetes Dashboard 的 YAML 文件，如代码 18-116 所示。

代码 18-116

```
$ wget https://raw.******.com/kubernetes/dashboard/v2.4.0/
aio/deploy/recommended.yaml
```

如代码 18-117 所示，编辑 recommended.yaml 文件。

代码 18-117

```
kind: Service
apiVersion: v1
metadata:
  labels:
    k8s-app: kubernetes-dashboard
  name: kubernetes-dashboard
```

```
  namespace: kubernetes-dashboard
spec:
  ports:
    - port: 443
      targetPort: 8443
  selector:
    k8s-app: kubernetes-dashboard
```

如代码 18-118 所示，更改代码，采用 NodePort 方式访问 Kubernetes Dashboard，并指定一个 nodePort 属性。

代码18-118

```
kind: Service
apiVersion: v1
metadata:
  labels:
    k8s-app: kubernetes-dashboard
  name: kubernetes-dashboard
  namespace: kubernetes-dashboard
spec:
  type: NodePort
  ports:
    - port: 443
      targetPort: 8443
      nodePort: 30002
  selector:
    k8s-app: kubernetes-dashboard
```

如代码 18-119 所示，使用 kubectl apply 命令部署 Dashboard。

代码18-119

```
$ kubectl apply -f recommended.yaml
namespace/kubernetes-dashboard unchanged
serviceaccount/kubernetes-dashboard unchanged
service/kubernetes-dashboard configured
secret/kubernetes-dashboard-certs unchanged
secret/kubernetes-dashboard-csrf unchanged
secret/kubernetes-dashboard-key-holder unchanged
configmap/kubernetes-dashboard-settings unchanged
role.rbac.authorization.k8s.io/kubernetes-dashboard unchanged
```

```
clusterrole.rbac.authorization.k8s.io/kubernetes-dashboard unchanged
rolebinding.rbac.authorization.k8s.io/kubernetes-dashboard unchanged
clusterrolebinding.rbac.authorization.k8s.io/kubernetes-dashboard unchanged
deployment.apps/kubernetes-dashboard unchanged
service/dashboard-metrics-scraper unchanged
deployment.apps/dashboard-metrics-scraper unchanged
```

创建一个名为 dashboard-adminuser.yaml 的文件，如代码 18-120 所示，创建一个 Dashboard 用户来登录。

代码18-120

```yaml
apiVersion: v1
kind: ServiceAccount
metadata:
  name: admin-user
  namespace: kubernetes-dashboard
---
apiVersion: rbac.authorization.k8s.io/v1
kind: ClusterRoleBinding
metadata:
  name: admin-user
roleRef:
  apiGroup: rbac.authorization.k8s.io
  kind: ClusterRole
  name: cluster-admin
subjects:
- kind: ServiceAccount
  name: admin-user
  namespace: kubernetes-dashboard
```

使用 kubectl apply 命令应用配置文件创建用户，如代码 18-121 所示。

代码18-121

```
kubectl apply -f dashboard-adminuser.yaml
```

如代码 18-122 所示，获取用户的 token 信息用于登录。

代码18-122

```
$ kubectl describe secrets -n kubernetes-dashboard admin-user
Name:         admin-user-token-rnbkf
Namespace:    kubernetes-dashboard
```

```
Labels:         <none>
Annotations:    kubernetes.io/service-account.name: admin-user
                kubernetes.io/service-account.uid:    49ca275a-b5d4-4456-b4d7-
46ac9b1561da

Type: kubernetes.io/service-account-token

Data
====
ca.crt:     1099 bytes
namespace:  20 bytes
token:
eyJhbGciOiJSUzI1NiIsImtpZCI6IjBCNFpWaXZoOGRIS3VrSWg4a01qR21kZHlsb1lWNE1LUD
Y0bzA4NjR1SVUifQ.eyJpc3MiOiJrdWJlcm5ldGVzL3NlcnZpY2VhY2NvdW50Iiwia3ViZXJuZ
XRlcy5pby9zZXJ2aWNlYWNjb3VudC9uYW1lc3BhY2UiOiJrdWJlcm5ldGVzLWRhc2hib2FyZCI
sImt1YmVybmV0ZXMuaW8vc2VydmljZWFjY291bnQvc2VjcmV0Lm5hbWUiOiJhZG1pbi11c2Vy
XRva2VuLXJuYm1Iiwia3ViZXJuZXRlcy5pby9zZXJ2aWNlYWNjb3VudC9zZXJ2aWNlYWNjb3VudC9zZXJ2aWNlLWFjY291
bnQubmFtZSI6ImFkbWluLXVzZXIiLCJrdWJlcm5ldGVzLmlvL3NlcnZpY2VhY2NvdW50L3Nlc
nZpY2UtYWNjb3VudC51aWQiOiI0OWNhMjc1YS1iNWQ0LTQ0NTYtYjRkNy00NmFjOWIxNTYxZGE
iLCJzdWIiOiJzeXN0ZW06c2VydmljZWFjY291bnQ6a3ViZXJuZXRlcy1kYXNoYm9hcmQ6YWRta
W4tdXNlciJ9.N4rW8bb7Bu_BKR40-cpd2sYKcSuG8SanmreKvjU855lEhKNkEkdTyjBeWhIj2h
WYq6d_vSRwh_AbttyvTy9nvBU4WvyLjlphTk7T1SNU85QjxQiu0yNl_niQOXkbpNUV4Jskh563
9YlCagRzTO2qiu35vTIWpW4kLqdvrphlKR2zkx2wO8Ilpj9LlglEERKLlkHbD1TIwjveL7Msna
qlfbXOMafHZaOAoZORdr8vFNpYba8uOk8CzTJts5gTVeCZABzjZXmap9i3pM9T641-OT1sigx_
O0uCAIpcZZvwAvcC4U0nELZiOtmvS_BiWRcuzxF5SuRFfb_Whv5fkXkj6g
```

如图18-20所示，通过Kubernetes Dashboard Token登录。

图 18-20 通过 Kubernetes Dashboard Token 登录

图 18-21 所示为 Kubernetes Dashboard 主页。

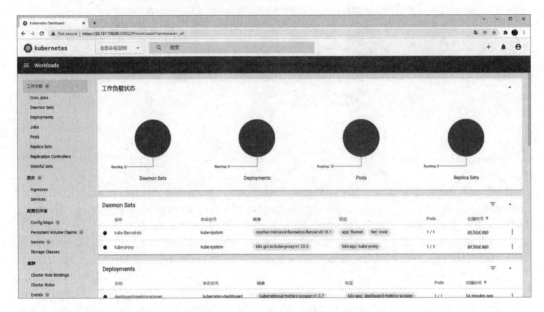

图 18-21　Kubernetes Dashboard 主页

18.7.3　部署 .NET 应用

创建一个 Kubernetes Deployment 对象来运行一个应用，可以在 YAML 文件中描述 Deployment。如代码 18-123 所示，YAML 文件使用了 DockerHub 仓库中的 hueifeng/aspnetapp Docker 镜像。

代码18-123

```
apiVersion: apps/v1
kind: Deployment
metadata:
  name: myapp-deployment
spec:
  selector:
    matchLabels:
      app: myapp
  replicas: 2
  template:
    metadata:
```

```
    labels:
      app: myapp
  spec:
    containers:
    - name: myapp
      image: hueifeng/aspnetapp
      resources:
        requests:
          memory: "32Mi"
          cpu: "100m"
        limits:
          memory: "128Mi"
          cpu: "500m"
      ports:
      - containerPort: 80
```

如果想要一个单节点的 Kubernetes，则删除这个 Taint 才是正确的选择，如代码 18-124 所示。

代码18-124

```
kubectl taint nodes --all node-role.kubernetes.io/master-
```

如代码 18-125 所示，使用 kubectl create 命令创建一个 Pod。

代码18-125

```
$ kubectl create -f myapp-deployment.yaml
deployment.apps/myapp-deployment created
```

使用 kubectl get 命令检查运行这个 YAML 文件的状态是否与预期的一致，如代码 18-126 所示。

代码18-126

```
$ kubectl get pods -l app=myapp
NAME                                READY   STATUS    RESTARTS   AGE
myapp-deployment-7c64869cbf-bdc76   1/1     Running   0          5m
myapp-deployment-7c64869cbf-crpm7   1/1     Running   0          5m
```

可以使用 kubectl describe 命令查看一个 API 对象的细节，如代码 18-127 所示。

代码18-127

```
$ kubectl describe pods myapp-deployment-7c64869cbf-bdc76
Name:         myapp-deployment-7c64869cbf-bdc76
Namespace:    default
Priority:     0
Node:         centosvm/172.18.0.5
Start Time:   Thu, 27 Jan 2022 13:59:33 +0000
Labels:       app=myapp
              pod-template-hash=7c64869cbf
Annotations:  <none>
Status:       Running
IP:           10.244.0.7
IPs:
  IP:         10.244.0.7
Controlled By: ReplicaSet/myapp-deployment-7c64869cbf
...
Events:
  Type    Reason     Age    From                Message
  ----    ------     ----   ----                -------
  Normal  Scheduled  100s   default-scheduler   Successfully assigned default/myapp-deployment-7c64869cbf-bdc76 to centosvm
  Normal  Pulling    99s    kubelet             Pulling image "hueifeng/aspnetapp"
  Normal  Pulled     57s    kubelet             Successfully pulled image "hueifeng/aspnetapp" in 41.747279s
  Normal  Created    57s    kubelet             Created container myapp
  Normal  Started    57s    kubelet             Started container myapp
```

在使用 kubectl describe 命令返回的结果中，可以清楚地看到这个 Pod 的详细信息，如它的 IP 地址等。另外一个值得特别关注的部分就是 Events。

使用 kubectl exec 命令可以进入这个 Pod，如代码 18-128 所示。

代码18-128

```
$ kubectl exec -it myapp-deployment-7c64869cbf-bdc76 -- /bin/sh
```

如果需要从 Kubernetes 集群中删除整个 MyApp Deployment，则可以直接通过命令执行，如代码 18-129 所示。

代码18-129

```
$ kubectl delete -f myapp-deployment.yaml
```

18.7.4 公布应用程序

如代码 18-130 所示，定义 Service 服务。

代码18-130

```yaml
apiVersion: v1
kind: Service
metadata:
  name: myapp-service
spec:
  type: NodePort
  selector:
    app: myapp
  ports:
  - port: 80
    nodePort: 30010
    protocol: TCP
    targetPort: 80
```

如代码 18-131 所示，通过 kubectl create 命令，以 YAML 文件为基础创建一个 Service 服务。

代码18-131

```
$ kubectl create -f myapp-service.yaml
```

如代码 18-132 所示，通过 Service 服务的 IP 地址 10.111.168.204，就可以访问它所代理的 Pod。

代码18-132

```
$ kubectl get svc myapp-service
NAME            TYPE       CLUSTER-IP        EXTERNAL-IP   PORT(S)        AGE
myapp-service   NodePort   10.111.168.204    <none>        80:30010/TCP   8m3s
```

NodePort，顾名思义，在所有的节点上开放指定的端口，接下来就可以通过 http://ip:30010 查看网站。图 18-22 所示为使用 Kubernetes 部署的网站。

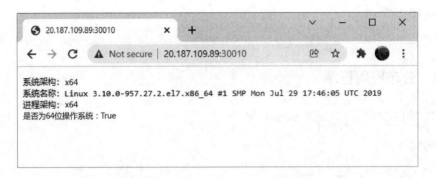

图 18-22　使用 Kubernetes 部署的网站

18.8　小结

本章介绍了 .NET 跨平台部署和容器化部署。通过学习本章，读者可以将 .NET 应用程序上云。在云原生时代，避免不了的是上容器云。本章以 Docker 这个容器工具为基础展开介绍，利用 Docker 可以让应用程序快速部署容器云，同时利用 Docker Compose 进行容器的编排，而利用 Docker Swarm 管理 Docker 集群可以将 Docker 打造为一个容器云平台。在 18.7 节中，笔者以 Kubernetes 为例，介绍从 Kubernetes 的安装到 Kubernetes Dashboard 可视化界面的安装，以及将应用程序部署到 Kubernetes 中，使其具备对外访问的能力。读者阅读完本章后，基本上就可以掌握应用程序快速迈向容器云的使用。